Thomas Scott Lambert

# Systematic Human Physiology, Anatomy and Hygiene

Second Edition

Thomas Scott Lambert

**Systematic Human Physiology, Anatomy and Hygiene**
*Second Edition*

ISBN/EAN: 9783337365356

Printed in Europe, USA, Canada, Australia, Japan

Cover: Foto ©berggeist007 / pixelio.de

More available books at **www.hansebooks.com**

SYSTEMATIC

HUMAN

# PHYSIOLOGY, ANATOMY,

AND

## HYGIENE:

BEING AN ANALYSIS AND SYNTHESIS OF THE
HUMAN SYSTEM,
WITH PRACTICAL CONCLUSIONS.

———◆◆◆———

### Many New and Complete Illustrations.

———◆◆◆———

BY

T. S. LAMBERT, M.D.

SECOND EDITION.

NEW YORK:
WILLIAM WOOD & CO.,
61 WALKER STREET.
1866.

# CONTENTS.

## INTRODUCTORY REMARKS.

## PART I.

### SYSTEMATIC GENERAL ANALYSIS.

# INTRODUCTORY REMARKS.

"What a piece of work is a Man! How noble in reason! how infinite in faculties! in form and moving how express and admirable! in action how like an angel! in apprehension how like a god!"

———◆◆◆———

XENOPHON, speaking of the studies that Socrates advised people to pursue, says:

"He earnestly recommended those who conversed with him to take care of their HEALTH, both by learning whatever they could respecting it from men of experience, and by attending to it, each for himself, throughout his whole life, studying what food or drink, or what exercise, or what clothing, was most suitable for him, and how he might act in regard to them so as to enjoy the best health."

Socrates thought health attainable. He was right; for though partly dependent on inherited constitution and the care taken of childhood, its preservation and improvement are to a great degree under the control of each person.

Socrates also thought health of such great importance in daily life, that no person should wait for his own experience to instruct him, but should avail himself of that of others. Again he was right; for health is a means as well as an end, to be used and enjoyed as well in the earlier as in the later periods of life.

In those days, the uncertain experience of even many persons cast but a feeble light, where now the developments of science completely illuminate the laws of health. Had Socrates lived in these days of the maturity of Anatomy, Physiology, Chemistry, and kindred branches, then in infancy or even unborn, with what earnestness would he have urged that every child should be taught the golden laws of Hygiene, be trained to practise them, and be instructed in their why and wherefore, that he might practise them intelligently, and of course more thoroughly and satisfactorily! ~

But there is a very marked and important distinction between health and a condition best adapted for use and appearance; though the latter must of course include the former. Indeed, within the boundaries of health there is a wide latitude for the activities of the different parts of the body and the display of taste and refinement in its culture. Laws, therefore, which govern the use of the various parts of the body, and promote personal appearance—laws of physical perfection—are evidently required. As they are obtained from the same sources, at the same time, and are often the same as the laws of health, they are classed under the same head.

Hygiene is not, therefore, and should not be, restricted as its primary meaning would signify; for it teaches not only how to keep the body (1st) in health, but (2d) in the best condition for desirable use and appearance.

As Addition enters into every arithmetical operation, so is Hygiene elementary in every pursuit, being of hourly utility to every person and in every possible circumstance in life— really most useful to the most healthy; and as if the Divinity had intended that what is so essential to man's welfare should be neither difficult to teach nor to learn, the mind has been so constituted and the body so constructed that a sufficient knowledge of Hygiene, if properly presented, can be easily acquired, and remarkably early; for,

In the first place, the mind has the greatest "natural curiosity" to learn the structure and use of its own body: it is the child's earliest wonder, and the old man's latest care.

In the second place, no other subject has such close practical relations to life, which alone would make it intensely interesting.

In the third place, no other subject can so fully illustrate cause and effect, so clearly show the why and wherefore, so readily exhibit evidences of rational, ingenious design,—a study above all others pleasing to the youthful mind.

In the fourth place, no other subject can be so abundantly and readily illustrated, either by objects themselves or by drawings of them, the body itself furnishing many illustrations.

In the fifth place, no other subject admits of so simple, clear, and complete a classification and arrangement, analysis and synthesis—which may be illustrated pictorially, grouped in tables and synopses, and in other striking ways—arresting and winning the attention, producing vivid and lasting impressions, and suggesting inductions and deductions; thus enabling the pupil easily to learn, remember, and apply his lessons.

For instance: Plates 6 and 7 present an epitome of Anatomy, having the figures arranged in such order as will impress the mind with a correct succession of ideas, and instantly suggest important facts and classifications; such as,

That various parts of the body are similarly constructed: for example,

The {
Walls of the HEAD,
Neck,
Walls of the Trunk,
Upper Extremities,
Lower Extremities,
} are similarly constructed of {
Skeleton,
Muscles,
Nerves,
Skin,
Blood-Tubes;
}

That there are two cavities shown by the Skeleton; one of the Head, the other of the Trunk;

That there are two centres, each connected with all parts of the body; one the political capital, the Brain, by means of the Nerves, Fig. 3; the other the commercial capital, by means of the Blood-tubes, Fig. 6, Plate 6;

That there are two great groups of organs in the body; one Mental, Plate 6; the other Blood-making (Sanguific, Sanguificatory, or Sanguinificatory), Plate 7. (See also table page 399.)

This leads to the observation that every child should study Hygiene, properly presented, because it will have a favorable influence upon his mind, habituating him to look for a cause for every effect; to perceive that design is essential to the best results; to analyze and synthetize; to classify and arrange systematically ; to induce and deduce—in other words, to think correctly.

Life, to be most useful, should be full of executed purposes. A true life may be called a great purpose compounded of many subsidiary ones. Hence the mind is constituted, and should be trained, to purpose, plan, and execute.

Now, in the Divine mind there was a purpose, to execute which the human body of coördinate and subordinate parts was most exquisitely planned. In its study there must, therefore, be a starting-point and a conclusion necessarily related, and permitting between them a most rigid analysis and synthesis.

Besides, so much is this study in accordance with the constitution and necessities of the mind, that, as nothing improves it more, so nothing pleases it more ; and, "to please is the first step toward instruction."

What, then, can be more normal than the study of Divine purposes, plans, and executions, as written in the human body, and exhibited by the properly arranged study of the human system ?

This study is also a constant hymn to the Deity. His own works praise him silently and profoundly, impressing the mind far beyond what can be done by any words of man, and rendering any special clauses and chapters unnecessary.

"As the twig is bent the tree is inclined."


Life should be computed not only by the lapse of time between birth and death, but also by what is done and experienced during that time. Hygiene prolongs life in a twofold manner, by adding to the number of its days, and by increasing their efficiency, both mental and physical. In this view, the effect of this study, universally and properly pursued, would be beyond belief. The aggregate length of human life would be much increased; pupils would study more, but not so hard; teachers would teach more, and with less exhaustion; farmers, mechanics, merchants, professional men, and men of science would achieve and enjoy greater results than now. This would be better than adding so much to our population; for while all good results would be correspondingly increased, the expenses would be diminished. Many of the same laws, also, not being restricted to man, but equally applicable to animals, (indeed, in the single item of properly providing for cattle, in the State of New York alone, the economy may not be stated, as it would appear incredible,) there is an additional reason for the study.

Let every teacher, and all others interested in the education of children, give due heed and full weight to the arguments adduced:

That health can be improved and preserved;

That the body is one of the tools every one must use in any avocation or enjoyment, and, for perfect use, must be understood and kept in perfect condition;

That Hygiene, properly pursued, induces correct methods and habits of thought;

That Hygiene is a great economist.

Thoughts and arguments like these, which should induce each pupil to study and each teacher to teach this subject, have influenced the author to write this work. True, many books, and good ones too, have preceded this. Nearly thirty years ago Dr. Charles A. Lee, the pioneer in this good work, produced a text-book which at this day ranks as one of the best. The names of a score who have written since might be worthily mentioned, for every one has exhibited much merit, and some least known, the most.

But the study is not yet as universal, not yet as attractive, as is desirable; and whoever can, or thinks he can, should add his mite to the best of efforts.

1*

Something valuable, it is thought, can be added to what is already of great intrinsic merit, by both illustration and presentation. This is made conclusive by the frequently and urgently expressed desire of teachers, well qualified to judge, that for their own and pupils' use the subjects should be presented in book-form, according to the methods that had given satisfaction when orally used before them and their classes.

These methods are neither so novel nor so extraordinary as to render useless any previously acquired familiarity with the subject; they merely make it more familiar by presenting known facts and ideas in a new light and with new features, in groups and relations suggesting new thoughts, removing obscurity, and exhibiting a unity and completeness highly satisfactory.

The art of producing the desired and gratifying result does not consist in the use of a single idea or one means only. The first and most essential idea to be kept in mind is, that, if possible, the reasons for the existence of any part should be pointed out before its structure is described; for, if the necessity for it, and what ought to be, is first shown, the description of the structure and its adaptation to use will be much more interesting and very easily understood.

Physiology naturally precedes Anatomy in the order of constructive thought. Purpose is always antecedent to construction, which is governed by and in accordance with purpose. This is the Creative idea and mode of thought, and also the most natural and delightful to the human mind. It is the key to the great charm that always abides in the study of Natural Theology.

Functions are not performed because certain parts exist, but conditional parts exist because certain functions must be performed: function is therefore always the object in view as the result; and how to perform it perfectly, with the greatest economy of time, space, and material, is the sub-necessary question.

Unfortunately, though the general purposes and the modes and results of action of the various parts of the body are known, in many cases it is otherwise when details are considered; and the process of discovery that describes structures, and conjectures or endeavors to deduce functions therefrom, is the only course that can be taken. When, however, a purpose and the action necessary to accomplish it is understood, they should be explained first; then the properties, structure, and arrangement adapted to fulfil the purpose should be suggested, and the student encouraged to construct mentally a part in accordance with the necessities of the case. Comparing his result with the reality, he will be delighted to observe how well he has executed his own imaginary work—will be surprised at the common-sense manner in which every portion of the body has been made, and cannot but admire in the highest degree the exquisitely perfect adaptation of every part to all, and more than all that he would have required.

Thus, not only is interest excited and acquisition made rapid, but brevity is likewise gained; for irrelevant characteristics may be omitted when description is only to correspond to purpose, while if it is unknown, every particular must be given.

A second idea, arising from the first, but not of secondary importance, is, that the subjects should be studied under divisions and subdivisions made in accordance with the purposes or uses for which the different parts of the body were designed; the parts being classed, not so much according to similarity as community of use, and so as to show their relations to each other, to the whole body, and to the mind.

When this is done, there will be exhibited in Physiology, Anatomy, and Hygiene, an analysis and synthesis, a systematic arrangement, more natural, obvious, and complete, than can be found in any other subject. (See page 399.)

Classification should not be artificial, arbitrary, accidental, fanciful, trifling, or superficial, but should exhibit the Divine purpose in organizing the human system, which in itself is practical knowledge.

By classification according to uses, it will be perceived and
appreciated that the body is a unit, constructed of correlative
and subordinate parts, but all having a unitary purpose; that,
commencing with man, composed of mind and body, we can descend, by a regular gradation of subdependent relations, through
Members, Groups, Apparatuses, Organs, Textures, and Tissues,
(with the addition of fluids,) to the primary elements constituting the substratum of all parts, from which, by synthesis, the
steps can be retraced till we find every subordinate part culmi-
nate through its superior, in being a servant to the mind, for
the use and development of which the body is designed—the
mind and body acting and reacting on each other in every circumstance in life. (See the Chart of Analysis and Synthesis, in
Appendix.)

It must be evident that, by the use of this natural and legiti-
mate classification according to use, the labor of acquisition,
retention, and application, will be in a very noteworthy degree
abridged.

This naturally leads to the third idea, which is, that a general
and brief analysis of the uses and construction of the body
should be given as an introduction to the synthetical study of
details, as then the necessity for them and their uses will be
more readily apprehended, since they are useful and necessary
only as they are subordinate to the great purpose for which the
body exists; and their value will be more appreciated, and they
will become more interesting, as their relations to those purposes
are more clearly perceived. It may be added that the different
parts of the body are so intimately associated with, and act and
react so constantly on, each other, that a general knowledge of
all is necessary to the complete understanding of any one.

If, however, any teacher thinks that details should be studied
first, he can adopt this plan by commencing with the synthetic
view. And in many instances it will be best for him to do this,
and conclude with the General Analysis, or he may have the

class merely read the General Analysis during class-hour the first time going through the book.

In the fourth place, it may be suggested that blackboard exercises should be conducted in connection with recitations. The tables or synopses should be built, or grow up, line by line, as the ideas that underlie them are unfolded, and should be allowed to remain from day to day before the pupil, and be frequently reviewed or rehearsed. Each table can be transferred, for convenience of preservation, to a large sheet of paper or of cloth. But when such sheets, exhibiting the table in full, are possessed, they should not be used till made up on the board by the pupil.* Tables should also be constructed with initials only, and in several forms, and be made, if convenient, with colored chalks, pencils, inks, or paints, and with letters of different shape and size, in order to make the distinctions of the classifications and their suggestions evident to the eye.

This leads to the fifth idea, or means: The classifications should be represented by objects when possible; by models, by pictures and drawings, and by outlines and sketches on the blackboard. All such illustrations are very impressive and lasting in their effects. Many classifications of parts of the body can be illustrated by corresponding parts of animals, which always proves exceedingly interesting to pupils. By *illustrations* and *classifications* something more is meant than the illustration of single and isolated parts; it means a systematic illustration of all the classifications so far as possible—making the illustrations by themselves a complete epitome or abstract of relations. Every teacher will of course understand, without a hint, the advantage and importance of illustrating individual parts.

---

* Sheets exhibiting the most important tables, in colored letters of various forms, can be had of the Publishers of this work.

The order in which illustrations are used is of no small importance: it should always be in accordance with the action, or use of parts, and have reference to the class of ideas to be presented. Plate 6 is an example of an order perfect for the intended purpose: thus, Eye—Nerve—Brain, or, Brain—Nerve——Eye, would each be a proper order in its peculiar case, as the eye acts upon the nerve, and the nerve upon the brain, the nerve connecting the eye and brain; but Nerve—Eye—Brain, or, Nerve—Brain—Eye, would never be a proper order for mentioning or illustrating those parts of the body, as the eye is not between the nerve and brain, nor is the brain between the nerve and eye, either in position or action. If plates are suspended in the recitation or other room, their order should be that in which the parts represented act upon or relate to each other, so that relations and uses as well as construction should be suggested.

Under the preceding six headings are exhibited the most distinctive methods for producing the result that so many teachers have been pleased to admire and to desire to produce; if used with other appropriate means, as complete success as that which has gratified them may be assured. It is not perfect. It is an improvement. Teachers can add to it, as by increments it has grown during many years' experience, now by a new reflection, now by a new truth, now by some suggestion of a teacher, and not unfrequently by the timely question of a pupil. These methods are only one convenient part of the apparatus to be used in teaching Hygiene.

In addition, it may be suggested that all ideas presented should as far as possible be illustrated by facts within the observation of the pupils or drawn from the teacher's own experience, by anecdotes, and in any other way that will arrest attention or excite interest; and especially should every practical idea be enforced by illustrations of the necessity for and the results of its application. (See Appendix C.)

It is above all things important that the teacher be persuaded, and persuade his pupils to believe the fact, that this subject is an easy one to learn. The body is not that complex apparatus it at first seems to be. This impression is induced by the manner in which its parts are packed, and as it were interwoven, in order that they may occupy a small space and present a neat appearance. But when the body is properly resolved into its constituent elements, they will be found few in number, easily comprehended; and a student will quickly learn how to build them up and take them down, and the correct philosophy of all the circumstances that work their harm or good.

The subject sometimes appears difficult and tedious on account of the many details and technical expressions with which it is encumbered. Most of these are entirely unnecessary in a work for popular use.

Let teachers observe and inculcate the very marked distinction, in this respect, between popular and professional use. The latter requires a knowledge of a thousand details and a corresponding number of terms, while professional writings so abound in technical expressions, that the sooner a professional student becomes familiar with them the better. When familiar with them, and most of his knowledge has been obtained in part through their use, it will be more difficult for him to express himself otherwise, than to translate from one language to another. Hence the multiplicity of such expressions in works intended for general use. It is better for the popular student to have the translation made, however inconvenient it may be to the writer; for the expressions are often awkward, and always sound pedantic outside of professional circles. It will be difficult for him to understand their true force, and he has not the time to make himself familiar with them. The various details, and the few terms necessary or of any advantage to him, will be so evidently convenient, that they will neither be, nor appear to be, burdensome. Even the use of these had better be avoided in the general presentation, or, so to speak, plan, of new ideas, to either the popular or the professional student; as ideas are usually, though not always, more easily obtained through language that is familiar, though imperfect, than when the meaning of it is also to be learned. (See Appendix E.)

Such an explanation as that just made will often remove a very natural prejudice from the minds of parents, and secure the favorable opinion of any professional man who would not be likely to think a few weeks of much use in his studies, or that they could be easily comprehended by young persons. That labor is not lost which will render this subject agreeable to every person. It is worthy of patience and painstaking. Not the wise, but the ignorant, need instruction, and ignorance must be removed, not offended nor opposed.

But any difficulties in connection with this subject will readily melt away under the warm enthusiasm of a teacher zealous to discharge all his duty, and ambitious to take a high rank, in one of the most useful and deserving professions.

This work was prepared—

(1st) Because the study of Hygiene will promote health, prolong life, improve personal appearance, render the action of body and mind more efficient, and much increase the sum of human happiness.

(2d) Because, though the subject is well presented in many books, and well taught through their use, improvements upon them are desirable and attainable. A little improvement in a subject of such importance is a gain exceeding the outlay of means; even without improvement a new book will refresh and invigorate a teacher, and it will be profitable to any pupil to have more than one work on this subject.

(3d) Because many persons, well qualified to judge, have decided that its distinctive features are superior, and desired to have a text-book prepared in accordance with them; and

(4th) Because the publishers were desirous of presenting to teachers and the public a work illustrated and executed in a style worthy of the subject.

The ideal plan of the work is—

(1st) To show, when possible, the necessity for a part, and the properties and construction required, before describing it as it exists.

(2d) To present a brief general analysis (retraced, a general synthesis) of the uses and construction of the whole body, preparatory to the complete study of details (synthetic, retraced analytic).

(3d) To classify parts according to their uses, co-ordination, correlation, and subordination.

(4th) To illustrate classifications as well as their constituent parts.

(5th) To exhibit relations by tables, synopses, and blackboard exercises.

(6th) To present new ideas in familiar language, and supply the appropriate terms afterward.

(7th) To lead the student himself, from sustaining facts, to make immediately those inductions, deductions, and inferences, that, with the instructions of experience, constitute practical Hygiene; to observe the constant relations between the Mind, Body, and External World, that he may not think time spent in providing for physical welfare, and in improving his surroundings, is taken from mental advancement, but may know that the Body must be kept in a perfect condition in order that he may possess the highest capabilities.

(8th) To throw into an Appendix, models of blackboard exercises, charts, cuts, and other matter, which, though relevant, explanatory, and useful, would appear to encumber the text and disturb its unity ; while the cuts, beautifully printed upon tinted paper, will be a very useful and attractive feature.

(9th) To frame, as far as possible, the questions at the bottoms of pages, so that they will be in part composed of the capitalized words of corresponding paragraphs, thus making the questions strikingly topical, and not liable to objections.

These improved methods, in connection with the admirably illustrative cuts furnished by the publishers, will, it is hoped and believed, assist in habituating students to correctly exercise and develop their thinking powers, and enable them to become more pleasantly, rapidly, and completely acquainted with the laws of Hygiene than previous opportunities have allowed.

The work is addressed to those desirous of self-improvement, and willing to take the pains necessary for it; who value themselves as having minds as well as bodies, the former of which can be as much improved by mental as the latter by physical gymnastics; who think that man in his rich endowment of immortality is ennobled above all comparison with earthly distinctions, and worthy of his own highest respect and care; who desire to learn, therefore, not merely to recite a lesson, but to obtain knowledge, and, feeling the value of this acquisition, are willing to invest the necessary labor :—also to teachers who esteem the "laborer worthy of his hire," and demand it, yet enjoy an additional rich reward in doing that which will constitute the germs of exalted habits in future generations through widespread communities, and are ever ready to impart knowledge to those desirous of learning the laws of Hygiene.

In the same spirit, nothing will give the author higher pleasure than to assist, personally or by letter, teachers and students in acquiring or disseminating a knowledge of the Laws of Hygiene; and he would solicit the favor of their acquaintance, or the continuation of it, upon the ground of community of opinions, feelings, and interests.

# SYSTEMATIC

### H U M A N

# PHYSIOLOGY, ANATOMY, AND HYGIENE.

## PART I.

### GENERAL ANALYSIS.

———•••———

## CHAPTER I.

### Systematic Analysis of Man.

*Composed of Mind and Body.*

1. ANTHROPOLOGY, from *anthropos*, a man, and *logos*, a discourse, treatise, or science, is the name given to a department of science that treats upon the entire nature of man. (See Appendix A.)

---

The capitalized words commencing a paragraph in the text usually indicate topics, and a word or two more will, with them and an interrogation point, form a question. What is, with "ANTHROPOLOGY," and an ? will bring out the whole matter of the first paragraph. Teachers can readily, therefore, ask questions without reference to those at the bottom of the page. In those, the dash before, between, or after words in questions, is to be understood as meaning the capitalized words of the paragraph correspondingly numbered. In some questions the capitalized verb must be repeated, and the order of the words changed. The Teacher may ask such questions on Ap. as is judged proper. It should be read in class. A Blackboard (B—d) should always be present at recitations; if not, a slate or paper should serve the same purpose.—Have you read "Introductory Remarks?" What is the purport of them? 1. What is —? Have you read Ap. A?

2. MAN IS COMPOSED of Mind and Body; therefore Anthropology must be divided into two sub-sciences.

ANTHROPOLOGY . . . . . { Psychology.
{ Biology.

3. PSYCHOLOGY, from *Psyche*, the soul, and *logos*, is the name of that science that treats upon the nature of the human mind, and the methods of developing its powers.

4. BIOLOGY, from *Bios*, life, and *logos*, is the most correct name of that science that treats upon physical life; that is, life exhibited by vegetables, animals, and also by man, but not including his mental life.

BIOLOGY . . . . . . . { Human.
{ Animal.
{ Vegetable.

5. HUMAN, ANIMAL, AND VEGETABLE BIOLOGY, are the names given to the three divisions of that department; they correspond to the three general forms in which physical life is exhibited. (See Ap. A. *g.*)

6. COMPARATIVE BIOLOGY, is the name given to a science the object of which is to compare the three kinds of life, and the various parts of vegetables, animals, and man, so as to make out analogies, discern uses, and draw instructive conclusions in respect to the use of parts.

7. EACH DIVISION OF BIOLOGY is subdivided into PHYSIOLOGY, ANATOMY, HYGIENE, *Pathology, Therapeutics, Materia-Medica,* and *Surgery* each classed as Human, Animal, Vegetable, and Comparative, according to application. (See Ap. B.)

BIOLOGY { Human,
{ Animal,
{ Vegetable, } { PHYSIOLOGY,
ANATOMY,
HYGIENE,
*Pathology,
Therapeutics,
Materia-Medica,
Surgery,* } } POPULAR.

} *Professional.*

---

2. Of what is — ?   Write Div. of Anthro. on B—d.   3. What is — ?   4. What is — ?   Write Div. of Biol. on B—d.   5. What are — ?   What does Ap. A. *g.* state?   6. What is — ?   7. How is — ?   Write Div. and Subdiv. of Biol. on B—d.

8. PHYSIOLOGY, or *Bio-Dynamics*, treats upon the purposes, uses, functions, actions, properties, results, and relations of the various parts of living things.

9. ANATOMY, or *Bio-Statics*, treats upon the structure, viz., color, size, form, surface, position, and composition of living things.

10. HYGIENE treats upon Health, what will improve and preserve it, what will impair and destroy it; and also upon the means best adapted to produce and preserve a desirable condition for use and appearance in the various parts of living things. It also draws in a few practical facts from its professional co-divisions. Human Hygiene treats also of the mind, so far as its condition is dependent on the body.

11. PHYSIOLOGY, ANATOMY, AND HYGIENE may be treated in either a professional or a popular manner: their co-divisions, treating upon the effects of disease, its cure, the nature and use of medicines, and manual operations, are professional.

12. THE PROFESSIONAL STUDY of Anatomy, Physiology, and Hygiene, includes all their details and modes of expression, requiring years of time in connection with the study of their co-divisions, and, for skilful application in curing disease, much experience and disciplined observation.

13. THE POPULAR STUDY of Physiology, Anatomy, and Hygiene, includes only so much of them as is of use to all classes of persons: the whole of Hygiene, much of Physiology, and less of Anatomy. Few details are required; few technical terms, and not one technical expression (that so much abound in the professional study); very little of the mere mechanical, very little of drudgery: whatever is curious, whatever is interesting, whatever is instructive or practical, is appropriate. It is brief, and at once matures its fruit for use in every-day life.

---

8. — treats upon what? 9. — treats upon what? 10. Upon what does —? 11. How may —? 12. What does — include? 13. What does — include? What is appropriate to the study of Hygiene?

14. THE MODES OF TREATING Human Physiology, Anatomy, and Hygiene are numerous; indeed, there hardly seems to be any method, or approach to unanimity of treatment, as in other sciences; but they may all be classed as the *disjunctive* and the *systematic,* which will correspond to professional and popular use.

METHODS . . . . . . . . $\left\{\begin{array}{l}\text{Disjunctive.}\\\text{Systematic.}\end{array}\right.$

15. BY THE DISJUNCTIVE OR DISCRETE METHOD, Human Anatomy, Physiology, and Hygiene are treated upon separately, and their divisions even independently; it is not therefore adapted to popular, but to professional use; if not the better for that use, there are at least reasons why it is adopted, and may be with propriety, especially if the popular precede the professional study, as ought always to be the case.

16. BY THE SYSTEMATIC OR CONCRETE METHOD, Human Physiology, Anatomy, and Hygiene are treated as the topics of a subject so intimately and systematically related throughout its divisions and subdivisions, that, for popular use, neither can, independently, be fully and practically discussed.

By this method, as far as possible, the structure or Anatomy of parts is treated as secondary to their relations and uses, or Physiology, on account of which the peculiarities of their structure are required, and the best condition for which, it is the province of Hygiene to point out.

By this method the body is treated as a system of sub-systems, well denominated *The Human System*—a whole composed of dependent and sub-dependent parts, but all working together harmoniously, their proper and only correct classifications being thus naturally indicated by their uses and relations.

17. *Inf.*—USES AND RELATIONS are, therefore, the

keynote to the *systematic method*, and to the interest, brevity, practical character, and completeness that distinguish it.

18. ANALYSIS, from two Greek words, signifies the separation of a compound into its constituent elements.

19. By ANALYSIS, ACCORDING TO THE SYSTEMATIC METHOD, the *Human System* is resolved into *Members, Groups, Apparatus, Organs, Tissues* and *Fluids, Anatomical Elements* and *Chemical Elements*, each a subdivision of that which precedes it, the whole exhibiting a perfect hierarchy of parts.

Body $=$ Members $=$ Groups $=$ Apparatus $=$ Organs $=$ Fluids & Tissues $=$ Anat. Elements $=$ Chem. Elements.

20. SYNTHESIS, from two Greek words, is the opposite of Analysis, and signifies the uniting together of elements to form a compound.

21. By SYNTHESIS, ACCORDING TO THE SYSTEMATIC METHOD, from the thirteen essential, fundamental, *Chemical Elements*, through the successive steps of constructing *Anatomical Elements, Fluids* and *Tissues, Organs, Apparatus, Groups,* and *Members*, the *Human System* is built up a complete unit, all its parts working systematically together for a common purpose, the development of the mind. (See page 399.)

Chem. Elements ; Anat. Elements ; Fluids & Tissues ; Organs ; Apparatus ; Groups ; Members ; Body.

SYSTEMATIC METHOD . . . . . . $\begin{cases} \text{Analytic.} \\ \text{Synthetic.} \end{cases}$

22. *Inf.*—SYSTEMATIC ANALYSIS AND SYNTHESIS thus triumphantly prove that man is composed of mind and body, the latter secondary—the servant, yet an instrument essential to the development and activity of the mind, and also the active channel through which the mind exerts influences upon, and receives them from, the great world external to the body.

$$\text{MAN} \dots \dots \dots \dots \begin{cases} \text{MIND.} \\ \text{BODY.} \end{cases}$$

**23.** *Inf.*—THE WELFARE OF MAN requires a knowledge of, and the training of, the entire MAN—not of the mind alone, not of the body alone; but the education of the former, and the exercise of the latter.

$$\text{TRAINED MAN} \dots \dots \dots \begin{cases} \text{Educated Mind.} \\ \text{Exercised Body.} \end{cases}$$

**24.** *Inf.*—THE ENTIRE WELFARE OF MAN also requires the proper arrangement of the external world, that it may act favorably upon the body, and through it upon the mind. (See Ap. D.)

$$\text{WELFARE OF MAN} \dots \dots \dots \begin{cases} \text{Mind Educated.} \\ \text{Body Exercised.} \\ \text{World Arranged.} \end{cases}$$

**25.** *Inf.*—SYSTEMATIC HUMAN PHYSIOLOGY, ANATOMY, AND HYGIENE, presented *Analytically* and *Synthetically*, is, according to the meaning given of those words, the appropriate title to a work that proposes, with distinctness, brevity, and completeness, to treat upon whatever in the Human System is curious, interesting, or practical, for popular use.

| Analytic, Synthetic, | Systematic, Popular, | Phys. Anat. Hyg. | Human | Psychology, Biology, | ANTHROPOLOGY. |
| --- | --- | --- | --- | --- | --- |
| | Disjunctive, Professional, | | | | |

(It is not necessary to introduce in the title the word popular, as the idea is expressed in systematic; neither is it necessary to refer in it to the mind, nor to the External World, since they are only treated incidentally, as intimately related to the body, as beneficent or injurious influences, or, as exhibiting the rationale of its structure and action.

Neither is it necessary in the title to express the ideas of Induction, Deduction, or Inference, as Analysis and Synthesis essentially include those ideas, especially if the Analysis and Synthesis is developed by a natural progress from one division to another, which is a point of prime importance.) (See Ap. F.)

---

23. What does — require? What has a trained man? Write table on B—d. 24. What does — require? Read Ap. D. Write table of Welfare. 25. What is —? Write synthetic table on B—d. Read Ap. F.

# CHAPTER II.

## Systematic Analysis of the Body.

### *Members.*

26. The Mind and Body are united in a manner unknown. It is one of the mysteries of science, not as yet to the slightest degree fathomed. As how the mind acts or is acted upon is not known, no reason can be given why it must be associated with any peculiar structure, nor why the peculiar structure is adapted to its purpose.

27. The Mind is not associated directly with all parts of the body; for large portions of it may be diseased or removed without immediately, if at all, affecting the powers of the mind.

28. The Seat of the Mind must be connected with all parts of the body, because the influences of the mind are exerted upon all parts.

29. The Brain or the Heart must be the seat of the Mind, since they are the only parts of the body connected with all the rest—the heart by means of the blood-tubes, and the brain by means of the nerves. But the heart and blood-tubes have one office to perform, the · circulation of the blood; and they are also under the influence of the brain and nerves.

Fig. 43, on the following page, is a plan of the Brain, Spinal Cord, and Nerves, supposing the body to be transparent and no parts visible, except the nervous system, which is not in a plane, but extends backward and forward, terminating in the skin as well as other parts throughout the body. Though the branches appear to be numerous, not one in a thousand of

26. How are —? 27. With what is —? 28. With what must —? 29. Why

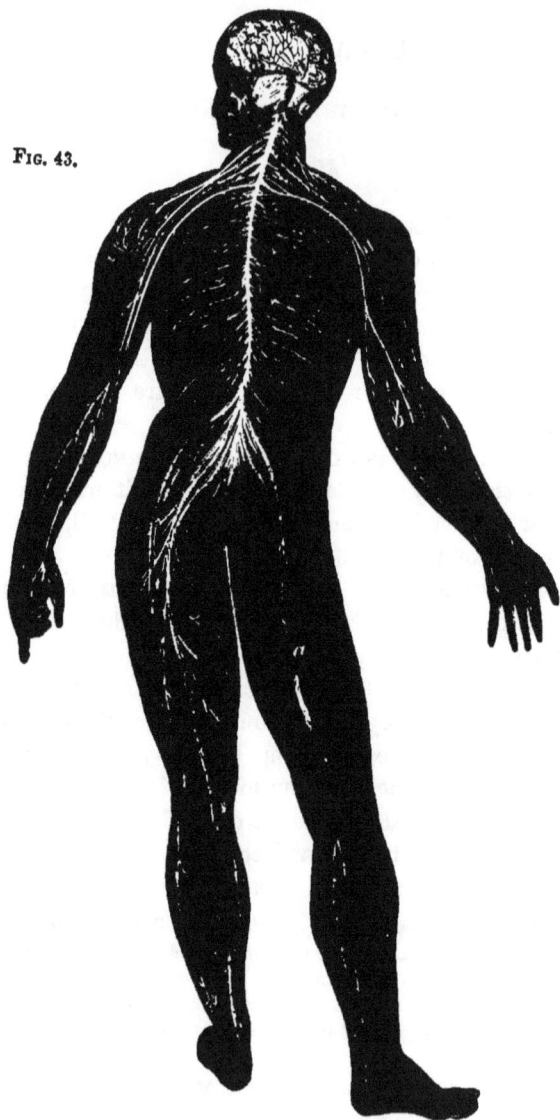

Fig. 43.

the minute nerves can be shown. The same is the case with the blood-
tubes of Fig. 44. There is not a portion of the body as large as a pin's
head that does not have both a nerve and a blood-tube commencing in it.

---

Describe Fig. 43. Where do the branches of nerve *a* seem to commence?
Where the branches of nerve *b* ?

Fig. 44 illustrates one class of blood-tubes, called arteries, commencing at H, the heart, and dividing and subdividing, according to the members, until all parts of the body are reached, where they terminate in the next class. The capillaries are a network of hair-like tubes, too small to be seen by the naked eye, into which the arteries pour their contents. They are more numerous in the brain than in any other part. From the capillaries the veins commence (see Pl. 5), and uniting together, and also receiving the contents of the fourth class of tubes, they at last open into another part of the heart, from whence the arteries lead out. The fourth class of tubes, called lymphatics (see Pl. 5*), commences in every part *except the brain and nerves,* and open into the veins, and thus their contents find their way to the heart.

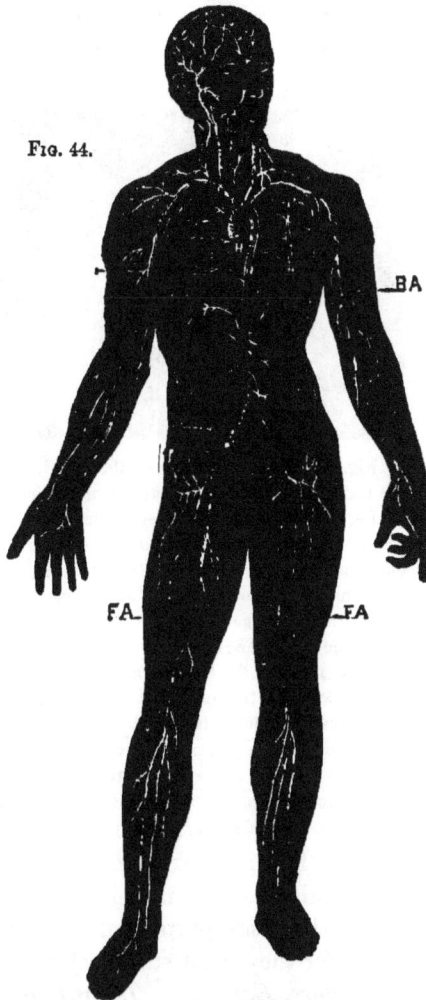

Fig. 44.

BA

FA          FA

FIRST MEMBER OF THE BODY.

30. THE DEVELOPMENT OF THE MIND, therefore, requires the use of that beautiful collection of similar parts, called the Brain, Brains, or Mental Ganglia, in the hidden recesses, and by means of which, in the most

Describe Fig. 44. What member of the Body is now to be treated upon? 30. What does — require?

mysterious manner, the Mind receives impressions, en-
joys emotions, meditates, and wills.    (Ap. G.)

FIG. 45.                    FIG. 46.

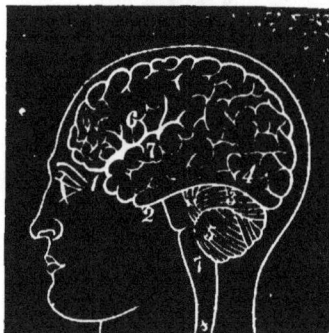
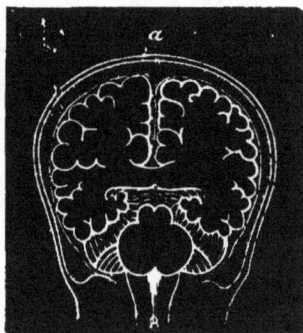

Fig. 45 shows the surface of the left Brain, 1 to 6, in the cranium, with
the facium and three of its organs of observation added.

Fig. 46 represents a perpendicular section of the Head, just back of
the line of the ears.  The external line represents the skin, the second
one and the dots represent the cranium, the third one the lining of the
cranium, at *a* following down the fissure between the right and left cerebrum, and back again, the two becoming one by adhering together, and
called the fal*x*.  The fourth or irregular line is intended to represent the
uneven surface of the large Brain, each one of the eminences being called a
convolution, and the indentations between them anfractuosities, some of
which are quite deep, as represented, the sides lying against each other.
Let the brain of a pig, or some other creature, be examined to illustrate
the touch, the texture, and general appearance; the convolutions are not
so numerous nor so prominent, nor the anfractuosities so deep, as in the
case of the human Brain, yet they give a good general idea of it.

31.  IT WILL NOT BE DIFFICULT, the intimate relation
of the mind and brain being thus determined, to previse
what other uses, parts, and structures it requires in the
body, for they must stand related to the brain as *Agents,
Protectors, Executors, Telegraphers, Examiners, Report-
ers, Orators, Purveyors, Chemists, Cooks, Calorifiers,
Distributors, Porters, Lacqueys, Scavengers,* &c., in fact
quite an establishment, all subsidiary to the Brain, and
willingly acknowledging its mastery.

32.  THE EXQUISITELY DELICATE STRUCTURE OF THE

---

What is the purport of Ap. G ?   Describe Fig. 45.  Fig. 46.  81.  What will — to
previse ?  32. What does — require ?

BRAIN requires the most considerate protection from blows not merely, but from jars even, and also from the slightest inequalities of temperature.

33. THE GANGLIA OF THE BRAIN find ample protection in the Walls of the Head, in which also are lodged the organs of observation needed by the Mind in exploring the External World, and by which it is warned of dangers yet at a distance—the upper and back parts of the wall being covered by the warmth-preserving hair.

34. THE FRAME OF THE HEAD is divisible into those parts that constitute the lower and front part, called the *Facium*, in which are the Organs of Observation (organs of sense); and the upper and back part, called the *Cranium*, and which encloses the brain.

35. THE ORGANS FOR OBSERVATION may with propriety be classed with the Contents of the Walls, as they are special to the head, their like not being needed or found in any other part of the body.

36. THE SPECIAL ORGANS of the Contents of the Walls are $\left\{ \begin{array}{l} \text{Ganglia.} \\ \text{Ear.} \\ \text{Eye.} \\ \text{Nose.} \\ \text{Mouth.} \end{array} \right.$

37. THE EAR, EYE, NOSE, AND MOUTH MUST BE CONNECTED with the central Ganglia, as they are, by white pulpy cords called Nerves, and all the special organs must be supplied with blood through tubes interwoven throughout. These are called general organs, because they exist in various parts of the body.

38. THE WHOLE CONTENTS of the Walls of the Head are $\left\{ \begin{array}{l} \text{Special} \left\{ \begin{array}{l} \text{Ganglia.} \\ \text{Ear.} \\ \text{Eye.} \\ \text{Nose.} \\ \text{Mouth.} \end{array} \right. \\ \text{General} \left\{ \begin{array}{l} \text{Nerves.} \\ \text{Blood-tubes.} \end{array} \right. \end{array} \right.$

39. THE PROPER WALLS OF THE HEAD ARE COMPOSED of, 1st, a strong *Framework*, called the Skull or Skeleton of the Head;

FIG. 47.

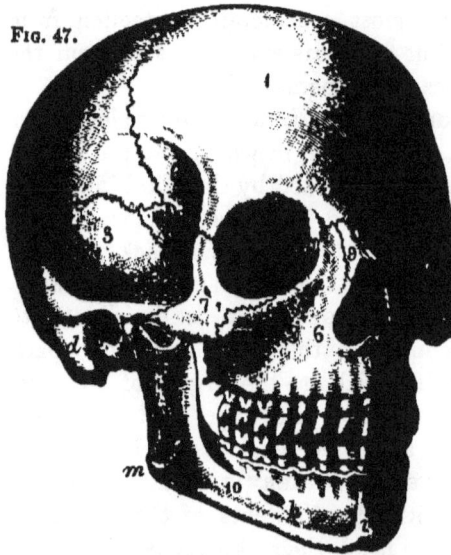

Fig. 47 represents the Skull; 1, 2, 3, 4, is the Cranium, and all the parts below and in front are the Facium, (except such as are seen in the socket of the eye,) including the part marked d, in which the inner ear is lodged: in earliest life it is a distinct bone, and in some animals always remains so.

2d, Means for producing motion, called muscles, (lean meat, as it is commonly called,) having the power of shortening or contracting under proper influences, and which can be felt working under the skin of the face when chewing or other facial motions are made;

FIG. 48.

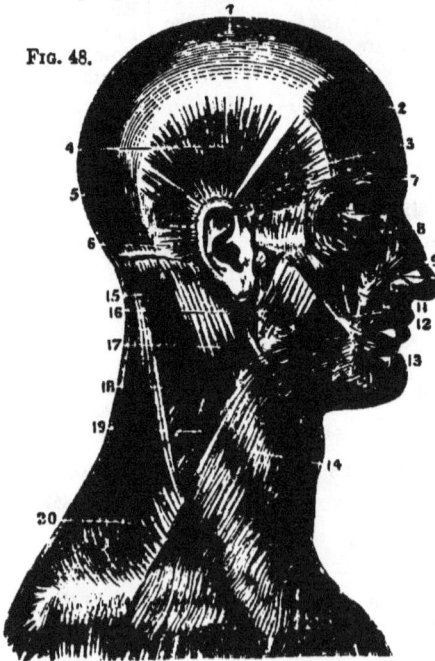

Fig. 48 represents the muscles of the right Head; 2, 5, move the scalp; 3, 4, 5, act on the ear; 7, the circular (orbicularis) muscle that closes the eye; 17, "the neck-cord," draws down the point back of the ear toward the upper end of the breast bone; 16, masseter, used and easily felt in chewing; 8 to 13, muscles that act on the nose and mouth. (For full description, see Synthesis.)

3d, The muscles must be connected with the central Ganglia by nerves;

4th, The whole must be covered with skin, also to be connected with the central Ganglia by nerves; and

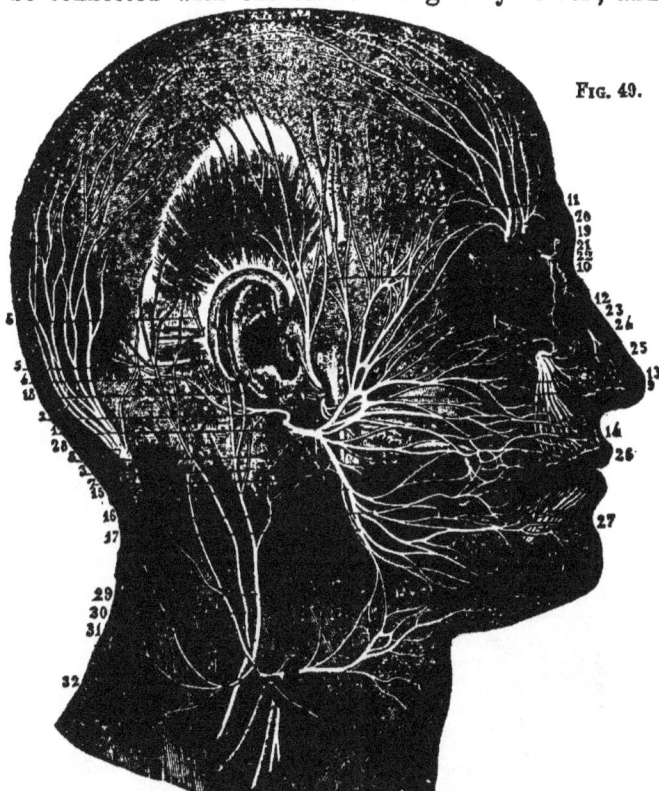

FIG. 49.

Fig. 49 is a beautiful representation of the superficial nerves of the right Head and Neck, which will be particularly interesting to those troubled with neuralgia of the face, since they can appreciate the fact that if a main branch of the nerves is affected, the pain may seem to be at many points, though most of the nerves shown are motory; 1, being the *Facial*, through which the muscles of the face are controlled by the will, and the avenue through which they are acted upon by the emotions and expressions caused. The bunch of nerves, 25, *Infra-orbital*, and 27, *Inferior-dental*, extending to the upper and lower lips, are especially noticeable.

Describe Fig. 47. Of what are the Walls of the Head composed, 2d? 3d? 4th? Describe Fig 48. Describe Fig. 49.

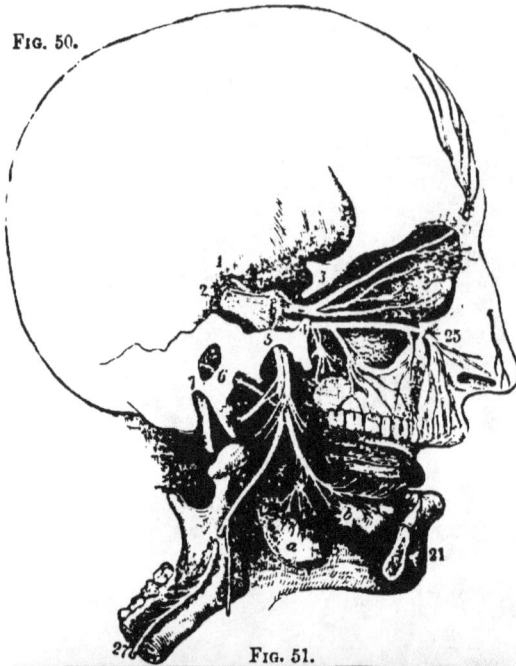

FIG. 50.

Fig. 50 represents a part of the facium removed so that the course of certain nerves, the *Tri - Facial*, may be traced. Branches are represented as commencing in the teeth, tongue, nose, eye - socket, and forehead, and the trunks extending into the Cranium at 1, 2 (25, 27, see Fig. 49).

FIG. 51.

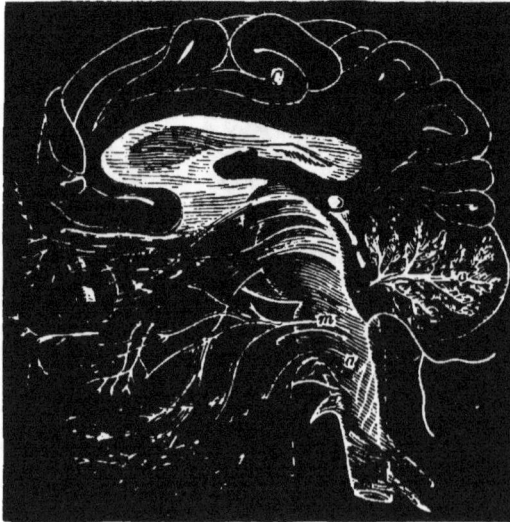

Fig. 51 represents the connection of nerves with the inferior parts of the right Brains, *C, c,* and *medulla oblongata, m, o,* which have been divided from the left ones on the middle line of the Head. S, 1st and 2d spinal nerves.

Describe Fig. 50. Have you compared 25 and 27 with the same numbers in Fig. 49? Describe 51.

5th, All the organs of the Head-Walls must have blood-tubes interwoven through them; or in tabular form,

40.   THE GENERAL ORGANS of the Head-Walls $\left\{\begin{array}{l}\text{Skeleton.}\\\text{Muscles.}\\\text{Nerves.}\\\text{Skin.}\\\text{Blood-tubes.}\end{array}\right.$

FIG. 52.

Fig. 52 represents a section of the skin, the right half removed; superficial Blood - tubes of the Right Head;   the darker network, veins; the beaded net-work the lymphatics, connected below the ear with small organs, called the lymphatic glands.

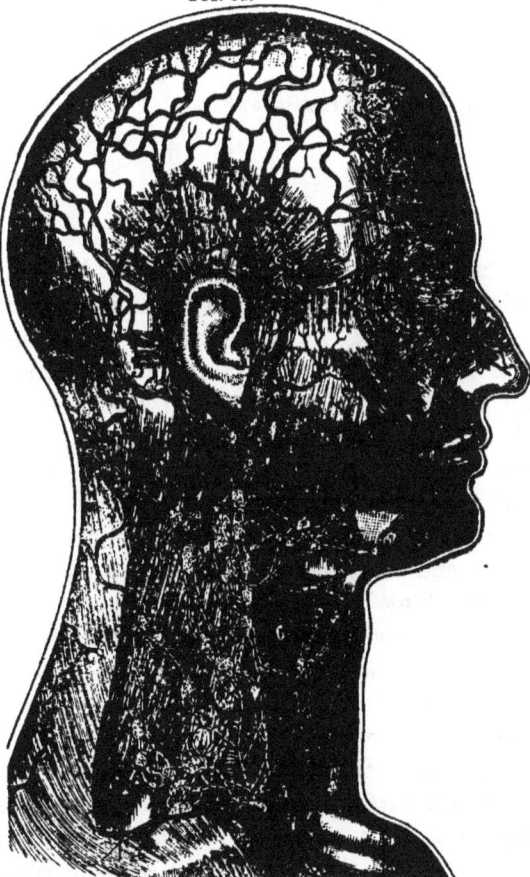

41.   THE ORGANS OF THE WALLS OF THE HEAD are all general, their like being found in various other parts of the body.

---

5th.   What parts of the Head-walls have Blood-tubes ?   Why ?   Describe Fig. 52.   40. What are — ?   41. Where are — found ?

2*

42. UNITING , THE ORGANS OF THE CONTENTS OF THOSE OF THE WALLS shows that the whole Head is composed of but ten different kinds of Organs, five of them special and five general, just equalling the ten digits.

43. ORGANS OF THE HEAD

Special, (*right hand begin with thumb.*)
- Ganglia.
- Ear.
- Eye.
- Nose.
- Mouth.

General, (*left hand close with thumb.*)
- Skeleton.
- Muscles.
- Nerves.
- Skin.
- Blood-tubes.

44. THE HEAD, composed of Walls and their Contents, may well be called the capitol of the body, since the mind is there enthroned in the midst of its legislative halls, its courts, its audience-rooms, and its executive chambers, fortified as in a citadel, with its picket-posts for observation on the very outworks. It is also truly the *Head-member* of the body, containing within itself the hints of what is required of all the rest. The Ear and Eye suggest the desirableness of information, the nose speaks of the importance of air, the mouth argues for water and food, the hair illustrates the need of clothing.

45. THE USES OF THE HEAD, not its position nor its structure, give it preëminence; indeed, its uses give it its position and structure.

46. LOCOMOTION OF THE HEAD is the next thing essential to the performance of its uses, for it must be carried up to the mountain-tops and down into the bowels of the earth, that the mind may search the truths of nature and be nourished; it must also be moved from place to place, that it may be itself nourished; it must also be moved without jar or shock, and be able to turn its face upward to the sky, downward to the earth, or around toward any part of the horizon.

42. What does — prove ? 43. Explain table of —. 44. What may the Head be called ? 45. What do — give ? 46. Why is — necessary ?

47. THREE MEMBERS of the Body are necessary to allow all that the Head requires in regard to motion.

## SECOND MEMBER OF THE BODY.

48. THE NECK PERFORMS very similar offices in reference to the head as the trunk does in the same regard; but in some of its offices the trunk is so different from the neck, which is no part of the head, that it is best to consider the neck as a distinct member.

49. THE USES OF THE NECK are, to elevate the head above the shoulders and allow it considerable latitude of motion, to assist in carrying it without jars, to move it vertically, as in affirmative nodding, to rotate it, as in negation, and to move it from side to side.

50. THE NECK MUST BE COMPOSED of a framework or skeleton of several pieces movable upon each other; of muscles that by contraction move the head; of nerves; of skin, and of blood-tubes interwoven throughout; being the same kind of organs as the general organs of the head, only differing in use, form, and position.

THE NECK . . . . . . . . . . ⎰ Skeleton.
⎱ Muscles.
⎱ Nerves.
⎱ Skin.
⎱ Blood-tubes.

## THIRD MEMBER OF THE BODY.

51. THE TRUNK (it should never be called the body, of which it is only a part) has a double relation to the Head, that of supporting it and that of supplying it with blood.

52. THE TRUNK IS DIVISIBLE, like the Head, into two classes of parts, the walls and their contents, corresponding to the two classes of ideas or double relations to be found in it.   (See Pl. 7, Fig. 2.)

---

47. For what are — necessary?  48. What does —?  49. What are —?  50. How —?  Write table of—?  51. What has —?  52. How —?  Describe Fig. 2, Pl. 7.

FIG. 54.　　　　　　　　　　　FIG. 55.

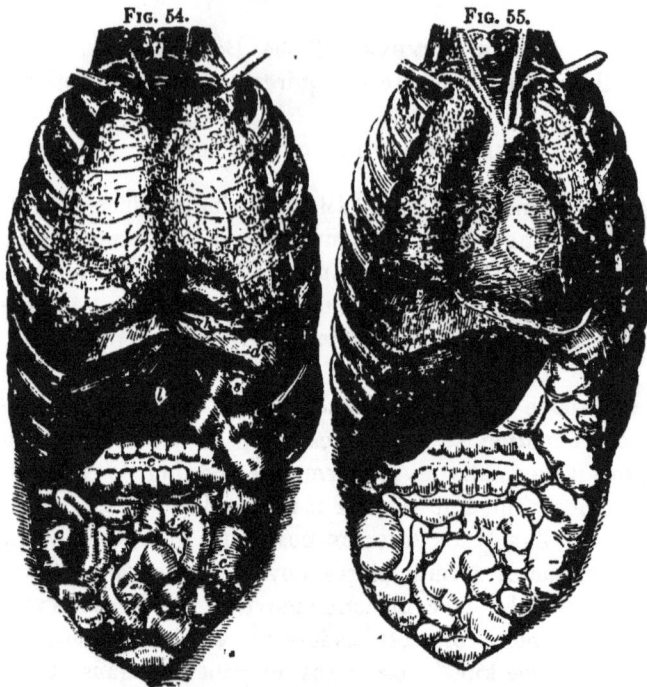

Fig. 54 represents the front part of the Trunk-walls removed, exposing their contents. Above in the neck, T, the Thyroid gland, is seen, on each side of which an artery stretches up from the Heart below, hidden except a small part, H, by the Lungs, L L, on each side, upon which the outline of the removed ribs is shown by the dotted lines. D, Diaphragm, arching quite across the body, as shown in Fig. 60. L, Liver, S, Stomach, nearly empty, c c c, colon, I I I. small Intestines. The ribs are cut farther back above than below the diaphragm.

Fig. 55 is like Fig. 54, except that the ribs of the chest are cut still farther back, and the lungs made to recede and expose the Heart and large vessels more to view.

These beautiful cuts are taken from Bennett's magnificent work on practical medicine, being copied by him from the excellent work of Sibson. They truly represent the position of the organs.

53. THE SUPPORT AND MOVEMENTS OF THE HEAD require that the Trunk-walls shall be composed of four classes of parts, one fixed or permanent, upon which the others can move, one an elastic support; the third must consist

Describe Figs. 54, 55. Which part of the Figs. is called the chest? What organs are shown in it? 53. What do — require?

of levers to assist in producing motions more easily, and the fourth of the muscles that act on the levers.

54.   THE TRUNK-WALLS ARE DIVISIBLE into the Pelvic, the Columnar, the costal (from *costa*, rib), and the abdominal.

$$
\text{TRUNK-WALLS} \ldots \ldots \ldots \left\{ \begin{array}{l} \text{Pelvic.} \\ \text{Columnar.} \\ \text{Costal.} \\ \text{Abdominal.} \end{array} \right.
$$

55.   THE PELVIC PORTION OF THE TRUNK-WALLS HAS FOR ITS FRAMEWORK a very notable broad ring called the Pelvis, composed of several pieces so bound together that they are immovable upon each other and are as a unit.

56.   THE PELVIS is that part of the framework of the Body that is fixed in relation to all the rest, and about which all the rest move; beneath it the lower extremities walk, and above it the back is flexible; it is the dead-point in the motions of the body, and in its want of motion is only comparable to the cranial part of the head.

FIG. 56.

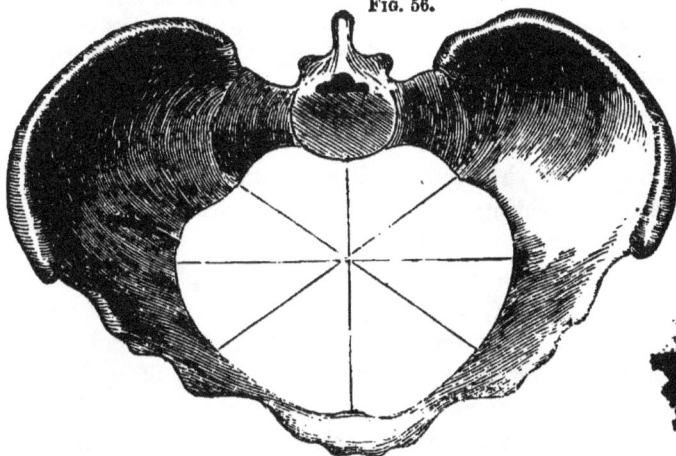

Fig. 56 represents the Pelvic Ring seen from above, the upper margin of which is several inches above the lower.

57.   THE PELVIC RING or bowl is not placed horizon-

---

54. How —?   Write table.   55. What has —?   56. What is —?   Describe Fig. 56.   57. How is — placed?

tally nor vertically, but at an angle to the direction of the spinal column and of the lower extremities; and from the irregularities of its form, has a favorable influence in preventing jars from passing from the lower extremities to the spinal column. (See Fig. 58.)

FIG. 57.

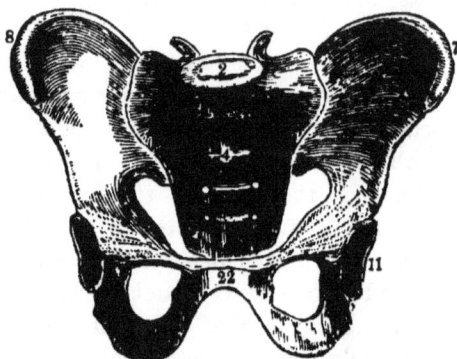

Fig. 57 represents a front view of the Pelvic Bowl, more vertical than when in its natural position. 1, 2, the upper surface of the sacrum, on which the spinal column rests; 4, body of the sacrum; 22, point of union between right and left pelvis; 7, 8, side ridge of hip-bones; 11, 11, sockets for thigh bones.

58. THE COLUMNAR PORTION OF THE TRUNK-WALLS HAS FOR ITS FRAMEWORK an exceedingly flexible, elastic column, called the spinal column, from the number of spines or projecting points on its back parts, that serve as short levers in controlling its motions.

59. THE SPINAL COLUMN is a series of springs, twenty-four in number, including those of the neck, of an elastic gristle or cartilage, separated from each other by bones called vertebræ, the size and thickness of which, like those of the springs or cushions between them, differ in different parts of the column. They have short arms extending out on each side of the back of their bodies or main part, with branches up and down to form joints with those above and below, and unite together

Describe Fig. 57. 58. What has —? What is the most conspicuous property of the gristle of the ear or nose? 59. What is —?

FIG. 60.  FIG. 61.  FIG. 59.

FIG. 58.

FIG. 62.

FIG. 63.

FIG. 64.

FIG. 65.

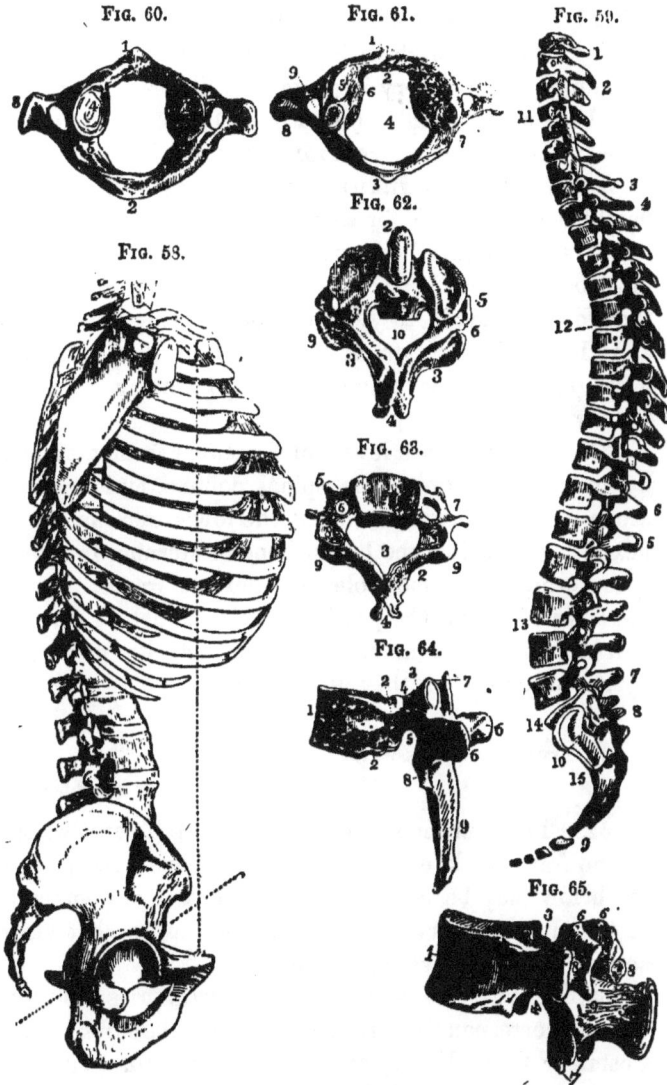

Fig. 58, framework of trunk. Fig. 59, spinal column; 1, atlas (Fig. 61, upper, Fig. 60, lower view of same); 2, dentatus (Fig. 62, upper view); 2 to 3, other cervical vertebræ (Fig. 63, upper view of); 4 to 6, dorsal vertebræ (Fig. 64, side view of one); 5 to 7, lumbar vertebræ (Fig. 65, side view of one); 8 to 9, sacrum; below 9, coxcyx.

behind to form the spinous processes or spines, thus leaving a hole through each and throughout the spinal column, forming a very important, well-protected canal, called the spinal canal. (See Figs. 60 to 63.)

60. The Costal portion of the Trunk-walls must have for its framework a strong yet slightly flexible set of levers; for though the springs or intervertebral substances of the spinal column will yield, they must be quite firm in order to support their burdens, and therefore not made to yield without considerable force; and the head is so much elevated that it acts with a great purchase or leverage, and long levers are required to stay it.

61. The Ribs are levers of a peculiar form, attached by movable joints to the spinal column, and curving around and downward: all but the lower ones are very flexibly attached to the breast-bone or sternum in front, thus forming the movable and flexible framework of the sides of a box, the passive *elastic* character of which will be found of as much use as the activity conferred by its muscles when they are added.

FRAMEWORK OF TRUNK-WALLS . . {
Pelvis.
Spinal Column.
Ribs.
Sternum, or breast-bone.

62. The Abdominal portion of the Trunk-walls has no framework, except as the ribs above and the pelvis below may be considered as such, but is composed of the muscles, nerves, skin, and blood-tubes, that close across from the pelvis up to the ribs.

63. Three Cavities exist within the Trunk-walls, corresponding to the Pelvic, Abdominal, and Costal or Chest-Walls, and similarly named. The former two are not divided internally, both being usually called as one, the abdominal; the two latter are parted by what is called the Diaphragm.

64. The Diaphragm is a thin, active partition, chiefly muscular, attached by its margin to the lower border of the chest, and arching upward, as seen by the figure of it, Fig. 1, Plate 7, also Fig. 66.

Fig. 66.

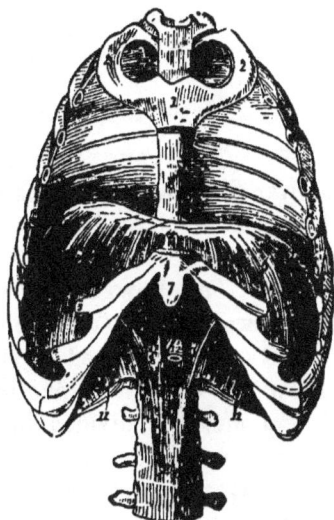

' Fig. 66 represents front walls of chest removed, 2 being upper ribs attached to 1, the upper portion of sternum, of which 7 is the lower portion, to each side of which the gristly ends of the lower ribs are attached. 3, the spinal column. 4, the centre, and 5, 6, the side arches of the Diaphragm.

65. The Movements of the Trunk-walls next require muscles, which must be connected with the brain (so, indeed, all parts of the Trunk-walls must be) by means of nerves, and then the whole must be covered by the skin, and interwoven by the blood-tubes, making up the five general organs, similar, except in size and form, to those of the head and neck.

TRUNK-WALLS {
Skeleton,
Muscles.
Nerves.
Skin.
Blood-tubes.
}
TRUNK-CAVITIES {
Pelvic.
Abdominal.
Costal.
}

66. The Blood-making or Sanguificatory Organs, forming the Contents of the Trunk-walls, require exactly the space afforded by the cavities of the Trunk-walls, and exactly the same structure that they have, so that, if they had been made for the purpose of enclosing, protecting, carrying, and assisting the Contents, nothing could have been added, removed, nor changed. The

elasticity of the spinal column, on which the contents are hung, the mobility of the ribs, and the pressure of the abdominal walls, are remarkable illustrations of the double utility of the Trunk-walls.

### FOURTH MEMBER OF THE BODY.

67. THE TRUE MEMBERS OF LOCOMOTION are the Lower Extremities, composed of two levers, having their lower ends enlarged to furnish a better support on the ground.

68. THE LOWER EXTREMITIES MUST BE FURNISHED with three joints : (1st) the upper ones, by which the extremities are attached.to the pelvis, must be rotary, to allow the levers to move in every direction; (2d) the levers must be divided in the middle by means of a hinge-joint backward, to allow the lever to be readily raised over an obstacle, or progression to be made up or down an incline, or upon steps; (3d) near the lower extremity of the lever it must be divided by another joint of considerable mobility, to allow the broader part to rest securely on every kind of surface.

69. THE LOWER EXTREMITIES ARE COMPOSED of three parts, the upper and lower leg, and the foot; and are furnished with three main joints, the hip, the knee, and the ankle.

70. THE SIMPLE USE OF THE LOWER EXTREMITIES is to cause motion, and each of their divisions is composed of a Framework or Skeleton, Muscles, Nerves, Skin, and Blood-tubes.

LOWER EXTREMITIES . . . . . . . . {
Skeleton.
Muscles.
Nerves.
Skin.
Blood-tubes.

71. BY THE FLEXIONS OF THE LOWER EXTREMITIES, very much of the jar which in walking acts upon the

---

67. What are —?  68. With what must.—?  69. Of how many parts are —?
70. What is —?  Write table of —  71. What is gained —?

foot is prevented from acting on the head. When unexpectedly a person steps down upon the ground with the limbs straight, a shock is felt that a bent or flexed state of the limbs would have prevented. The irregular form of the pelvis and the hip-joint, not being directly under the spinal column, nor in its plane, as well as the curvatures of the spinal column, assist in the same result of preserving the brain from receiving jars.

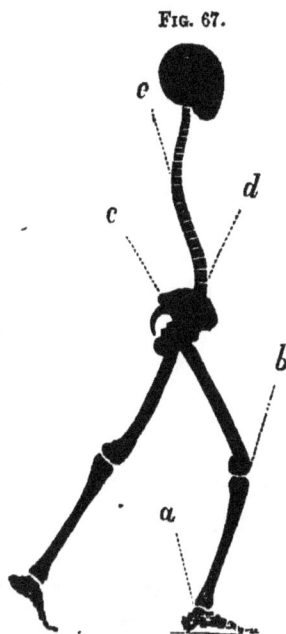

FIG. 67.

72. THE CENTRAL PORTIONS OF THE BONES ARE CONSTRUCTED either as canals, or as cells, like those of a sponge, filled with marrow, which serves to make the framework lighter, as a stock of food in time of need, and especially to deaden the jars that are passing upward to assault the brain, and annoy it.

Fig. 67 is a plan to represent how the force with which the foot strikes the ground is sent off in the directions $a$, $b$, $c$, $d$, $e$, so that but little reaches the head. Those are but a few of the points of dispersion.

73. *Illus.*—If in sickness, or from exposure to starvation, the marrow of the bones is exhausted, each step, unless taken with great care, jars the brain very unpleasantly.

74. UNDER THE BRAIN, and between it and the skull, there is a sponge-like cushion, the spaces of which are filled with a watery fluid; while a portion and continuation of the same extends down around the spinal cord, or large bundle of nerves that nearly but not

Describe Fig. 67. 72. How are —? 73. What Illus. to show value of marrow? 74. What is —?

quite fills the canal throughout the spinal column.

75. WHEN THE SPINAL COLUMN IS FLEXED, the central canal is lessened, and a part of the fluid within is instantly pressed up into the cushion beneath the brain, holding it firmly, as in a vice, but with the gentleness of an infant.

76. THE PROTECTIONS OF THE BRAIN from jars, found in the three members, the neck, trunk, and lower extremities, are: (1st) the elastic springs, cushions, gristles or cartilages of the spinal column; (2d) the fluid of the spinal canal; (3d) the marrow in the bones; (4th) the gristle of the joints; (5th) the curvatures of the column, pelvis, and the bones of the extremities; and (6th) the bending of the limbs at the joints.

FIG. 68.

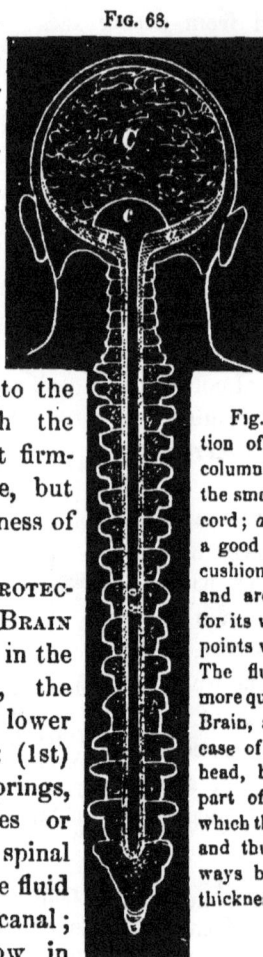

Fig. 68 is a plan of a section of the Head and spinal column. $C$, the large brain, $c$, the small brain, and $s c$, spinal cord; $a a$ is intended to convey a good idea of the fluid-filled cushion beneath the Brain and around the spinal cord for its whole length, except at points where nerves leave it. The fluid will always move more quickly than the less fluid Brain, and of course will, in case of rapid motion of the head, be impelled into that part of the cushion against which the brain will be thrown, and thus the cushion will always be thickest where the thickness is most needed.

FIFTH MEMBER OF THE BODY.

77. THE MIND NEXT REQUIRES means for handling the various objects of its regard. These means are furnished in the upper extremities.

78. THE UPPER EXTREMITIES are levers of a somewhat more complex structure than the lower extremities,

---



75. What effect —? Describe Fig. 68.  76. What are - ?  77. What does — ? 79. What are — ?

as a greater variety of motions and more extended ones are required; but the general character of the two must be similar.

79. The Upper Extremities are composed of the shoulders, the upper and lower arms, and the hands.

80. The Shoulders are not a part of the trunk, but are movable upon the upper part of it, in order to bestow more extended motion upon the arms and hand, that must serve the mind by great latitude of motion at times. They are hinged through the inner end of the collar-bone to the upper end of the breast-bone, which is the centre of a partial circumference, through which the shoulder can move.

81. The Upper Arm must, like the upper leg, be fastened above by a rotary joint, so as to have motion in every direction.

82. The Lower Arm, or the forearm, must be fastened to the upper by a hinge-joint, allowing upward or forward motion, just the reverse of that of the lower leg.

83. *Inf.*—The similarity of the upper and lower extremities was not accidental, nor a necessity, but is owing to the similarity of purpose; for in any case of dissimilarity of purpose there is a corresponding dissimilarity of structure in them.

84. The entire length of the Forearm must be a joint, adapted to allowing a rotary motion of the hand, which could not be gained by any construction of the wrist-joint; but by simply rolling the lower end of one bone of the lower arm over the other, the hand being attached to the former is carried with it and made prone.

85. The Hand is attached to one of the bones of the lower arm by a compound joint, allowing less motion than the rotary, and not restricting it as much as the hinge.

86. All the Divisions of the Upper Extremities,

---

79. Of what are - ?   80. What are — ?   81. What must shoulder joint be ?   82. How must the — be fastened ?   83 What is Inf. in regard to — ?   84. Why must — ?

being intended merely for motion, are composed of Skeleton, Muscles, Nerves, Skin, and Blood-tubes, similar to the general organs of the other members.

UPPER EXTREMITIES . . . . . $\left\{\begin{array}{l}\text{Skeleton.}\\\text{Muscles.}\\\text{Nerves.}\\\text{Skin.}\\\text{Blood-tubes.}\end{array}\right.$

### SIXTH MEMBER OF THE BODY.

87. IT WILL BE CONVENIENT TO THE MIND, besides the means thus far described, to have a special way of communicating its thoughts to others, and of seeking information that it may use while all the other parts are busy.

88. THE LARYNX is the sixth member of the body. It is the upper portion of the windpipe, often called Adam's Apple; is the seat of the voice, that is, where the full tones are produced. It is frequently called the vocal organs.

89. IT WOULD SEEM TO MANY, that the Larynx is a part of the neck, and so it might be termed; but its office is very distinct from that of any other part of the neck, and though anatomically small, and not by a cursory observation to be distinctly honored as a member, Physiology at once recognizes its eminent merits, and exalts it to its true position, while Hygiene confirms and Mental Philosophy applauds the decision.

90. The LARYNX MAINTAINS ITS DIGNITY as a member by requiring for its composition, Skeleton, Muscles, Nerves, Skin, and Blood-tubes.

LARYNX . . . . . . . . . . . $\left\{\begin{array}{l}\text{Skeleton.}\\\text{Muscles.}\\\text{Nerves.}\\\text{Skin.}\\\text{Blood-tubes.}\end{array}\right.$

91. THE FORM AND CONDITION OF THE STRUCTURE OF THE BODY are thus, by the uses required of them, shown to be not an accident, but the result of utilities;

---

85. How is — ? 86. How are — composed? Write table of —. 87. What will —? 88. What is — ? 89. What —? 90. How does — ? Write table of —. 91. Of what are

and the more the uses of any animal's body shall approach those of man, the more nearly will the form and structure approach the Human.

92.   THE BODY IS DIVISIBLE into, or composed of, six members, because the six, and no more, are necessary for the perfect development of the mind.

BODY . . . . . . . . . . $\left\{ \begin{array}{l} \text{H, HEAD.} \\ \text{N, Neck.} \\ \text{T, Trunk.} \\ \text{lx, lower extremities.} \\ \text{ux, upper extremities.} \\ \text{l, larynx.} \end{array} \right.$

93.   THREE OF THE MEMBERS are essential to life; hence in the table their names are printed with capital, the other three with lower-case, initials.

94.   THE SIX MEMBERS OF THE BODY are equally indicated by a superficial view of it, or by a synthetic construction of the five general Organs and the Contents of the Head and Trunk-walls.

95.   THE SKELETON is peculiarly and distinctly divisible into that of the Head, the Neck, the Trunk, the lower extremities, the upper extremities, and the larynx; though the latter is not shown in connection with the skeleton of the other members, since it is not jointed to any part of them. If shown in the neck, it must be disconnected; therefore it may better be shown apart. (See Plate 1, Fig. 1.)

96.   THE MUSCLES ALSO MAY BE CLASSED as those of the Head, the Neck, the Trunk, the lower extremities, the upper extremities, and the larynx, according to the motions they are adapted to produce. (See Plate 2, Fig. 1.)

97.   THE NERVES cannot be classed as distinctly as the Skeleton and Muscles according to the six members, since most of the nerves are connective between the brain and all other parts of the body, and those connecting the more distant parts must extend through the

---

— the result? 92. Into what is — ? Write table of —. 93. Why are three members in different letters? 94. What indicates — ? 95. How is — divisible? Describe Fig.

nearer parts, and in their course, as they approach, they unite in the same sheath, forming larger bundles, in which the fibres are so small as not to be distinguishable; and it is impossible to determine the origin of each when thus united. (See Plate 3, Fig. 1.)

98. THE NERVES ARE OF DIFFERENT KINDS and for different uses, and therefore a better classification than that by members can usually be made.

99. FOR SOME PURPOSES, TO CLASS THE NERVES by the members in which their outer ends are found is very advantageous, as it would always be if their course was distinct from each member to the brain.

100. THE SKIN MAY BE CLASSED, as it usually is, by members, with great propriety, though it is merely a covering in its relations to the parts beneath; as to the mind, any of its parts may be named by anything that will distinguish them. (See Fig. 4, Plate 7.)

101. THE BLOOD-TUBES, being, like the nerves, of a connective character, cannot very distinctly be classed by the members, yet to an extent this is always done, and with advantage; and the branching of the blood-tubes indicates distinctly the six members. (See Plate 5.)

102. THE CONTENTS OF THE HEAD AND TRUNK-WALLS will be respectively located in the Head and Trunk.

103. THE USES BY ANALYSIS, the structure by Synthesis, and common observation, except as to the larynx, would correctly divide the Body into its six members.

104. THE USES OF THE MEMBERS EXHIBIT the necessity for, and the structure of them exhibits the fact of, a remarkable similarity and simplicity of structure in the larger parts of the body; four entire members and a considerable portion of the other two, differing only in the size, form, and position of *only five* different kinds of organs, their substantial and vital characteristics being

the same in all the members; they therefore require the same treatment in all the members. The following table exhibits this important practical idea more clearly:

$$
\text{BODY} \ldots = \left\{\begin{array}{l}
\text{HEAD-walls,} \\
\text{Neck,} \\
\text{Trunk-walls,} \\
\text{lower extremities,} \\
\text{upper extremities,} \\
\text{larynx,} \\
\text{CONTENTS of Head-walls.} \\
\text{Contents of Trunk-walls.}
\end{array}\right\} = \left\{\begin{array}{l}
\text{Skeleton.} \\
\text{Muscles.} \\
\text{Nerves.} \\
\text{Skin.} \\
\text{Blood-tubes.}
\end{array}\right.
$$

105. *Inf.*—To LEARN THE ESSENTIAL CHARACTERISTICS OF ONLY FIVE KINDS OF ORGANS is to become acquainted with, and understand how to provide for the welfare of, much the larger part of the body.*

---

* The Teacher can now present the sheets exhibiting the Tables of this chapter, if he have them, or cause the pupils to produce them, and also suspend Anatomical Plates, outline diagrams, or whatever will be illustrative, bearing in mind that some truths become impressed by the constant action of their illustrations; and on the other hand, that the minds of youth are always awakened by something new. Therefore arrange and rearrange the old, and keep always something in reserve, to be brought out little by little at each exercise, especially if it is a review. It is a lamentable fact that pupils often study through a book on Hygiene, making good recitations at each lesson, yet in a short time know very little of the subjects they have studied. This is for want of a linked chain running through the whole, and binding the parts together. If this chain is fixed in the student's mind, it is not necessary for him to try to remember the details, as they will suggest themselves in connection with the appropriate link of the chain whenever he reviews the subject. To make ideas practical, they must, so to speak. become incorporated in the mind, which does not occur until they have been again and again presented before it in various aspects; after they have once been thus fairly introduced, they never cease to exert their leavening influences. It is of so great importance that a pupil should be familiar with the ideas of the correct classification and relations of parts, that it is hardly possible for him to analyze and synthesize the divisions over which he passes too many times. The object is to have him, when through the work, thoroughly understand the subject, not only what exists, but the rationale of it as far as possible; and where that cannot be, to know precisely where the line of clear knowledge is, and where he must begin to search for that yet undeveloped. Let the Teacher make sure, therefore, that every pupil in the class understands not only that there are six members, but why there can be neither more nor less than six. Let him also be assured, that each pupil understands what parts are composed of the five general organs, and why there can be neither more nor less than five. The relative location of the five general organs, which are at the surface, which the deepest, and which in the midst of the others, should be made so familiar to the student by varied questions upon the structure of members, that the Body shall be before his mind's eye like a transparent structure.

99. Can nerves ever be classed by members? 100. How may —? 101. How may Blood-tubes be classed? 102. Where are — located? 103. How would — divide the body? 104. What do —? Write table of —. 105. — is what?

# CHAPTER III.

## GROUPS OR MECHANISMS.

*Mental and Sanguificatory.*

106. THE HEAD INCLUDES the five kinds of general organs in addition to the five kinds composing its contents.

107. *Inf.*—To UNDERSTAND THE HEAD is to understand the essentials of all the rest of the Body except the contents of the Trunk-walls.

108. As THE HEAD-WALLS are wholly composed of the five general organs, and are also of use in protecting and serving the brain in common with the Neck, Trunk-walls, lower extremities, upper extremities, and larynx, those members may be said in one sense to be the HEAD-walls extended and modified in form, size, and by position, to meet the exigencies of the case.

109. THE QUESTION MIGHT BE ASKED, Why, if the Trunk-walls are composed of the five general organs, and are so perfectly adapted to their contents, may not the other parts be the extension of the Trunk-walls as well as of the HEAD-walls?

110. ANOTHER CLASSIFICATION MAY BE SUGGESTED as follows:

$$\text{BODY} = \begin{cases} \text{CONTENTS OF HEAD-WALLS} \\ \text{Contents of Trunk-walls} \end{cases} \text{\& or } + \begin{cases} \text{Skeleton.} \\ \text{Muscles.} \\ \text{Nerves.} \\ \text{Skin.} \\ \text{Blood-tubes.} \end{cases}$$

---

106. What does —? 107. What is it —? 108. What are the Head-Walls extended? 109. What —? 110 What —? Write table.

and the question asked, with which of the centres do the other parts belong?

111. THE PARTS FORMED OF THE FIVE GENERAL ORGANS evidently should belong and be classed with the centre that controls them by its will, and not, except indirectly, to that which only feeds them; they therefore belong with the contents proper of the Head-walls.

112. A CORRECT CLASSIFICATION, then, will be as follows:

$$\text{BODY} \cdot \cdot = \left\{ \begin{array}{l} \text{Contents of HEAD-WALLS} \quad \text{\& or} + \\ \text{Contents of Trunk-walls} \end{array} \right. \left\{ \begin{array}{l} \text{Skeleton,} \\ \text{Muscles,} \\ \text{Nerves,} \\ \text{Skin,} \\ \text{Blood-tubes;} \end{array} \right.$$

or,

$$\text{BODY} = \left\{ \begin{array}{l} \text{Contents of HEAD-WALLS} \quad \text{\& or} + \\ \text{Contents of Trunk-walls} \end{array} \right. \left\{ \begin{array}{l} \text{HEAD-WALLS,} \\ \text{Neck,} \\ \text{Trunk-walls,} \\ \text{lower extremities,} \\ \text{upper extremities,} \\ \text{larynx;} \end{array} \right.$$

or, grouping together,

$$\text{BODY} \cdot \cdot = \left\{ \begin{array}{l} \left. \begin{array}{l} \text{CONTENTS OF HEAD-WALLS} \\ \text{HEAD-WALLS,} \end{array} \right\} \text{HEAD} \\ \text{Neck,} \\ \text{Trunk-walls,} \\ \text{lower extremities,} \\ \text{upper extremities,} \\ \text{larynx,} \\ \\ \text{Contents of Trunk-walls,} \end{array} \right. \begin{array}{l} \\ \\ = \text{G'roup,} \\ \\ \\ \\ = \text{G'roup;} \end{array}$$

or, the same in another form,

$$\text{BODY} \left\{ \begin{array}{l} \text{HEAD+Neck+Trunk-walls+lower ex.+up. ex.+larynx=G'roup} \\ \text{Contents of Trunk-walls} \quad \cdot \cdot \cdot \cdot \cdot \cdot =\text{G'roup;} \end{array} \right.$$

or, by initials,

$$\text{BODY} = \left\{ \begin{array}{l} \text{H + N + T-w + l ex + u ex + l} \cdot \cdot \cdot \cdot = \text{G'} \\ \text{C of T-w} \cdot \cdot \cdot \cdot \cdot \cdot \cdot \cdot \cdot = \text{G';} \end{array} \right.$$

or, introducing the classification by members,

$$\text{BODY} = \text{H+N+T+l ex+u ex+l} = \left\{ \begin{array}{l} \text{H+N+T-w+l ex+u ex+l} \quad = \text{G'} \\ \text{C of T-w} \cdot \cdot \cdot \cdot \cdot = \text{G'.} \end{array} \right.$$

113. BY A CORRECT CLASSIFICATION the members of the body are beautifully arranged by division into

---

two groups; the first or prime group, much the larger, has the honorable office of directly serving the mind, and therefore may be called the Mental Group, while the second, or secondary as its name implies, waits on

FIG. 69.　　　　　　　　　　　　　　FIG. 70.

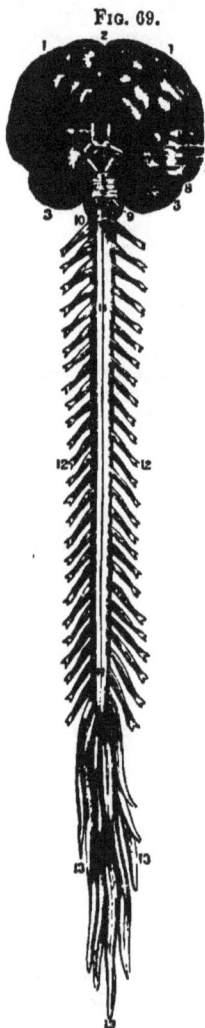

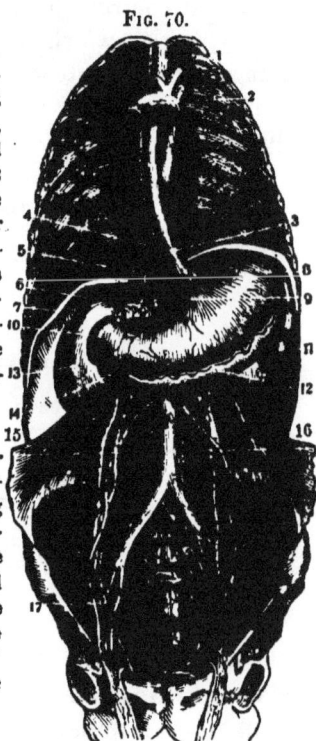

Fig. 69 represents a front view of Brains, Spinal cord and nerves, from 2 to 13, each side being so similar that their double character is evident, especially if compared with the organs illustrated by.

Fig. 70 represents Œsophagus 1, leading into the stomach 9; the large artery 2; all these parts are single, while the walls *x* of the Trunk are double.

the first by supplying it with the life-giving blood, and may hence be called the Sanguific, Sanguificatory, or blood-making and supplying.

114. ANOTHER PROOF of the correctness of this classification, and of the importance of the prime group, is that it is double, which the second is not.

115.  It is evident that parts of the mental group must be double; a person could not walk without two lower extremities, and all the parts that control them, the nerves and parts of the brain inclusive, must be double.  So also it must be with the upper extremities. All will allow that there should be two ears and two eyes for the purpose of hearing and seeing well; but the same persons will not as readily notice that there are two noses, because they are near together, and are spoken of as single, while practically each nostril is a nose.  In fact, all the mental class of organs are, and of right ought to be, double; and the seeming exceptions confirming the rule will be accounted for under synthetic details.

Fig. 71 represents a section across the neck, to illustrate the double and single character of parts belonging to the two groups.  11, 12, 13, the Thyroid Gland, Windpipe or Trachea, and Œsophagus or meat-pipe, are single, and belong to the 2d group; all the other numbers belong to the 1st group, are double, and are muscles, except 14, Blood-tubes ; 23, vertebræ ; s c, spinal canal.

FIG. 71.

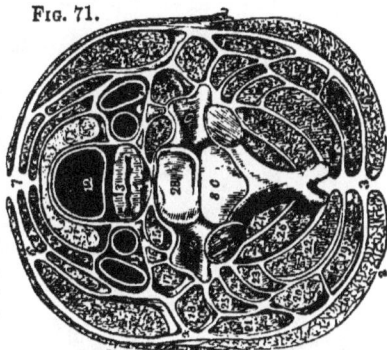

116.  It may appear that some of the contents of the Trunk are double; they are not, properly speaking. The Lungs are called right and left, and so are the kidneys; but both of each are required to effect their functions, and the whole blood and body suffers if either is deficient in its duties.  Anatomically they may be spoken of as double, but Physiologically they are considered as single and composed of two halves.  It is not so with the organs of the first group; one eye does not enlarge when the other is diseased, as one lung sometimes does, nor will the ear enlarge as one kidney will.  Each lung and

each kidney might, therefore, be most properly called a half.

117. THE ENTIRE MENTAL Group is double; and each part is to be spoken of as Right and Left. There are two Heads, the right and left, composed of two brains, nearly separated by a deep fissure. The Cranium is divisible into the right and left. The Ears, the Eyes, the Noses, the Mouths are double so far as regards sensation. The Muscles are right and left; and although those around the mouth are named as of the upper and lower lips, or usually as the orbicularis oris (circular of the mouth), as if a single muscle swept around the mouth, they are divided by a thin tendon at the mid-

FIG. 72.

Fig. 72 represents the muscles of the face, all of them evidently being double, except 13, which is also double, the tendon at the middle line of the upper and lower lips, perfectly though thinly, separating the right from the left. If a person's right or left facial muscles are paralyzed, he will be able to use only the other muscles of the face including that of both lips from the middle line, and by a little practice any person can exhibit the action of the lips of one side in talking while the other is quiet. Fig. 49 also shows that each group of muscles has its own nerves.

dle line. The Nerves also are double. The skin at its surface is continuous, but when we pass down to its nerves it is an organ of sensation, and composed of the right and left, the line of division being most exactly marked in some cases of paralysis, on one side of it sensation being perfect, and on the other altogether wanting. The same is true of the Neck and Walls of the Trunk; the Skeleton, Muscles, Nerves, Skin, and Blood-tubes of those parts are exactly double.

117. How —? Name parts in detail that are double. What is said of muscle of lips? Describe Fig. 72. What other parts of members are double?

118. USUALLY THE RIGHT GROUP is the stronger, this being most noticed in the hand, the person is called right-handed, but he is also right-footed, right-faced, and stronger on that side throughout the body.

119. THE FACE most remarkably shows the predominating character of one side or the other, usually the right side, which from more active use exhibits more expression, and is the "view" usually preferred by those artists who take pictures. It is seldom that the features are well balanced, and when that is the case, the person will use one hand as well as the other, excepting the effect of practice, the whole body being well balanced.

120. THE LARYNX being also double, and the right and left frequently not perfectly balanced, the voice will seldom be perfectly smooth and harmonious. Neither the right nor the left larynx can be used by itself, which is the case with other parts, but each is nerved distinctly, and often inharmoniously; so that though, Anatomically, it might with great propriety be spoken of as composed of halves, Physiologically and practically it is double.

121. The following table, or initial synopsis, will be correct, as showing what parts are Right and Left:

$$R + L$$
$$B = H + N + T + l\,x + u\,x + 1 \begin{cases} H + N + T\text{-}w + l\,x + u\,x + 1 = G' \\ C \text{ of } T\text{-}w \quad . \quad . \quad . \quad = G'' \end{cases} \quad R \& L$$

122. THE CLASSIFICATION OF THE HUMAN MECHANISM INTO TWO MECHANISMS, one secondary to the other, is important, as showing the primary object of the Body to be the development of the Mind, and that the formation of blood, which all the world esteems most practical, is so, after all, only as secondary to a greater object. (See Ap. H.)

123. BY SYNTHESIS, the Right M, Mental, Mentory, or ' Group, or Mechanism, may be constructed by taking the right skeleton of the Larynx, applying the right

---

118. What is — comparatively? 119. — shows what? 120. What is said of —
121. What does table show? Write table. 122. — shows what? Purport of Ap. H

muscles, nerves, skin, and blood-tubes, and following this
by a similar construction of the Right Upper and Lower
Extremities, Right Trunk-Walls, Neck and HEAD-Walls,
placing in the last their Right CONTENTS. The Left
Group or Mechanism may be constructed in the same
way. Then unite the Right and Left Mechanism on the
middle line, and into the cavities of their Trunk-Walls
pack the close-fitting Contents of Trunk-Walls, the B.
Blood-making, sanguific, sanguificatory, or *II* Group. To
the CONTENTS of the HEAD-Walls add the Mind, and the
passive potential Man is complete! Let the External
World act through his body upon his mind, and excite
its reaction, and give him air, water, and food, from
which to produce blood, and he becomes mentally and
physically an active man.

(It is not intended that the student shall conclude
that the right and left G'roups are entirely independent
of each other, and without any relations. The skin of
the two is continuous,—so are the vertebræ of the skeleton; and the nerves from a part of each eye extend to
the opposite Brain, as must be the case for perfection of
sight. Nerves, under the name of commissures, from
each spinal cord and brain extend into the other. Yet
the general idea is correct and very valuable toward giving a correct idea of the structure of the Body.)

124.   How ACTION AND REACTION between the mind
and world, and the making of Blood, take place, it will
be our duty in the next chapter to begin to unfold, by
an analysis of the two mechanisms into the requisite apparatus for performing all the functions required by the
Mind, and into those necessary for perfecting and circulating blood. The last clause exhibits the fact, that in
reference to Blood, two sections will be needed, one upon
making it, and the other upon circulating it through
both Groups.

---

123. Construct Man from Groups —? How shall he become active? 124. What
will be unfolded in the next chapter?

# CHAPTER IV.

## SYSTEMATIC ANALYSIS OF GROUPS OR MECHANISMS.

### APPARATUS.

### SECTION I.

### *Mentory Apparatus.*

125. To DEVELOP THE MIND, TO MAKE BLOOD, AND TO CIRCULATE IT, several distinct operations, called functions, must be performed. It is one operation or function for the mind to receive an influence, and another for it to exert one. To take air and to take food for making blood are distinct functions.

126. EACH FUNCTION REQUIRES an Apparatus, which is the collective name of the parts that successively, or combined, perform a function.

127. THE NUMBER OF APPARATUSES into which each mechanism can be divided will depend upon the number of operations or functions required for the development of the Mind, the making of blood and the circulation of it.

128. THE OPERATIONS OF THE MIND may be included under four classes, Sensational, Emotional, Intellectional, and Motional; that is, by which it experiences Sensations, as of sight, hearing, warmth, hunger, pain, health, &c.; Emotions, as of joy, grief, anger, &c.; Intellections, as thoughts, &c.; and the power of producing Motion. (See Ap. I.)

---

Subj. of Chap.? How many Sec.? Subj. of 1st Sec.? 125. What must be performed —? 126. What does —? 127. On what will — depend? 128. How class —?

3*

58    GENERAL ANALYSIS.

129. The Mentory Apparatus must, like the Functions, be of four kinds, Sensatory, producing *Sensations;* Emotory, *Emotions;* Intellectory, *Intellections;* Motory, *Motions;* or, in tabular form,

| M. Mentory or Mental, } Apparatus . . | { Sensat-ory (-ion). Emot-ory (-ion). Intellect-ory (-ion). Mot-ory (-ion). |

SECTION II.

*Sanguificatory Apparatus.*

130. Blood is made from Air, Water, and Food: there must be an Apparatus to receive the air and pass it into the Blood, and another to operate upon the food; the water is of such a character that it can pass in through either of the two. The former is called the Respiratory, and the latter the Digestory Apparatus, and their Functions are called Respiration and Digestion.

131. Impurities must be eliminated or removed from the blood; therefore an Eliminatory Apparatus will be required, its Function being Elimination.

132. The Elements of the Blood, and what it gathers as it passes through the body, must be worked over in various ways, or modified according to the requirements of the case: hence there must be parts that, taken together, may be called the Modificatory Apparatus; its Function is Modification.

133. All the Requirements for Blood-making or Sanguification can be fulfilled by four kinds of Apparatus, in tabular form as follows:

| Sanguificatory Apparatus . . = | { Respirat-ory (ion). Digest-ory (ion). Eliminat-ory (ion). Modificat-ory (ion). |

129. Kinds of —? Write table. Subject of 2d Sec.? 130. Of what —? 131. How —? 132. What must be done to —? 133. How fulfil? Write table.

## SECTION III.

### Circulatory Apparatus.

134. BLOOD MUST BE CIRCULATED in order to be of avail, and it has been seen that tubes are provided for this purpose. There must be, therefore, an Apparatus for the circulation of Blood, called the B. (Blood) Circulatory Apparatus; its Function, B. (Blood) Circulation.

135. THE CIRCULATION OF BLOOD MUST BE CONTROLLED, increased, or diminished, as the necessities of parts require; and as all other controlling influences in the body are nervous, so is this. For the sake of euphony the parts controlling the circulation of blood may be called the N. (Nervous) Circulatory Apparatus.

136. CIRCULATORY APPARATUS . $= \begin{cases} \text{B. Circulat-ory (ion).} \\ \text{N. Circulat-ory (ion).} \end{cases}$

137. THE CIRCULATORY APPARATUS is, in fact, to an extent, a part of the Mentory and of the Sanguificatory Apparatus, as Blood-tubes are found in all of them. In the Mentory the tubes are right and left, as the apparatus is; but in the sanguificatory, as its parts are. (Fig. 48.)

138. BOTH KINDS OF CIRCULATORY APPARATUS MAY BE INCLUDED with each, which will give six to each Group, though there are only ten in all, as follows. (See Ap. J.)

$$\text{BODY} = \text{MEMBERS} \begin{cases} G' = \begin{cases} \text{MENTORY APPARATUS.} \cdot \cdot \begin{cases} \text{Sensat-ory (ion).} \\ \text{Emot-ory (ion).} \\ \text{Intellect-ory (ion).} \\ \text{Mot-ory (ion).} \end{cases} \\ \text{CIRCULATORY APPARATUS} \begin{cases} \text{N. Circulat-ory (ion).} \\ \text{B. Circulat-ory (ion).} \end{cases} \end{cases} \\ G'' = \begin{cases} \text{SANGUIFICATORY APPARATUS} \begin{cases} \text{Respirat-ory (ion).} \\ \text{Digest-ory (ion).} \\ \text{Eliminat-ory (ion).} \\ \text{Modificat-ory (ion).} \end{cases} \end{cases} \end{cases}$$

139. THE USES OF EACH KIND OF APPARATUS will determine their varieties, and of how many and what kind of parts or organs each apparatus consists. That Analysis will be the object of the next chapter.

# CHAPTER V.

SYSTEMATIC ANALYSIS OF APPARATUS.

ORGANS.

SECTION I.

*Sensatory Organs.*

140. "LIFE DEPRIVED OF SENSATIONS AS USEFUL AS THOSE OF HEARING, is a kind of premature death; the deaf man is necessarily a dumb man, and who can compute his loss? his never-sleeping guard that warned him of a thousand dangers is dead; and now the tread of the midnight thief, the scream of the drowning child, and the mutterings of the coming storm, fall on his ear as vainly as the tear of sorrow on the brow of death; who can compute his loss? the sweet echoes of the valley, the voice of friendship, the hallelujahs of the Sabbath, and the loud artillery of heaven, are alike condensed into barren nothingness, and in the very excess of stillness he loses all the pleasures of solitude."—*Le Cat.*

141. LE CAT shows by the above eloquent description of the Ear, not only its importance, but by analogy that the mind is indebted to Sensatory Apparatus for a knowledge of danger, of philosophy, and of all the enjoyments the External World is adapted to produce.

142. THE PHYSICAL MEANS BY WHICH SUCH GRAND RESULTS ARE PRODUCED are wonderfully simple and few. Waves of air, colors and direction of light, odorous properties, savory properties, temperatures and presence of objects, and density of objects, requiring only a cor-

responding number of kinds of apparatus to utilize them fully.

143. DIFFERENT OBJECTS PRODUCE different kinds of waves of air; different colors of light, from different positions; have different odorous or savory properties, or temperatures, or density; for if they are alike in all other respects, they will be in different positions, and light will come from them in different directions; and if the mind be furnished with the few kinds of apparatus necessary to perceive those differences and similarities, it can learn all it can know about external objects.

144. THE MIND ALSO REQUIRES a knowledge of the state of the various parts of the Body; and that must be gained by means of Sensations, produced through appropriate apparatus.

145. TWO CLASSES OF SENSATORY APPARATUS ARE NECESSARY; the one called objective, external, special, or the senses, because it deals with objects external to the Body, is specially adapted to its purpose, and is the evident cause of sense; the other, called general, internal, or subjective, because it exists generally throughout the internal parts of the Body, and because through it observations are made upon the condition of the various internal parts of the subject himself.

SENSATORY APPARATUS . . . . . . . . $\begin{cases} \text{External.} \\ \text{Internal.} \end{cases}$

146. The number and kinds of organs or parts necessary to compose each variety of Sensatory Apparatus can be best appreciated after noticing the means necessary to cause a Sensation.

147. SENSATION is the name of a peculiar effect produced by a part of the Brain on the Mind, of which it is conscious.

148. How THE BRAIN ACTS upon the Mind in the production of a Sensation is not known or conjectured.

149.　To PRODUCE A SENSATION, there is necessary a portion of Brain, called a Sensatory Ganglion, active to such a degree and in such a way as to excite in the mind a sensation.

150.　THE SENSATORY GANGLIA ARE CONNECTED only with the Mind, with other parts of the Brain, and with the Sensatory Nerves, which are in one sense the white part of the Ganglia, or continuations of it, reaching or extending out, in the form of minute pulpy cords, into all parts of the body, and almost to its very surface. (Pl. 3.)

151.　THE ACTIVITY OF A SENSATORY GANGLION, and consequently a sensation, must be caused either,

1st.　*By the action of the Mind* upon the Ganglion, causing reaction upon itself; which is doubtful.

2d.　*By the action of other parts of the Brain.*

3d.　*By the action of the Ganglion itself.* This may easily be the case, for blood is constantly circulating through it, and changes are thus constantly taking place in the ganglion, and it must be constantly active, if not to the degree to cause sensation.

4th.　*By the action of the sensatory Nerves.*

152.　THE *immediate* CAUSE OF SENSATION is always a Ganglion, but that which excites the ganglion to sensatory activity, no matter through what steps, is commonly called the cause of the sensation and of the activity of the Ganglion; therefore it is caused,

5th.　*By the action of the general organs*, in which the outer extremities of the general sensatory nerves are found; the organ exciting the nerve, and the nerve exciting the ganglion; thus by two steps its activity is excited.

6th.　*By the action of the special organs* in which the special sensatory nerves commence outwardly. In this case through two steps the ganglion is excited.

7th.　*By the action of external objects*, through the

---

149. What is necessary —? 150. How —? 151. How must — 1st? 2d? 3d? 4th? 152. What is —? How is sensation caused? 5th? 6th? 7th?

special sensatory organs, upon their nerves, that in turn excite the ganglia, formed in part of their inner extremities. In this case by three steps the activity is excited.

8th. *By the action of objects* exerting an *influence* upon the organs. In this case there are four steps between cause and activity; for example, an object throws light upon the eye, through it affecting its special nerve, that acts upon its ganglion.

153. THE ORDER OF THE PRODUCTION OF SENSATION may be represented as follows:

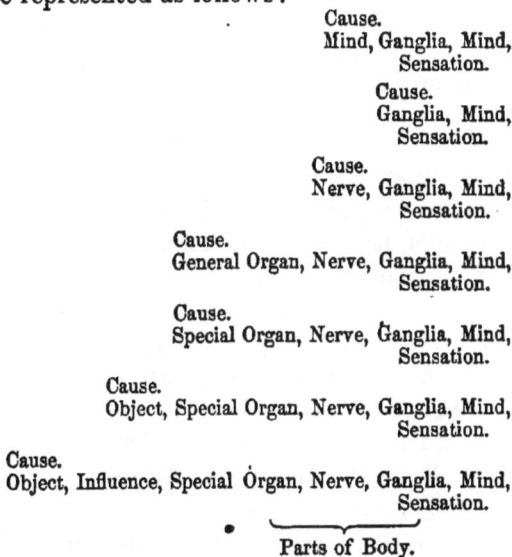

Cause.
Mind, Ganglia, Mind,
Sensation.

Cause.
Ganglia, Mind,
Sensation.

Cause.
Nerve, Ganglia, Mind,
Sensation.

Cause.
General Organ, Nerve, Ganglia, Mind,
Sensation.

Cause.
Special Organ, Nerve, Ganglia, Mind,
Sensation.

Cause.
Object, Special Organ, Nerve, Ganglia, Mind,
Sensation.

Cause.
Object, Influence, Special Organ, Nerve, Ganglia, Mind,
Sensation.

Parts of Body.

154. IN THE PRODUCTION OF A SENSATION IT IS EVIDENT that one, two, or three parts of the body are always concerned; they are called Sensatory Organs, or Organs of Sensation.

155. A COMPLETE SENSATORY APPARATUS IS COMPOSED of three kinds of Sensatory Organs; thus,

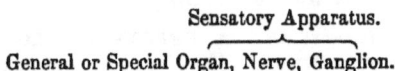

Sensatory Apparatus.

General or Special Organ, Nerve, Ganglion.

8th? 153. Write table and explain. 154. What —? 155. Of what —? How is sensation caused?

156. The General Organs of the Body, in which the outer ends of the sensatory nerves exist, do not require any peculiar adaptation to their sensatory office or function; all that is necessary is to have the nerves commence in, and thus connect every part with, the Ganglia; then, whatever the condition of the organ, a corresponding effect can be produced in the nerve, and a corresponding activity in its Ganglion, and a corresponding sensation.

157. *Illus.*—If any part be in health, it causes a delightful sensation; if diseased, an unpleasant sensation; if it require food, a sensation of hunger; if drink, a sensation of thirst; if too warm or too cold, appropriate sensations; and thus of all other states, through scores of varieties and hundreds of shades.

158. A very natural question would be, If all the sensations are produced in the head by the activity of the Ganglia, how can it be determined that they are caused in one part rather than another?

159. To determine in what part any sensation has been caused, it is only necessary to have the mind believe, when it has a sensation, that it was caused at the outer end of the nerve through which the activity of the Ganglion has been caused. This is the case, and the act is called perceiving the sensation.

160. *Illus.*—When a person strikes the elbow, or funny bone, he says he feels a sensation in his little finger, but he does not; he strikes a nerve (*a*, Fig. 43) between the skin and bone of the elbow, which nerve commences by numerous branches in his little finger, and extends to the Ganglia. When struck, this nerve excites the Ganglia and causes a sensation that the mind refers to the part in which the outer end of the nerve is; that is, the Mind perceives the sensation in the finger, or as if it was in the finger.

161. *Illus.*—When a person by long sitting compresses the nerve (*b*, Fig. 43) between the seat

and the haunch-bone, he causes a sensation that the Mind perceives in the foot and lower leg, where the branches of the nerve commence, and he says his foot is asleep, when in fact there is nothing the matter with the foot; nor will rubbing the foot or lower leg, or anything else, except removing the pressure, do any good.

162. *Illus.*—IF AN ARM BE CUT OFF (*a*, Fig. 1, Pl. 3) below the elbow, or the leg below the hip (*b*, Fig. 1, Pl. 3), and the elbow struck or the nerve at *x* compressed, precisely the same result will be produced; the sensation will appear to be where the nerves naturally commence, and the individual, through his sensations, will appear to have a hand and foot. If he is asked with his eyes shut to point out where his sensations seem to be, he will accurately indicate where his foot or hand would be, both in distance and direction; and as the stump is turned, so will it seem to him that the foot or hand changes position.

163. DISEASES OF THE GANGLIA OR NERVES SOMETIMES PRODUCE the most excruciating pains, which will be referred to or perceived in parts perfectly sound.

164. *Illus.*—A LADY REQUIRED her physician to extract eight teeth, that, as she insisted, "ached." But she was not relieved, for the teeth were sound, and the disease existed in the nerves or Ganglia.

165. *Illus.*—IN THE DISEASE NEURALGIA, Nerve-pain, or tic-douloureux, applications to the parts apparently affected do not usually remove the pain, because these parts are not usually the seat of the disease.

166. PAIN PRODUCED BY PARTS DISEASED is not always referred to them, but sometimes to others; why, is not known.

167. *Illus.*—PAIN PRODUCED BY SOME DISEASES OF THE LIVER is frequently referred to a location under the shoulder.

168. *Illus.*—PAIN PRODUCED BY INCIPIENT HIP-DISEASE is frequently perceived in the knee.

169. *Inf.*—IT IS BY NO MEANS CERTAIN that because a part is painful it is diseased, or that pain is even present, for it may seem to be in a part that is removed.

---

162. What —? 163. What do —? 164. What did —? 166. How is —referred? 167. 1st Illus.? 168. 2d Illus.? 169. What —?

**170.** *Inf.*—IT WILL REQUIRE MUCH EXPERIENCE oftentimes to decide upon the kind and locality of disease from the character of pains that a patient will describe, and the places where he will perceive and locate them.

171. THE ORGANS AT THE OUTER EXTREMITIES OF THE SPECIAL SENSATORY NERVES have special sensatory uses, need special names, and are called Organs of Sense as a collective name. (Observe that this expression differs from Organs of Sensation and from Sensatory Organs. Organ of sense is the technical name of one kind of the Organs of Sensation.)

172. THE ORGANS OF SENSE, with one exception, are found at the surface of the Body, ready to catch the influences that objects around can exert directly or indirectly.

173. THE NUMBER OF KINDS OF ORGANS OF SENSE, and their essential characteristics, and of course the number of special Sensatory Apparatuses, can be determined by again noticing the characteristics of external objects, and comparing them with what is necessary to produce sensations.

ORGANS OF SENSE OF HEARING.

174. WAVES OF AIR HAVE three characteristics, *force*, *quality*, and *pitch*, and different objects produce waves that differ in some or all of these respects, some in an exceedingly minute, and some in a great degree.

175. FOR THE MIND TO LEARN THE CHARACTER OF OBJECTS and enjoy them, it is necessary it should have the means of distinguishing all the different degrees of force, quality, and pitch that objects produce.

176. EACH WAVE POSSESSES in some degree all its three characteristics; a single organ therefore must be so constituted as to receive a wave, and cause it to exert its three influences, through three kinds of parts, upon three kinds of nerves connected with three kinds of

ganglia, and the mind will perceive all the sensations of sound, which are called feeble, loud; sweet, rough; low, high; and by a thousand other names expressive of different degrees of force, quality, and pitch.

177. THE EAR, or Organ of Sense of Hearing, is constituted of three parts, called the external, middle, and internal ear.

FIG. 73.

Fig. 73 is a beautiful view of the Ear; 1, external, 2, middle, 3, inner ear; 13, a section of the air-tube, which section extends through the front of 2, and the middle of 19, a tube called Eustachian, leading to back part of nostril. If a person close the nose and mouth, and blow air from the lungs, it will press up that tube into 2, and produce a ringing. 14, bottom of air-tube, a tremulous membrane that, acting on nerves, determines *force ;* 22, semi-circular canals, that determines *quality ;* 24, cochlea, that determines *pitch ;* 18, bones of ear.

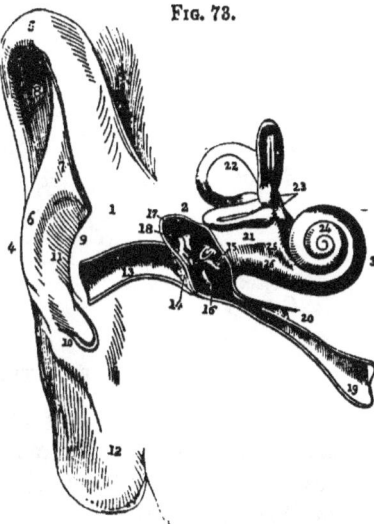

178. THE EXTERNAL EAR is adapted to receive a wave of air and transmit it to

179. THE INTERNAL EAR, COMPOSED of three parts, each acted upon by its corresponding influence, and connected by nerves with Ganglia. If all these parts are perfect, correct sensations of the three kinds will be produced; but if one should be imperfect by constitution or by disease, the sensations of that class will be imperfect or wanting. Thus a person may have correct ideas of the force of a sound and not of its pitch, &c., as is not unfrequently the case. The ears may be unlike, and

177. — how divided? Describe Fig. 73. 178. — how adapted? 179. How —? May a person have correct sensations of one class and not of the other?

thus the hearing through both imperfect, and perfect through one with the other closed.

180. The Ear is placed in the side of the head, where it will most readily catch each passing air-wave, and where it can be easily turned in any direction to assist in judging whence the wave came.

Thus in the simplest manner, and with the most exquisite delicacy, is the ear made useful to a wonderful degree, and will be found still more admirable when the details of this organ are wrought out.

181. The Apparatus of Hearing, through which waves of air act, is double, being right and left, and is composed, acted upon, and acts, as follows:

<pre>
            Apparatus of Hearing,
            Three kinds of organs.
 1      2        ┌──────────┐        6
Object, Waves of Air, Ear, Nerves, Ganglia, Mind.
Cause.           3    4        5    Sensations.
</pre>

#### ORGANS OF APPARATUS OF SIGHT.

182. Rays of Light are of many different colors and shades, and pass off in a radiating manner from the minutest points of all objects except those called black.

183. Similar objects throw off similar, and dissimilar objects dissimilar, colored rays, and therefore they may be the means of distinguishing the character of objects.

184. The minutest points of objects (which are in fact the real objects of vision) are in different positions, and therefore throw off rays in different directions, each in a direct line from its starting-point. Therefore, to know the direction of a ray of light is to know the direction of the object whence it came.

185. To learn the character and direction of objects, the mind requires means for distinguishing the color and direction of rays of light.

---

180. Where —? 181. How is — composed? Table? 182. What are —? 183. What do —? 184. Where are —? 185. What required —?

186. EACH RAY OF LIGHT has color and direction, therefore a single organ is required that shall produce two sensations.

187. A SINGLE RAY will produce a very feeble effect, therefore the organ to receive it must be so constructed that several rays from the same point shall act together and thus increase the effect.

188. SEVERAL RAYS RADIATING FROM THE SAME POINT have different directions, but if the organ be properly constructed, the effect of several acting upon the. same point in the organ will be the same as if one only acted with the same power as all of them. (See Pl. 8.)

189. THE SAME MEANS NECESSARY TO CAUSE THE RAYS FROM ONE POINT OF AN OBJECT TO ACT UPON ONE POINT IN THE ORGAN, WILL BE SUFFICIENT to cause the rays from thousands, yes, millions of adjacent points, to act upon adjacent points in the organ, and it will only be necessary to furnish it with the corresponding num-

FIG. 74.

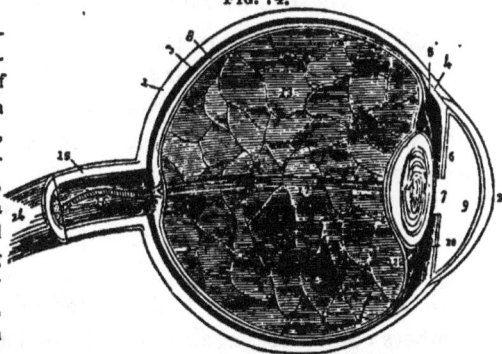

Fig. 74 represents a perpendicular section of the eye, from front to rear. 1, white of eye, extending round, and forming a strong spherical box, the opening in front filled by 2, a window; 7, the pupil, an opening in 6, the iris, muscles that contract or enlarge 7; 9, a space filled by a transparent fluid; 11, lens; 13, transparent substance, filling all back of 11 to 8, a multitude of nerves spread out to receive light coming back through 7; 15, the bundle of nerves 8, continuing back to brain. Make a hole in paper, place two candles before it, and the light will shine through in two directions; place a lens behind the hole, and at a proper distance the light will be brought to two points.

ber of nerves to enable the mind to receive millions of sensations of three kinds, color, and direction, at the same instant, and to take in a wide field of vision.

190.  THE EYE is an organ adapted to cause the rays of light, from millions of points without, to act upon millions of points within it, where commence nerves connecting with ganglia.  Through each nerve two sensations, one of color, the other of direction, are produced, and the mind is thus furnished with the data for further knowledge.

191.  THE EYE IS LOCATED in the front and upper part of the head, the motions of which, together with those of the eye, give all the extent desirable for vision.

192.  THE APPARATUS OF SIGHT must be double, right and left, and is composed, acted upon, and acts, as follows:

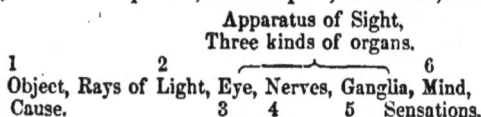

Apparatus of Sight,
Three kinds of organs.

| 1 | 2 | | 6 |
|---|---|---|---|
| Object, | Rays of Light, | Eye, Nerves, Ganglia, | Mind, |
| Cause. | | 3    4    5 | Sensations. |

#### ORGANS OF APPARATUS OF SMELL.

193.  ODOROUS PARTICLES are exquisitely minute, dissolved in the air, and differ as the objects do from which they are derived, and require only to be brought in contact with appropriate nerves, when each will excite its peculiar sensation.

194.  To DISTINGUISH ODORS, therefore, and learn through them the character of objects, it is only necessary that the air containing them shall sweep over a delicate surface near to which nerves of smell in great numbers commence, against which the laden air may at times be pressed with some force.

195.  THE NOSE is an organ admirably adapted to be the organ of smell.  The air loaded with odorous particles can be drawn through it, and over the delicate membrane with which it is lined, near the surface of which commence numerous nerves that extend to ganglia related to the mind.

---

190. What is —? 191. Where —? 192. Write table. 193. What are—? 194. What is necessary — ? 195. What is —?

**196. THE APPARATUS OF SMELL** is double, right and left, and composed, acted upon, and acts, as follows:

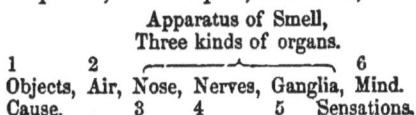

Apparatus of Smell,
Three kinds of organs.

| 1 | 2 | Nose, | Nerves, | Ganglia, | 6 |
|---|---|---|---|---|---|
| Objects, | Air, | Nose, | Nerves, | Ganglia, | Mind. |
| Cause. | | 3 | 4 | 5 | Sensations. |

FIG. 75.

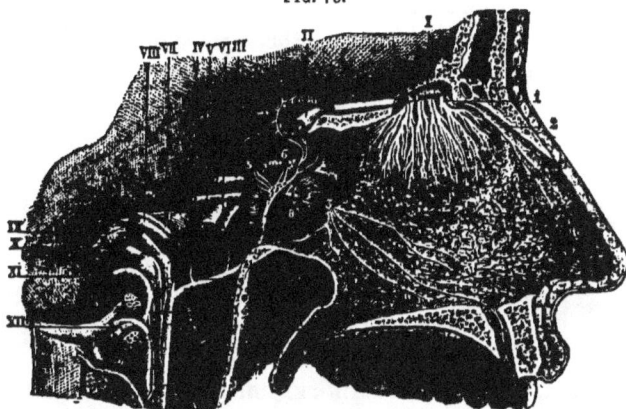

Fig. 75 represents a section of the nose near its partition, toward which the observer is supposed to look. The section is carried back into the cranium, and down into the mouth and throat. I is the Ganglion of smell, just above the roof of the nose, on which it rests, and through numerous holes in which the nerves of smell, 1, extend; 2 and 3 are divisions of nerves, one twig of 3 extending through an opening in the bony roof of the mouth to the skin lining it; *u* is the uvula, or tip of the hanging palate. The white line from it, up to 3, shows the termination of the partition, behind which the eye can look into the other nostril, and see *x*, the opening into the Eustachian tube, leading to the middle ear. (See Fig. 78.)

ORGANS OF APPARATUS OF TASTE.

**197. SAVORY PROPERTIES EXIST** in the minute particles of certain objects that must be dissolved for the exhibition of their savors, and applied to nerves, when each one will produce its appropriate sensation.

**198.** To DISTINGUISH ‾SAVORS, therefore, and learn

---

196. Table. Describe Fig. 75. What connections has the cavity of the nose according to the cut? 197. How do—? 198. What is necessary—?

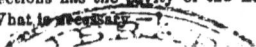

through them all that they can teach of the characters of objects, it is only necessary that a fluid in which savory objects are dissolved should be spread over a delicate surface, near to which commence numerous nerves of taste.

199. THE MOUTH is an organ admirably adapted to Taste. It is provided with an abundance of fluid for dissolving the savory objects, and also furnished with teeth to grind and prepare them for solution, and with a delicate membrane, near to the surface of part of which commence millions of nerves connecting with appropriate ganglia.

200. THE APPARATUS OF TASTE must be double, right and left, and is composed, acted upon, and acts, as follows:

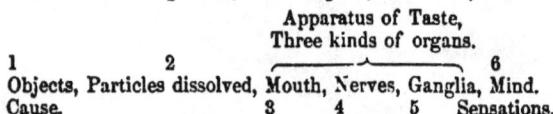

Apparatus of Taste,<br>
Three kinds of organs.

| 1 | 2 | 6 |
|---|---|---|
| Objects, | Particles dissolved, | Mouth, Nerves, Ganglia, Mind. |
| Cause. | | 3       4       5       Sensations. |

### ORGANS OF APPARATUS OF TOUCH.

201. TO DISTINGUISH TEMPERATURES AND THE PRESENCE OF OBJECTS, and thus learn all that it is possible to learn from them, requires merely the presence of nerves of Touch near to the surface acted upon.

202. THE SKIN is the necessary organ of this sense, and is constructed with nerves of Touch, commencing just below its surface, and extending thence to their ganglia.

203. THE LINING OF THE MOUTH AND NOSE, and other parts, are to be included in the skin of those parts, and as being the organs of the sense of Touch. (2, 3, Fig. 75.)

204. THE APPARATUS OF TOUCH is double, right and left, and is composed, acted upon, and acts, as follows:

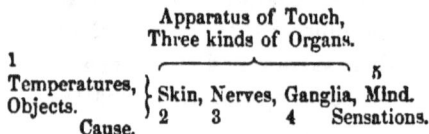

Apparatus of Touch,<br>
Three kinds of Organs.

| 1 | 5 |
|---|---|
| Temperatures, | Skin, Nerves, Ganglia, Mind. |
| Objects. | 2        3         4        Sensations. |
| Cause. | |

### ORGANS OF APPARATUS OF MUSCULAR SENSE.

205. DENSITY, RESISTANCE, GRAVITY, and the like properties of objects, always represent power, and are distinguishable by the different degrees of power requisite to oppose or overcome them.

206. To DISTINGUISH THE DENSITY, etc., of different objects, it will be necessary to construct organs capable of exerting power in various degrees, which shall produce corresponding effects on nerves commencing in the organs and extending to ganglia adapted to produce the appropriate and informing sensations.

207. A MUSCLE (a piece of lean meat), by contracting, is capable of exerting power; and if the muscle should be properly attached by its extremities to two parts of a framework, connected by a joint, its power could be applied to overcoming density, &c.; then if nerves should commence in every part of it and be differently affected by every degree of contraction, and extend to appropriate ganglia, a complete sensatory apparatus would be formed, as follows:

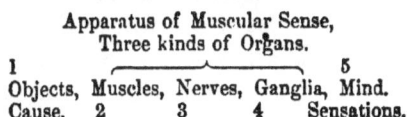

Apparatus of Muscular Sense,
Three kinds of Organs.
1　　　　　　　　　　　　　　　5
Objects, Muscles, Nerves, Ganglia, Mind.
Cause.　　2　　　3　　　4　　Sensations.

208. THE MUSCULAR SENSE, the simplest and least easily deranged, is yet the most useful of the senses, manifesting itself the earliest, being blunted the last, and serving under the most varied circumstances of life.

209. THE USE OF THE MUSCLES in distinguishing objects is only one, or one form, of the many uses to which they are adapted, and which the muscular sense assists in performing; its office in regard to objects being superadded or an accident to its prime motory office.

210. MUSCLES MUST CONTRACT to perform every mo-

tion required in or of the Body, and each minute contraction of each and all the muscles must be every
instant present to the mind: this result is a wonderful
use of the muscular sense.

211. ALL PARTS OF THE BODY ARE BALANCED, or their
motions adjusted to the requirements of any desired
motion, by means in part of the muscular sense.

212. THE MIND IS WARNED of danger when any
part of the body begins to fall, and to yield to a shock
it cannot overcome, by means of the muscular sense.

213. THE MIND IS TAUGHT the form of objects by
the muscles grasping them, or moving round them, and
also distances moved through by the whole body, or a
part of it, by means of data furnished by the muscular
sense, especially in combination with the data furnished
by touch.

214. THE POWER OF SPEECH IS DUE to the rapid action of very small muscles, the surprising and wonderful
control of which is in part dependent upon the faithful
action of the muscular sense, which reports what the
sound will be before it is produced, by reporting the
condition of the muscles which will control the voice.

215. A KNOWLEDGE IS ALSO REQUIRED BY THE MIND
of the condition of disease or health in the muscles,
their weariness, fatigue, and exhaustion, on one hand,
and the advantages of their repose and proper exercise
on the other, and for all this the muscular sense is sufficient.

216. THE MUSCULAR SENSE, THEREFORE, ABOUNDS in
enjoyment as well as utility more than any other sense.

217. THAT A MUSCLE MAY CONTRACT, and its power
be applied to the examination of objects, or any other
purpose, a nervous influence must be exerted upon it;
and as the degree of that influence must be greater or
less according to the contraction demanded, and must

also be induced by the mind, it may assist in determining the particular characteristics of examined objects, and thus be a part of the muscular sense, or rather an addition to it.

218. *Inf.*—TWO CLASSES OF NERVES MUST BE FOUND IN THE MUSCLES, one through which influences act on the muscle causing it to contract, and another through which the muscle, by contracting, produces its effect on the Ganglia.

NERVES RELATED TO MUSCLES . . . . . $\begin{cases} \text{Motory.} \\ \text{Sensatory.} \end{cases}$

Fig. 76 represents a section of the spinal cord, with 1, 2, the roots, as they are called, of nerves extending from it on each side. 1 is motory, 2 is sensatory; beyond 3 they are enclosed in one sheath, and no longer distinguishable. All the spinal nerves are similar. Where they arise from the spinal cord, they are seen to be numerous; they are equally so in their sheath.

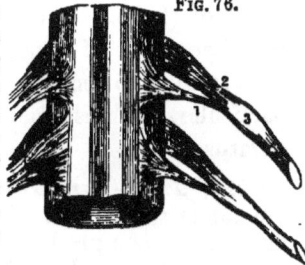

FIG. 76.

219. *Inf.*—NERVES are only channels for conducting influences.

220. *Inf.*—THERE MUST BE two kinds of Ganglia for the muscles, one to exert an influence through, and another to receive one from, the two kinds of nerves.

221. *Inf.*—THE MUSCLES are both sensatory and motory—they must be the latter in order to be the former —and the control of them as motory organs is dependent on their sensory action.

222. A REVIEW OF ALL THE SENSATORY APPARATUS will show that each kind is composed of three organs, as follows:

Organ of Sense; Nerve; Ganglion;

and that

223. THE SENSATORY APPARATUS is of two classes, one wholly devoted to the internal condition of the body, its organs of sense corresponding to all the organs of the body; the other devoted to gaining for the mind

---

218. What —? Table. Describe Fig. 76. 219. What are —? 220. How many kinds of Ganglia related to muscles? 221. What are —? 222. — shows what?

the data of knowledge of external objects, its organs of
sense being of six kinds, corresponding to the kinds of
apparatus to which they belong, as follows: (to be read
from right to left.)

SENSATORY APPARATUS
- General=Ganglia; Nerves; General Organs of Sense.
- Special=Ganglia; Nerves; Organs of Sense.
  - Ear.
  - Eye.
  - Nose.
  - Mouth.
  - Skin.
  - Muscles.

Parts of the Skeleton are also to be appropriated to
the Apparatus of Sensation, for, though in one respect
not essential to sensation, it has been constituted in some
particulars with especial reference to the Sensatory Apparatus.

224. *Inf.*—ALL PARTS OF THE BODY, however minute,
are connected with the Ganglia of the Brain by means of
nerves, and can produce an infinite number of sensations.

225. *Query.*—Is each different kind of sensation produced by different nerves, or can many kinds of sensation be produced through the same nerve?

226. MANY DIFFERENT SENSATIONS CAN BE CAUSED
THROUGH THE SAME NERVE, for in the use of the eye all
the different colors at different times act on all the
nerves, and are equally recognized by each.

227. *Query.*—ARE ALL THE NERVES ALIKE, being
only channels of different influences, or are the Ganglia
unlike?

228. THE SAME CAUSE PRODUCES DIFFERENT SENSA-
TIONS by acting on different nerves. The nerve of sight
pinched, pricked, or cut across, does not cause a sensation of pain, but of light. Electricity passed through
the nerves of muscular sense, of touch, taste, smell,
sight, and hearing, produces in each case the sensation
appropriate to the nerve. Therefore, either the nerves
or the ganglia differ from each other, probably the latter,

the nerves are probably alike, and were the nerve of smell exchanged for that of sight the result would probably be the same as now; or, if the termination of the nerve of smell could be transferred to the ganglion of sight (optic ganglion), an odorous particle acting through the nose would probably produce a sensation of light.

229. *Inf.*—THE NUMBER OF GANGLIA need not be so great as that of the nerves, nor greater than that of the kinds of sensations classed according to the causes producing them.

230.

| | | | |
|---|---|---|---|
| SENSA-TIONS | General | Pleasant & Unpleasant, } of | Appetites & their satisfaction. Satiety & Repose. Ennui & Exhilaration. Fatigue & Rest. Disease & Health (normal and abnormal). |
| | Special | Pleasant, Unpleasant, & Neutral, } of | Muscular Sense. Touch. Taste. Smell. Sight. Hearing. |

231. SIX KINDS OF GANGLIA ARE EVIDENTLY NECESSARY for the production of the Special Sensations, while for the General Sensations one kind may be sufficient, though it is probable that several of the Appetites have their peculiar ganglia.

232. APPETITES are the sensations that signify, strictly speaking, when the blood requires supplies of food—the demand for water being called thirst, and for air a sense of suffocation. But appetite is allowed to include a wider range than the demands for supplies for the blood. A sensation demanding the exercise of any part, having its origin in the condition and wants of the physical system, is called an appetite.

233. *Inf.*—APPETITE FOR FOOD DIFFERS from an inclination to gratify the palate, which is merely a desire to produce pleasant sensations of taste, even to the sacrifice of appetite.

---

229. Relative —? Write and describe table of sensations. 231. Why are —?
232. What are —? 233. From what does —?

234. *Inf.*—ALL PARTS OF THE BODY CRAVE exercise, especially if they are in the habit of being exercised, and this demand, with the succeeding one for relaxation, may well be called an appetite, since they must be treated like an appetite, the pleasant and unpleasant effects being the same.

235. DIFFÈRENT PARTS OF THE BODY ARE COMPOSED of different substances, and demand different kinds of food; and as the appetites have their seat or exciting cause in the part that needs the food, the appetites for food ought to differ, or crave different kinds of food.

236. *Inf.*—THERE OUGHT TO BE a muscular appetite, a nervous appetite, an osseous appetite, an appetite for calorific food, for food for the secretions, &c.

237. *Illus.*—IF A BONE BE BROKEN, very much food and of peculiar kinds should be craved; often it is thought because the man is not using his muscles, he is not doing much; he is—he is growing new bone, and will digest much food.

238. STUDENTS DO NOT ALWAYS LEARN TO DISTINGUISH a nervous appetite, entirely unlike a muscular or calorific one to which they have been accustomed, and more like a feeling of exhaustion or irritability, that experience soon teaches can be overcome by eating the right kind of food.

239. TEACHERS DO NOT ALWAYS distinguish, in the lassitude, uneasiness, and discomfort of their pupils, or even of themselves, the indications of an appetite for air.

240. *Inf.*—It is certainly very unfortunate for persons in so important relations as teachers and students, not to know what is the matter with themselves, nor what to do to improve their condition!

241. THE APPETITES ARE RELATED to sensations of Satisfaction, Satiety, Repose, Ennui, Exhilaration, Fatigue, Rest, Exhaustion, and Quiet; all of which may be produced through the same apparatus as appetites.

242. ABNORMAL SENSATIONS, or those of Disease, include all kinds of pains, those from injuries as well. In many instances the same diseases produce different sensations in different organs, and different diseases produce different sensations in the same part.

243. NORMAL SENSATIONS, or those of Health, are the constant inspiring source of enjoyment always produced by parts when free from disease. They do not require any special apparatus, nor do those of disease.

244. ALL THE GENERAL SENSATIONS are pleasant or unpleasant, to induce or prevent their repetition.

245. THE ORGANS OF THE SPECIAL SENSES EXCITE an appetite for exercise and repose, sensations of fatigue and rest; possess health and are diseased, like all parts of the body; and in the same manner promote enjoyment.

246. EACH OF THE SENSES has its peculiar activity, excited by external objects, and their mode of action will determine the character of the sensations.

247. *Illus.*—FISH BEFORE MEAT is palatable, after it, insipid. A Red color succeeding Green, Orange succeeding Blue, &c. are agreeable. Combinations of certain notes please, &c. (See App. K.)

248. CERTAIN SENSATIONS of the special class are neutral, having relation only to the acquisition of knowledge, and being indifferent to the physical system.

249. THE OBJECT OF LIFE being activity, and the repair of the body requiring its relaxation, alternate activity and relaxation ought to be the conditions productive of the highest physical enjoyments.

250. TO KNOW HOW TO ALTERNATE *the activities and repose of the various parts of the body directly, and by arranging objects to act favorably through the senses, is to possess the art of making life pleasant both to the possessor and to others.*

---

## SECTION II.

### *Motory Organs.*

251.　MOTIONS of several different kinds are required in the Body.

252.　CHEMICAL AFFINITY AND CAPILLARY ATTRACTION MUST OPERATE IN THE BODY as well as out of it, and produce movements in the particles over which they exert their sway.

253.　THE CILIA OF THE CILIATED CELLS MUST PERFORM their wonderful though unobtrusive movements at the surfaces where they are needed, all within the limits of microscopic observation.

FIG. 77.

Fig. 77 represents 2, cilia, very minute extensions of the surface ends of four cells, 1, very much magnified. The cilia have an exceedingly rapid vibratory motion, tending to move along any substance resting on them. The cause of their motion is not at all understood, nor even conjectured. 4, other cells, growing to take the place of 1. 5, basement membrane, from which 4 take their rise. 6, sub-tissue, containing Blood tubes, &c. 3, nucleus of cells.

254.　ELASTICITY COMES INTO PLAY very frequently in the production of motion, and without it the body could not exist.

255.　THE MOTIONS CAUSED BY AFFINITY, ATTRACTION, CILIA, AND ELASTICITY, are called non-nervous, because they are not directly influenced by nerves, and of course are not under the direct control of the mind.

256.　THE MIND REQUIRES the means of producing motion at its option in various parts of the body.

257.　THE REQUIREMENT OF THE MIND IS THREEFOLD. 1st. Various parts of the Body are to exhibit motions at the option of the mind. 2d. It follows that they must be connected with the centre of the Body, where the mind has been shown to be; and, 3d. The

231. What—does the Body require? 252 Does—? 253. What must—? Describe Fig. 77. 254. When does —? 255. What — called? 256. What does—? 257. What —?

mind must have means there from which to discharge an influence to act through the conducting medium upon the parts that are to directly cause the motion.

258. THE SIMPLEST FORM OF MOTORY APPARATUS imaginable, would be a Ganglion to produce an influence at the option or will of the mind, a nerve to conduct the influence, and a part or organ to receive the influence and shorten (contract) or lengthen (relax) accordingly.

259. AN APPARATUS COMPOSED OF ONLY THREE OR-GANS will seem to be too simple for the purpose of producing nervous motion, when the great number and apparently complicated motions in the Human Body are considered, some of them almost infinitely delicate, as in producing the shades of speech, and others as rude and as strong as the grasp of a madman.

260. *Query.*—Is it possible that every nervous motion, each mental expression and the beating of the heart, the winking of the eye and the inhalation of air, the voice of command and the whispered lullaby, the beckoning hand and the repulsing foot, the writhings of anguish and the merry laugh of joy, the ploughman's labor and the boy's play, are all performed by the action of an apparatus so simple as to be composed of only three kinds of organs?

261. IN FACT, ONLY GANGLIA, NERVES, AND MUSCLES are necessary for constituting a motory apparatus, and performing all the varied nervous motions in the Human Body not only, but also in the bodies of all animals; the swiftness of the bird, the strength of the beast, the wriggling of the serpent, and the hum of the insect, are due to the same three kinds of organs of a motory apparatus.

262. *Inf.*—THE MOST PERFECT AND WONDERFUL RE-SULTS are produced in the Divine architecture of the Human Body in the simplest manner.          `

263. THE REASONS WHY THE THREE KINDS OF MO-

---

258. What is —?   259. What will — seem ?   260. What is —?   261. What neces-sary for motion ?   262. How are — produced ?   263. What are — ?

4*

TORY ORGANS SEEM INADEQUATE, are because the results
of motion are so numerous, and likewise so diverse, that
at first glance it does not seem reasonable that they
should have a similar cause; but motion is the simple
idea in all the cases, and the direction or mode in which
it will be exhibited will depend upon,

264. THE DIFFERENCE IN SIZE, FORM, AND POSITION
of similar apparatus, which will account for all the results of nervous motions observable in the body.

265. *Size.*—SUPPOSE a minute oblong cylindrical
cell, shaped like a bead, to have the power of alternately
contracting, or shortening, and relaxing when nervous
influence is sent into it. This will be the element of all
nervous motion, and a sufficient number of cells properly
arranged will produce all the results desirable.

FIG. 78.

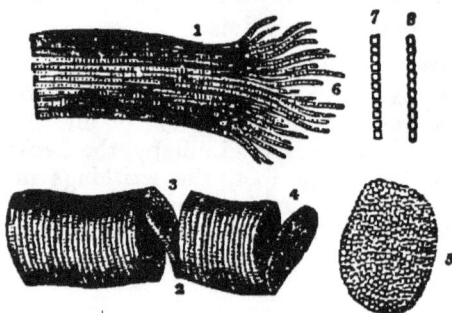

Fig. 78, a plan of
Muscle-Cells. 7, 8,
cells, end to end; 1,
cells end to end and
side by side; 6, same,
spread; 2, 3, 4, bun-
dle of cells, separable
into discs, of which
5 is one. The form
of the bundles is very
irregular in the mus-
cles. The size here
is very much magni-
fied.

266. IF CELLS BE CONNECTED, END TO END, they will
increase the extent of motion.

267. *Inf.*—IF ONE CELL CAN SHORTEN only to the
extent of one hundredth the thickness of a hair, a hundred cells would produce motion to the extent of a hair's
breadth, and a million of cells would produce a very
appreciable motion.

268. IF THE CELLS BE CONNECTED, SIDE BY SIDE,
they will increase the strength of motion.

269. *Inf.*—IF ONE CELL COULD NOT BEND A HAIR,

two cells might, and enough millions would exert all the strength desirable.

270. EXTENT AND STRENGTH OF MOTION ARE DETERMINED by the number of cells placed end to end and side by side.

271. *Inf.*—ONE HUNDRED CELLS WILL BE REQUIRED to produce at once the extent and strength of motion for either of which alone ten cells would suffice.

272. *Form.*—IF A SERIES OF CELLS IS FORMED INTO A CIRCLE OR OBLONG, their contraction will diminish the circle, and close or tend to close the orifice about which they are placed. (See muscle around the eyes in Fig. 71.)

273. IF RINGS FORMED OF CELLS BE PLACED BY THE SIDE OF EACH OTHER, a contractile tube will be formed, and by the successive contraction and relaxation of its rings, substances can be forced through it. (1, Fig. 70.)

274. IF CELLS ALSO EXTEND LENGTHWISE the tube, it will be shortened in length by their contraction.

FIG. 79.    FIG. 80.

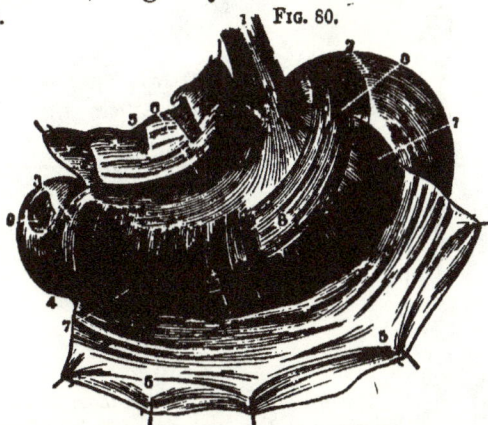

Fig. 79, *a*, represents the diagonal, *b*, the circular, and *c*, the longitudinal fibres, formed of muscle-cells, in the Œsophagus: that at 1, Fig. 80, opens into the stomach, of which 5 is the outer coat, dissected and turned back from 7, 8, 9, the muscular fibres, extending, as shown by the lines, in different directions.

270. By what —? 271. For what —? 272. What effect —? 273. What —? 274. What if —? Describe Fig. 79. Fig. 80.

275. IF RINGS OR PARTIAL RINGS OF CELLS BE SO IN-TERWOVEN AS TO form a pouch, its contents may be com-pressed, moved about, or expelled, according to the pe-culiar action of the rings.

276. MUSCLES MAY THEREFORE BE CLASSED as follows:

MUSCLES . . . . . . . . . . . { Direct. Circular. Pouched. }

277. *Position.*—IT IS EVIDENT that if a part to be moved by a direct muscle is small, only a few cells can be attached to it, so that if it is to be moved with much force, a modified arrangement must be made.

278. IF A SMALL, STRONG CORD, OR IF A MEMBRANE, IS ATTACHED to the part to be moved, the cells in any required number can be attached to it, and though they will be obliged to act at some disadvantage, yet their number can be correspondingly increased.

FIG. 81.          FIG. 82.

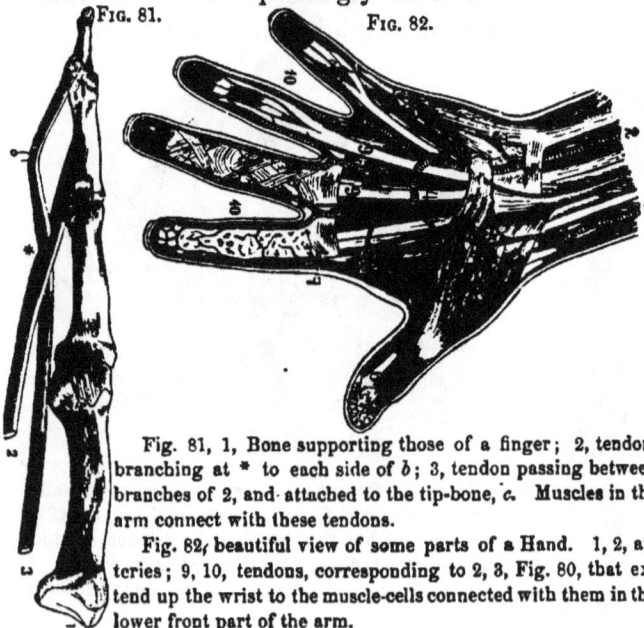

Fig. 81, 1, Bone supporting those of a finger; 2, tendon, branching at * to each side of *b*; 3, tendon passing between branches of 2, and attached to the tip-bone, *c.* Muscles in the arm connect with these tendons.

Fig. 82, beautiful view of some parts of a Hand. 1, 2, ar-teries; 9, 10, tendons, corresponding to 2, 3, Fig. 80, that ex-tend up the wrist to the muscle-cells connected with them in the lower front part of the arm.

275. What —? 276. How may —? 277. What —? 278. What —? Describe Fig. 81. Fig. 82.

279. THE BEAUTY AND CONVENIENCE OF A PART WILL OFTEN DICTATE that the cells by which it is to be moved shall not be placed directly in connection with it, but at a little distance, and act upon it through the medium of tendons.

280. VARIOUS RESULTS CAN BE PRODUCED by a succession of different contractions, or by combined contractions, neither of which could alone effect the object.

281. THUS HAS THE ONE ELEMENTARY MUSCLE-CELL the potential property that, properly applied, in connection with an appropriate framework, will produce all required nervous motions.

282. A FRAMEWORK IS NOT ESSENTIAL to a motory apparatus, for certain motions can be produced without a framework ; but,

283. THE COMPLETE EFFECTS OF MOTORY APPARATUS REQUIRE a framework light and strong, having a large surface, and composed with many joints of varying character.

284. THE FRAMEWORK AND THE MOTORY APPARATUS MUST BE MUTUALLY ADAPTED TO EACH OTHER, so that what is impossible in the Framework may be supplied by adaptation in the Motory Apparatus, and what is impossible in the Motory Apparatus may be supplied in the Framework.

285. THE MOTORY APPARATUS AND THE FRAMEWORK are, in one sense, parts of one whole, and not to be separated in action, as reasons for the construction of each are to be found in the necessities of the other.

286. THE CRANIUM, or framework, containing the brain, is the only part of the whole skeleton that is not constructed with reference to the production of motion.

287. JOINTS to the number of more than two hundred will be needed in the right and left skeleton, and their exquisite beauty, and the perfection of their adap-

tation, will surprise and delight the student when he studies their details.

288. ON THE OTHER HAND, the connecting nerves and the Ganglia that send influences through them must be adapted to the contractions, their combinations and successions.

289. HOW THE NERVE EXERTS ITS INFLUENCE upon the cell, what the essence of that influence is, or how it is generated in the Ganglia, is not known; whether it is of the same nature as the influence producing sensation is uncertain, but it is probably different.

290. THE NERVES THROUGH WHICH MOTION IS PRODUCED must be different from those through which sensation is caused, because both are produced at the same time.

291. IN THE BODY, GENERALLY, BOTH KINDS OF NERVES ARE FOUND in the same bundle, since their outward direction is the same; but as they approach their inner termination, their course being to different Ganglia, they separate and show their twofold character. (Fig. 76.)

292. THE CONTRACTION OF SOME PARTS BEING ASSOCIATED with several others, in the production of different motions, the nerves ought to associate them with several centres.

293. *Illus.*—THE CONTRACTIONS OF THE CHEST ARE USED in simple breathing, in sneezing, in speech, and in several other combinations.

294. THE MOTORY GANGLIA MUST BE DIFFERENT from the Sensatory Ganglia, since their offices are so different, and both are active at the same time.

295. THE MOTORY MUST BE INTIMATELY ASSOCIATED WITH THE SENSATORY GANGLIA, since it is often necessary that the causes of sensation should excite the muscles to activity without the apparent intervention of the mind.

288. What said —? 289. Is it known —? 290. What is said —? 291. What is said —? 292. What is said of —? 293. Illus. 294. Why? 295. Why —?

296. *Illus.*—A BAREFOOT BOY STEPS ON A THORN. His foot must be raised quicker than thought, and the same cause that excites a sensation of pain must cause the ganglion to exert an influence on the contractile muscles that raise the foot.

297. *Illus.*—IN PLAYING AN INSTRUMENT, it is desirable to have the familiar sounds of the oft-played tones influence the motions necessary to produce the succeeding ones, so that the piece may be played without thought.

298. THE GANGLIA OF MOTION DIFFER from those of sensation, and from each other, in size, form, and position, and somewhat in their internal structure; but the rationale of their structure or their action is but very slightly understood, nor can any one say how many of them there ought to be, or are.

299. IT IS PROBABLE NOT ONLY, BUT IT SEEMS NECESSARY, that there should be a ganglion for the production of each class or group of motions, so that they shall take place simultaneously; and the distribution of the nerves seems to prove this.

300. THE SMALL BRAIN, OR CEREBELLUM, has been supposed to be a Motory Ganglion, adapted to harmonize the movements of all parts of the body, and adjust or balance the action of all the muscles.

301. THE MOTORY APPARATUS is conclusively proved to be composed as follows:

MOTORY APPARATUS . . . . . . . $\left\{\begin{array}{l}\text{Ganglia.}\\\text{Nerves.}\\\text{Muscles.}\end{array}\right\}$ Skeleton.

302. THE MOTORY APPARATUS INCLUDES much the larger part of the body by weight; if the skeleton is included, not less than three fourths.

303. THE ELEMENTARY MOTORY APPARATUS is a minute portion of brain, a correspondingly minute nerve, and a muscle-cell, all of which, to be seen, must be exam-

---

296. What if —?    297. What said of —?    298. How do —?    299. What —? 
800. What is —?    301. How — composed?    302. What does —?    303. What is —?

ined with a microscope, and are simply repeated or multiplied billions of times, and properly arranged to produce the grand result, the Motory Apparatus.

Elementary Apparatus.

$$\text{MINUTE} \left\{ \begin{array}{l} \text{Part of Ganglion,} \\ \text{Nerve-fibre,} \\ \text{Muscle-cell,} \end{array} \right\} \times \text{by billions} = \left\{ \begin{array}{l} \text{Entire} \\ \text{Motory} \\ \text{Apparatus.} \end{array} \right.$$

304. *Inf.*—It cannot be difficult to compre-
hend the character, necessities, and methods for perfecting an apparatus so simple in its structure as the
Motory.

305. *Inf.*—To understand the Motory Apparatus is to understand a large part of the body.

306. The Motory Apparatus includes, in one
sense, a part of the Sensatory, as, for the perfect action
of the motory, sensations must be caused. The amount
of influence sent down to cause contraction might determine the intended motion, but any opposition would require greater exertion. The sense of Touch must also
be a guide in the production of many motions.

307. *Illus.*—A man who had lost the sense of Touch
for two years, when seen by the author, would drop anything from his hand except it was under his eye, by
which, also, he was obliged to guide his steps, observing
when his foot was on the ground.

308. Several of the Sensatory Ganglia are
doubtless in intimate relation with the Motory Ganglia,
influencing them to actions required for the safety or
convenience of the different parts of the body.

309. Influences constantly pour in upon the
mind through the Sensatory Apparatus, and influences
flow out through the Motory.

310. A complete circuit is formed by the Sensatory and Motory Apparatus, the mind being included on
the one hand, and the External World on the other, as
follows:

| M I N D | Ganglia; | Nerves; | Organs of Sense | Ear, Eye, Nose, Mouth, Skin, Muscles, | Skeleton. | Ex. World. |
|---|---|---|---|---|---|---|
|  | Ganglia; | Nerves; | . . | Muscles, |  |  |

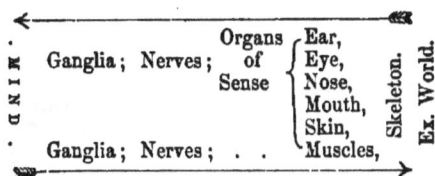

311. MUSCLES ARE INCLUDED among the organs of sense, and are the only organs that require to be connected with the mind by two kinds of nerves.

312. *Inf.*—SENSATION AND MOTION are the only functions needed by the mind, not only for superintending all the operations in the body, but all its relations to the external world.

313. *Inf.*—This is another exhibition of the manner in which beauty, simplicity, and completeness are exhibited in the construction of the human body.

### SECTION III.

### *Emotory Organs.*

314. EMOTIONS are functions, for producing which the mind requires the activity of Ganglia only.

315. THAT EMOTIONS RESULT FROM THE ACTIVITY OF THE GANGLIA is proved by the effects of ether, laughing gas, tickling, and other physical causes that excite emotions, despite the will of the mind that they shall not.

316. HOW THE MIND AND EMOTORY GANGLIA act upon each other, or how they should be classed, is a profound mystery, the solution of which, it is hoped, will be made ere long, and immortalize the discoverer.

317. THE EMOTORY GANGLIA ARE ALSO RELATED to the Sensatory on the one hand, and to the Motory on the other, through which they are constantly manifesting their activities, as illustrated in the following tables, in which the arrows indicate the direction in which the influence is exerted: the Mind acting on and being acted on by the Emotory Ganglia, is connected by two arrows:

Write table. 311. With what —? 312. What said of —? 314. What are? 315. How proved —? 316. What said of —? 317. How —?

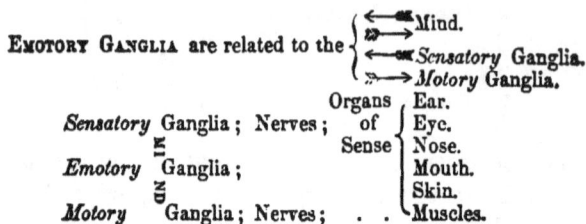

EMOTORY GANGLIA are related to the $\begin{cases} \longleftarrow \text{Mind.} \\ \longrightarrow \\ \longleftarrow \text{Sensatory Ganglia.} \\ \longrightarrow \text{Motory Ganglia.} \end{cases}$

$\begin{array}{ll} & \text{Organs} \\ \textit{Sensatory} \text{ Ganglia; Nerves;} & \text{of} \\ \textit{Emotory} \text{ Ganglia;} & \text{Sense} \\ \textit{Motory} \text{ Ganglia; Nerves;} & \end{array} \left\{ \begin{array}{l} \text{Ear.} \\ \text{Eye.} \\ \text{Nose.} \\ \text{Mouth.} \\ \text{Skin.} \\ \text{Muscles.} \end{array} \right.$

## SECTION IV.

### Intellectory Organs.

318. INTELLECTIONS are functions for the production of which the mind requires Ganglia only. The elements of facts it can receive through the Sensatory Apparatus, and its determinations can be executed through the Motory Apparatus.

319. THE INTELLECTORY GANGLIA MUST BE INTIMATELY RELATED to the Sensatory and Motory; they are also related to the Emotory, but not very intimately—the latter being more readily excited to activity in early life, and the former retaining their vigor till declining years, dying the last of all, in the perfectly healthy person.

320. IT IS VERY IMPORTANT to particularly observe the intimate relation that must exist between the Emotory and Intellectory Ganglia and the Motory; this shows how

321. HABITUAL EMOTORY AND INTELLECTORY ACTIONS INFLUENCE muscular action, so that the Emotory and Intellectory character of a person is sure to be exhibited.

322. *Inf.*—IF ANY PERSON THINKS HE CAN DECEIVE OTHERS in regard to his habitual character, he deceives only himself—a child or even a dog will detect him, at once.

---

Write tables. Explain them. 318. What —? 319. To what —? 320. What—? 321. What do —? 322. What is said if —?

323. *Inf.*—EMOTIONS MOST READILY AFFECTING EX-PRESSION in early life, and Intellections in later life, the respective ganglia by which they are produced must predominate at those periods.

324. The following tables illustrate the relations of the Intellectory Ganglia:

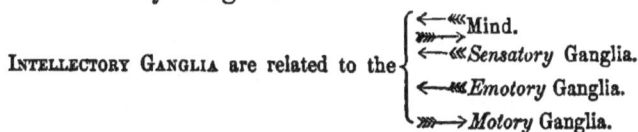

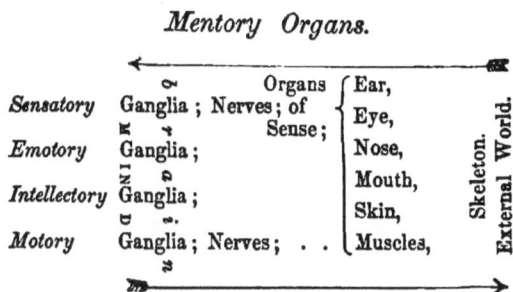

INTELLECTORY GANGLIA are related to the
- ←—*Mind.*
- ⟫⟫→ *Mind.*
- ←—*Sensatory* Ganglia.
- ←—*Emotory* Ganglia.
- ⟫⟫→ *Motory* Ganglia.

*Mentory Organs.*

325. THE MIND IS. ASSOCIATED with four kinds of Ganglia, all double, and collectively forming the brain or brains.

326. FROM THE BRAIN THE MIND REACHES OUT into the various parts of the body, by means of two kinds of nerves. By one it takes hold of the muscles only; by the other it draws in influences from every part of the Body, so that by both kinds of nerves it is, so to speak, everywhere present in the Body.

327. THE MIND SEIZES UPON THE EXTERNAL WORLD and makes use of it by means of six kinds of organs at the extremities of the nerves, supported and assisted by a skeleton.

328. THE MENTORY ORGANS INCLUDE nine kinds, as necessary to constitute them: *Ganglia* (four kinds, *S.,*

---

323. What is said —? 324. Explain table. Write and explain table. 325. With what —? 326. How does —? 327. How does —? 328. What do —?

*E., I., M.*), *Nerves* (two kinds, *S., M.*), *Ear, Eye, Nose, Mouth, Skin, Muscles,* and the *Skeleton,* to which must be added *Blood-tubes,* to keep the organs in good condition, making ten kinds of organs as the number that should be enumerated as Mentory Organs.

329. THE MENTORY ORGANS EQUAL the digits upon both hands, with which they may be associated; putting those which belong to the Head only, on the right hand, commencing with Ganglia on the thumb, then Ear, Eye, Nose, Mouth; Skin on the little finger of the left hand, Muscles, Nerves, Skeleton, and Blood-tubes on the left thumb.

### SECTION V.

### THE USES OF BLOOD.

### *Warming, Cooling, Excretion, Nutrition.*

#### BLOOD-CIRCULATING ORGANS.

330. EVERY ACTION OF THE MENTORY ORGANS— every Sensation, every Emotion, every Intellection, every Motion—is attended by, not only, but dependent upon, a change, called a decomposition, of some of the substance concerned in the action.

331. THE CHANGE OR DECOMPOSITION is such that the substance is no longer fit for use in that part—must be removed, and its place supplied with new material, in order that the part may be kept in a perfect condition for use.

332. THE DECOMPOSITION does not take place in any particular portion of the active part of an organ, but throughout the whole—now here, now there; so that in the course of time, and as the result of many actions, all parts of all the organs undergo a change, its rapidity depending on the amount of action.

---

829. What do —? Subject of Section? 830. What is said —? 831. What is —? 832. What is said — ?

333. *Excretion* is the name given to that part of
the process by which the useless substance is removed.

334. *Nutrition* is the name given to that part of the
process by which the organs are renewed.

335. THERE IS ANOTHER CONDITION ESSENTIAL TO
THE HEALTHY ACTION OF THE MENTORY ORGANS: they
must be kept at certain temperatures, for which

336. THEY MUST BE warmed, on the one hand, if
too cool, and cooled, on the other, if too warm; hence,

337. THE HEALTHY ACTION OF THE MENTORY OR-
GANS REQUIRES that provision be made for accomplishing
four objects or processes: Excretion, Nutrition, Warming, and Cooling.

338. ALL FOUR OBJECTS CAN BE ACCOMPLISHED by
a very simple arrangement.

339. Let minute tubes, smaller than hairs, and with
sides thinner than those of a soap-bubble, be interwoven
like a network through all the active parts of the organs,
and connected with two sets of tubes, one opening into,
and the other out from, the network; let these two sets
of tubes be also connected with a pump that shall receive
the contents of one and pour them out into the other;
then fill this piece of mechanism with a proper fluid, and
connect with it certain arrangements to maintain the
supply in a perfect condition, and the circuit is complete for supplying each of the four necessities of the
organs.

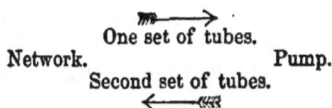

One set of tubes.

Network.                    Pump.

Second set of tubes.

340. THE CAPILLARY (hair-like) TUBES OR VESSELS
is the name of a set of microscopic tubes, that, in the
form of a network, are interwoven through the active
parts of all organs.

FIG. 83.           FIG. 84.

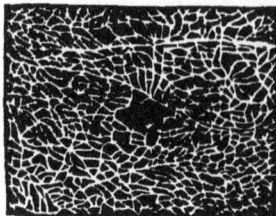 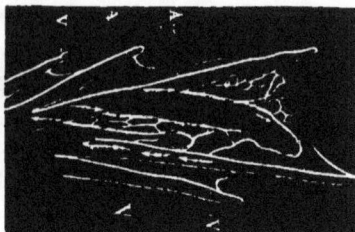

Fig. 83 represents one form of capillary network, very much magnified. Fig. 84, part of a frog's foot, magnified; dotted lines, capillary network; *A*, arteries; *V*, veins; arrows show direction of current.

341. THE VEINS are the names of tubes, continuous with the capillaries, that gradually unite together and lead their contents into the Heart.

FIG. 85.           FIG. 86.

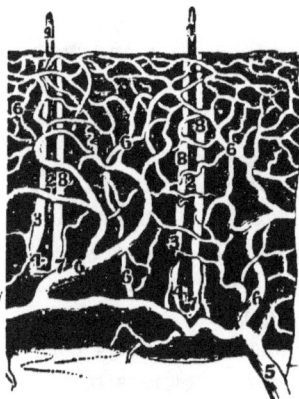 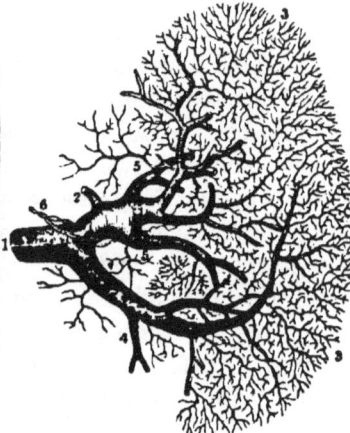

Fig. 85, a section of Skin; 6, small branches of veins connecting with capillary network and uniting to form the large branch, 5.

Fig. 86 represents the Veins of the Spleen uniting to form the trunk, 1.

342. THE HEART is a muscular pouch or pump that receives its contents, on one hand, from the veins, and, being furnished with a set of valves, throws out at each stroke what it receives, meantime, into

---

Describe Fig. 83. Fig. 84. 341. What are —? Describe Fig. 85. Fig. 86. What is the subject of Pl. 5? 342. What is —?

FIG. 87.

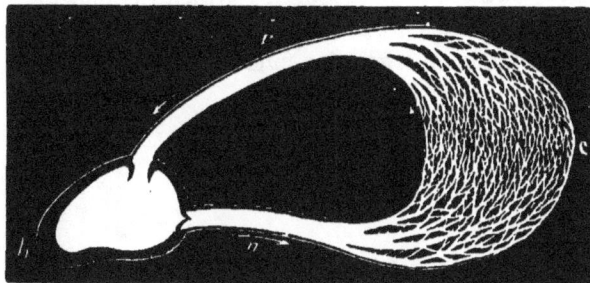

Fig. 87, plan of *h*, Heart, with closed valves, leading into *a*, Arteries; *c*, Capillaries, leading into *v*, Veins, leading through open valves into *h*, ready to contract, closing one and opening the other set of valves.

343. THE ARTERIES is the name of a set of tubes that commence from the Heart as a single trunk, and, dividing and subdividing, become continuous with the capillary network, and pour their contents into it.

FIG. 88.

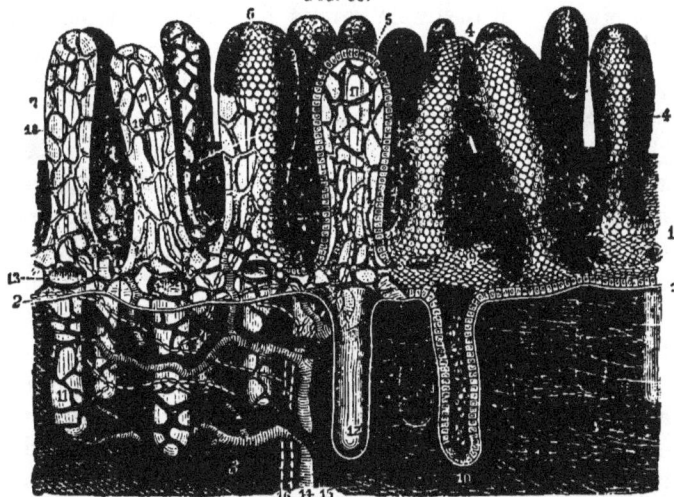

Fig. 88, view of a section of a minute portion of the intestinal canal. 14, Artery, its divisions connected with Capillaries 18 and 11, also connected with 15, a vein; thus Blood is poured through 14, 18, 15.

Describe Fig. 87. 343. What are —? Describe Fig. 44, page 27. Describe Fig. 88. Can the course of the blood be traced from 14 round to 15?

344. THE WHOLE CIRCUIT by which the four necessities of the organs are satisfied, is composed of only four different kinds of organs, as follows:

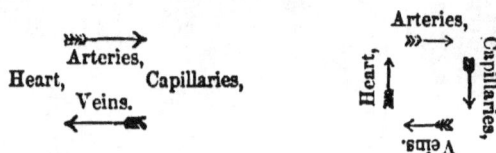

Arteries,
Heart, Capillaries,
Veins.

Arteries,
Heart, Capillaries,
Veins.

345. THE FLUID CALLED BLOOD, with which this circuit is filled, is not a simple fluid, but must be a mixture of four classes of material: 1st, The waste substance produced by the action of organs; 2d, the nutritive material adapted to replace the waste; 3d, the materials for producing heat; and, 4th, the means of cooling the organs.

346. THE MOST WONDERFUL ILLUSTRATION OF THE BEAUTIFUL SIMPLICITY OF THE HUMAN STRUCTURE, and the gaining of desirable results by unexpected means, that can be traced in the whole system, is this one simple arrangement, by means of a single fluid to produce complex and apparently unattainable results.

347. THE BLOOD COURSING THROUGH THE CAPILLARIES can gather every decomposing particle of substance and bear it away; and if in some part of the Blood-circuit modifying organs, and in another part eliminating organs, are located, the substance can either be fitted for some secondary duty, or cast out of the current of the blood, never more to enter it.

348. THE BLOOD CAN NOURISH THE ORGANS, if proper organs for replenishing the Blood are applied to its circuit, so that it can receive nutritious substances and pour them into the capillaries, through the sides of which organs can feed themselves as liberally as their exigencies demand. (See Figs. 90 and 100.)

349. THE BLOOD CAN KEEP THE ORGANS WARM, if

it be itself warmed; it can be, 1st, by the action of
the organs themselves, which always produces heat; and
2d, by receiving substances that are burned in it; also,
if it move around rapidly it will distribute as well as
produce heat more rapidly. This important result can be
gained by having its motion quickened, or its quantity
lessened by removing its water, for that does not diminish its heat-producing substance; the result will be still
greater if its motion is quickened and its water lessened
at the same time. Therefore, there should be connected
with its circuit an organ that can rapidly remove water
without evaporating it, or causing thirst; for if that exist,
water will be drank, and no more heat be produced than
before.

350. THE BLOOD CAN COOL THE ORGANS through
which it passes, if it is itself cooled, as it can be if it
receive substances that prevent the production of heat,
if it be largely diluted, and if it have associated with it
an organ that can take from it water, and evaporate it,
and at the same time cause thirst; for unless it exist and
induce drinking, the blood will not be sufficiently diluted.

351. IT IS EVIDENT that, as the different mentory
organs are composed of different substances, their action must excrete into and draw from the blood different
substances.

352. IT WOULD BE NATURAL TO SUPPOSE that different and corresponding organs would be required to eliminate the different excretions, as well as different organs
to supply the different substances necessary for nutrition,
warming, cooling, etc.

353. DIFFERENT CORRESPONDING ORGANS DO EXIST,
at least to a certain extent, for eliminating the substances
excreted by the action of the mentory organs.

354. THE KIDNEYS HAVE BEEN CONSTRUCTED with
reference to eliminating from the Blood the substances

excreted into it by the action of the branial and nervous organs.

355. THE INTESTINAL CANAL AND SOME OF ITS AP-PENDAGES HAVE BEEN CONSTRUCTED with reference to the elimination from the Blood of substances excreted into it by the action of the muscles.

356. THE LUNGS HAVE BEEN CONSTRUCTED with reference to eliminating from the Blood useless substances arising from the production of heat.

357. THE PERSPIRATORY GLANDS OF THE SKIN HAVE BEEN CONSTRUCTED with reference to removing water from the Blood when it is too warm. (When too cool, the water is removed by the kidneys.)

358. THE WHOLE SIMPLE PLAN BY WHICH EXCRE-TION, NUTRITION, WARMING, AND COOLING, ARE PRO-DUCED, is to have the single compound fluid, Blood, circulate from a Heart outward, through arteries, into and through capillaries, and return through veins—a cluster of organs by which it can be modified, purified, supplied with all needed substances, and cooled, being attached to some convenient part of the circuit, so that during

FIG. 89.

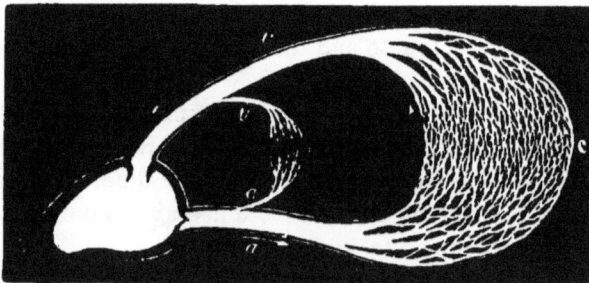

Fig. 89 is a plan like Fig. 87, except that $O$ represents an attached or-gan either for supplying to, or eliminating from, the Blood received through a branch of $a$. At each circulation some of the blood will pass through $O$, and if it should be multiplied the same would be the case with all.

---

355. How —? 356. How —? 357. How —? 358. What is —? Describe Fig. 89.

each circuit a portion of the Blood passes through each eliminating organ (after a sufficient number of circuits, all the blood will be subjected. to the action of each organ), while the blood also receives, during some or all of its circuits, its needed supplies.

359.  THE SIMPLE PLAN IS MODIFIED in different animals and in man to meet the minor exigencies of the case, but the general idea is the same wherever there is a circulation.

360.  THE ORGAN BY WHICH COOLING TAKES PLACE CANNOT BE PART OF A CLUSTER, as an extensive surface exposed to the air is essential to rapid evaporation, and this is found only in the skin.

361.  IT WILL BE EASILY UNDERSTOOD that the activities of different parts of the body must produce very different quantities, as well as qualities, of substances to be eliminated, and will require equally different quantities of supplies.

362.  IT WILL BE AS READILY INFERRED, that if more substance is to be thrown off or received through one organ than through another, there must be larger blood-tubes, or more of them, leading to the organs having most to do, and the number or size may be increased until all the Blood of the circuit shall be included.

363.  CASES EXIST IN THE HUMAN BODY in which the organs of supply or elimination require blood-tubes of all sizes, from the diameter of a pin up to that of the entire circuit.

364.  WHETHER THE BLOOD AT EVERY CIRCUIT MUST BE SENT THROUGH AN ORGAN, WILL DEPEND UPON, 1st, how fast the substance to be eliminated accumulates, and how long it can remain with safety in the blood; and, 2d, how large a supply the Blood can receive at each circuit, and how rapidly it is used.

365.  THE CIRCUMSTANCES HITHERTO MENTIONED do

---

359. How is —?  360. Why —?  361. What —?  362. What —?  363. What —?
364. — what?  365. What said of —?

not vary the simple plan of the Blood-circuit; it may
also be mentioned that the eliminating or supplying organs may be attached to either the arteries or veins, or
a part of them to each, without essentially varying the
simple plan.

366. IF THE FORCES THAT MOVE THE BLOOD THROUGH
A SINGLE CIRCUIT DO NOT SUFFICE to move it through
an eliminating organ with sufficient rapidity, another
Heart must be introduced, and the veins which lead the
Blood back must open into that Heart, from which another set of arteries must arise and lead to a set of
capillaries in the eliminating organ, from which veins
must lead to the first heart, as follows :

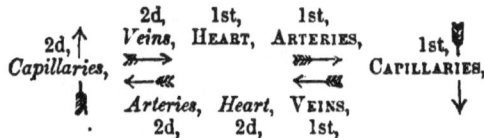

367. THE TWO HEARTS AND THEIR CONNECTIONS ARE
SPOKEN OF as forming a double circulation, called the
greater or systemic, and the lesser respiratory or pulmonary circulation, when in fact all the parts constitute
but a single circulation, and the greater should be called
the greater part of the circulation, and the smaller the
smaller part, etc. It is no longer a simple but a compound circulation.

368. THE TWO HEARTS SHOULD BE LOCATED by the
side of each other, and enclosed within the same external
covering; therefore externally they appear to be one
thing, and are, in fact, so called, viz., the Heart. For
the purpose of close packing, the tubes or vessels leading
into and out of them are so intertwined that great complexity at first appears, where, in fact, there is great
simplicity. (See pl. 5 and 27.)

---

366. What —? Write and describe table. 367. How —? 368. How —? Is
there complexity or simplicity in arrangement of Blood-tubes?

FIG. 90.

Fig. 90, plan of compound circulation; 1, 2, two Hearts, enclosed in one membrane, so that externally they appear as a unit; *a*, arteries leading out of 1 and branching upward and downward, and leading into *c*, capillaries, through which their contents can pass into *v*, veins that open into 2, from which *a''* lead into *c''*, capillaries, through which their contents can pass into *v''*, opening into 1. Thus a compound circuit is formed, so that when the Hearts alternately contract and relax, they can pour their contents through the entire course. The whole of the Blood in this case passes through the eliminating organs, *R*, respiratory, also marked lungs.

A cluster of organs is represented by 4, 5, 6, 7, 8, 9, of which those marked *d* are digestory, *m*, the spleen, modificatory, and *e*, the kidneys, eliminatory. These receive at each circulation a part of the Blood thrown out by 1 downward, and, after an uncertain number of circulations, each of them must have the opportunity of acting upon all the Blood, if it awaits their action. It is also noticeable that the Blood circulating through 4, 5, 6, 7, circulates through 8 before it passes into the veins.

1 is called the Left, Back, and Systemic Heart—its initials, *S H;* *a c* and *v* are called Systemic Arteries, Capillaries, and Veins, and have the respective initials, *S A, S C, S V;* 2 is called the Right Front, Pulmonary, or Respiratory Heart; *a''*, *c''*, *v''*, are called the Pulmonary or Respiratory Arteries, Capillaries, and Veins, having the initials, *P A, P C, P V,* or *R A, R C, R V.*

Describe Fig 90. How many arteries leading out from each Heart? How many veins back?

369.   In man the results of producing Heat and the like results to be eliminated are so great and so constant, that large corresponding eliminating organs are required to receive all the Blood each circuit; and an additional Heart and vessels, to drive the Blood around through those organs, will also be required.

Fig. 92.

Fig. 91, thumb of natural size, with the skin and nail transparent, to show the commencing lymphatics, a little magnified; a portion of the skin is removed at *a* to show the larger branches.

Fig. 92, a view of hand and arm, dissected to show, 7 to 12, superficial lymphatic branches; 13, two of several glands in the arm-pit; 11, another; 1 to 6, superficial veins.

Fig. 93, plan of lymphatics of the whole Body, including the digestory lacteals opening into the veins of the neck.

Fig. 93.

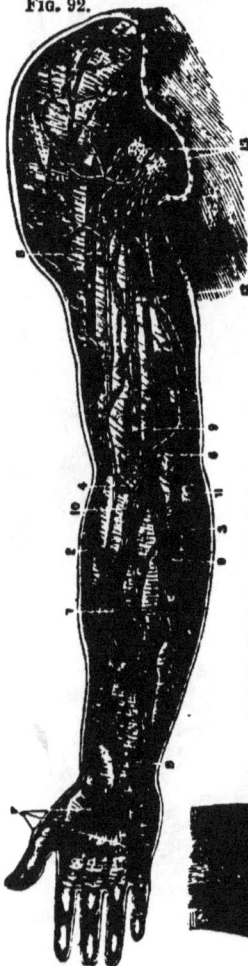

Fig. 91.

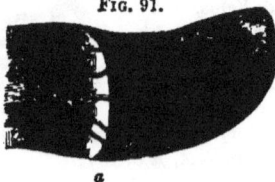

*a*

369. What are —?  Repeat the order of parts through which Blood circulates. Describe Fig. 91.  Fig. 92.   Fig. 93.

370. In man, Lymphatics are also required. They are a system of tubes and glands, commencing in every organ except the brain and nerves, and opening into veins. Why the veins cannot take up all the substance they contain is not known. It may be that it requires a peculiar modification, and, if all the glands necessary should be in one mass it would find no convenient place in the Body, and therefore they have been distributed for convenience of packing, and their tubes have been adapted to bring to them substances adapted to their action. They are attachments to the veins, and gather into them the lymph, modified by their action, therefore the

B. Circulatory Organs = $\begin{cases} \text{Hearts, Arteries, Capillaries,} \\ \text{Veins, Lymphatics.} \end{cases}$

## SECTION VI.

### *Nervous Circulatory Organs.*

371. It is easy to conceive that it will be very important, even essential, to have some means of controlling the circulation of Blood, now quickening, now slackening it, both as to the whole Body and as to each part through which it moves.

372. The Circulation must go on and be regulated during sleep, as well as during waking hours, and must, therefore, as well as for other reasons, not be directly under the control of the Mind.

373. There is no means of controlling the circulation except by ganglia (of which the case would require many) and nerves, and these must be indirectly connected with the Brain and Mind. (See Pl. 3*.)

374. This class of nerves is called the Ganglionic system, its ganglia being so numerous; the Sympathetic system, since it was supposed to weave together all parts

---

370. What required also —? What said of them? Write table.  371. What —?
372. When —?  373. What is said of the —?  374. What is —?

in sympathetic action; and the Organic system, to distinguish it from those nervous parts directly under the
influence of the mind, which are called the Nervous system of Animal life.

375.  IT IS ALSO THOUGHT that this same system su-
perintends the activities of all the organs of the Blood-
making class.

376.  IN THIS WORK, THE CONTROLLING GANGLIA AND
NERVES WILL BE CALLED the N. (nervous) Circulatory Organs, as being concerned and having their chief duty to
regulate the circulation of blood, and will in the tables
be placed next to the Mental Organs, as being connected
with them, and the connecting link between them and
the Sanguificatory Organs.

$$\text{N. CIRCULATORY ORGANS} \begin{cases} \text{Ganglia.} \\ \text{Nerves.} \end{cases}$$

SECTION VII.

*Respiratory Organs.*

377.  THE ANCIENTS FABLED that there were three supreme gods, the trinity of Pluto, Neptune, and Jupiter,
and their dwelling-places were respectively the earth,
the sea, and the air, over which they presided, Jupiter
being the most powerful and the supreme.  Thus did
they represent the solid, fluid, and gaseous elements of
nature, and their relative importance.  And, as nature
exhibits herself under the three forms of matter, so does
the body in this respect represent nature, the blood being
composed of the three forms; so do we also add to the
blood, matter under the three forms of food, drink, and
· air, while also the eliminations take place in the same
three forms: here also Jupiter is supreme, for the demands for air are more imperative than those for water
or food.  The last can be "stocked" in the blood in ad-

---

vance of need—so can the water, to a more limited extent; but air can only be stored for a moment's time, and with each returning circuit of the blood the life-giving influences of the air must be breathed upon it, and its poisonous burden drawn out, or the pulse fails to perform its now useless task, and animation will be suspended.

378. THE RESULTS OF PRODUCING HEAT and the like results, which are so great in quantity as to require that all the Blood at each circuit should be sent through an organ adapted to eliminate them, are gases dissolved in the Blood, or substances that very easily become gases by contact with pure air, and by its influence are in either case easily withdrawn from the blood; therefore,

379. THE RESPIRATORY ORGANS ARE REQUIRED, in order that the Blood and Air may be freely and rapidly brought within influence of each other.

380. TWO THINGS ARE THEREFORE TO BE PROVIDED FOR: one, the mode of bringing the Blood to the Air, and the other, the mode of bringing the Air to the Blood.

381. THE METHOD OF BRINGING THE BLOOD TO THE AIR has been described in a previous section. A Heart exists, ready to pour its contents into arterial tubes leading to a network of capillaries, through the sides of which air, if present, can act upon and be acted upon by the Blood, when it is ready to return to the other Heart, and be sent on its way.

382. IT ONLY REMAINS TO DEVISE some way by which Air shall be brought to act freely through the sides of the capillaries.

383. IF A SMALL SAC OR CELL BE CONSTRUCTED with exceedingly delicate sides, composed in part of the network of capillaries, with a tube through which Air can be drawn into the cell, the material of which has a certain amount of elasticity, it will be complete, and a minute lung will be formed.

---

378. What form have —? 379. Why —? 380. What —? 381. What is —? 382. What —? 383. What —?

5*

384.  A Lung is an exceedingly simple organ in its ideal or plan state.

385.  A real Lung is merely an aggregate of cells such as just described, the whole being covered in by a skin or membrane, called Pleura.

FIG. 94.

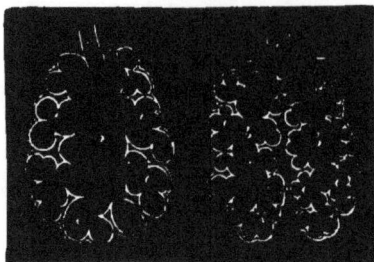

Fig. 94, plan of a very highly magnified cluster of air-cells, c, into which a allows the air to enter.

Fig. 95 represents the windpipe dividing into two branches, each of which subdivides in its corresponding lung, much more minutely than represented, and opens into clusters of cells, like Fig. 94. 1, 2, 3, a continuous membrane, forming surface of lungs.

FIG. 95.

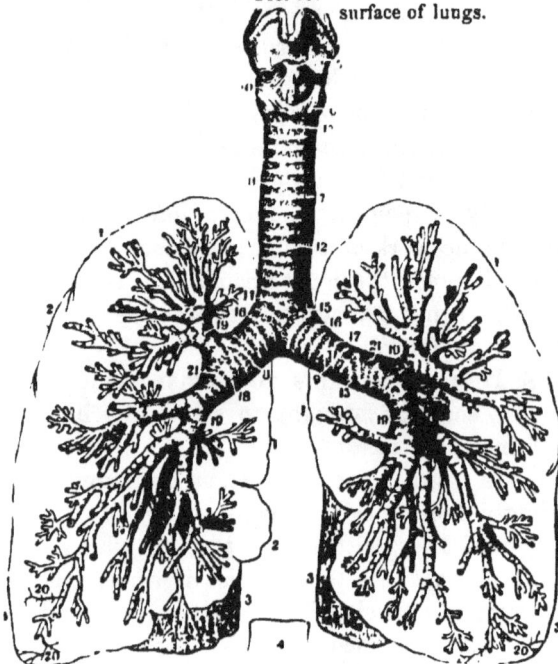

384. What —? 385. What is —? Describe Fig. 94.  Fig. 95.  Can the enlargement 5 be felt?  What its usual name?

386. THE NUMBER OF CELLS and the size of them will vary according to the size of the Lungs and the individual, and the amount of air that will act in a given time upon the blood will vary accordingly.

387. THE CELLS ARE COLLECTED, like the leaves on a tree, into clusters called lobules, and the lobules into clusters called lobes, and these again into the right and left (halves of) lung; the tubes from the cells, lobules, lobes, and lungs unite to form the trachea, as the trunk of a tree is formed by the uniting of the branches.

388. THE MODE OF FILLING AND EMPTYING THE CELLS is as simple as their structure.

389. IF AN ELASTIC BAG WITH AN OPEN MOUTH HAVE THE PRESSURE OF THE AIR REMOVED FROM ITS EXTERNAL SURFACE, the pressure of the air through the mouth will distend the bag; if then the external pressure is allowed to act again, it will balance the internal pressure, and the elasticity of the bag will throw out the air.

390. SUPPOSITION: Let 15 = External pressure.
     " 15 = Internal pressure.
     " 10 = Elasticity.

391. WHEN EXTERNAL PRESSURE, 15, AND ELASTICITY, 10, ACT TOGETHER, 25, they overcome Internal pressure, 15, and expel the air; take away External pressure, 15, and Elasticity, 10, is overcome by Internal pressure, 15, and the lung is distended.

392. To CAUSE BREATHING, viz., the passing of air into the cells and out again, all that is necessary is to contrive some means to remove the pressure of the air from and return it against the external surface of the lungs, which is the easiest thing in the world to do.

393. PLACE THE LUNG in an air-tight box, with the pipe that connects with the air-cells extending through and closely fitting an opening in the side of the box; this must also be so constructed as to be readily enlarged in all directions and again returned to its previous size.

394. WHEN THE BOX IS ENLARGED the pressure of the air, 15, is removed from the external surface of the Lung, and the internal pressure of the air, 15, correspondingly distends the Lung, overcoming its elasticity, 10; and when the box is being restored to its former size the external, 15, balancing the internal, 15, the elasticity, 10, of the Lung as surely drives out the air, and the inner surface of the box and the outer surface of the Lung will be in the gentlest contact with each other all the while that the box is enlarging and diminishing, and only by closing the pipe leading into the Lung can the box be made to press upon the Lung.

395. EACH CHEST IS such a box, most readily enlarged and diminished by means of the muscles and car-

FIG. 96.

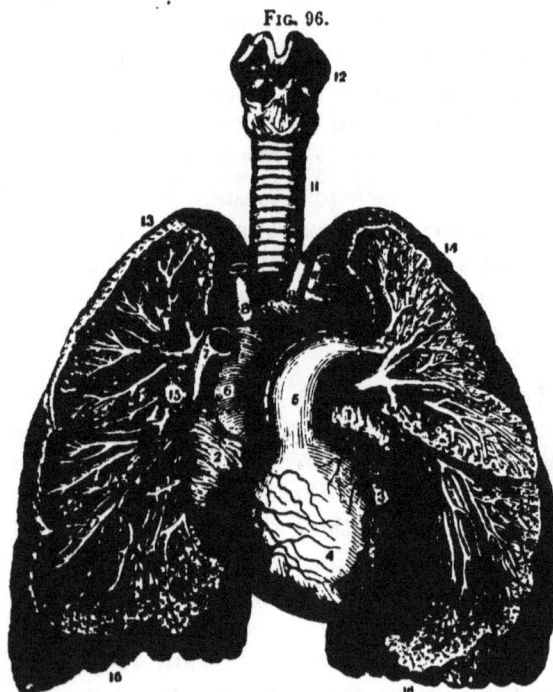

Fig. 96 represents the Hearts, 3, 4, between the lungs, distended as in life with the front parts cut away, to show the divisions of veins, arteries, and 12, the windpipe; 6, systemic artery; 7, systemic vein; 1, 2, right and left auricles.

394. What effect —? 395. What —? Describe Fig. 96. Is the Heart in the centre of the chest?

FIG. 97.

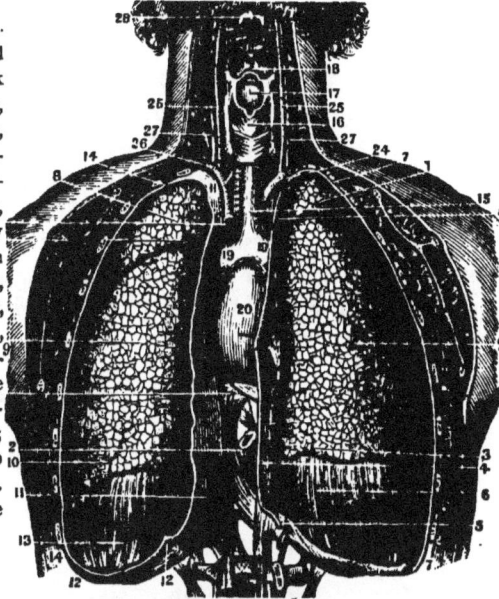

Fig. 97 repre-
sents the spinal
column, the back
parts of the ribs,
the œsophagus,
and the aorta dis-
sected away, ex-
posing 1, 2, 8, 9,
the Lungs, as they
would appear in
life and full of air,
filling their boxes,
formed by 4, 7, 11,
14. 6, 12, the Di-
aphragm; 20, the
Heart; 19, Divis-
ion of Windpipe;
17, opening into
it; 18, Tongue.
6, 13, arches of the
Diaphragm.

FIG. 98.

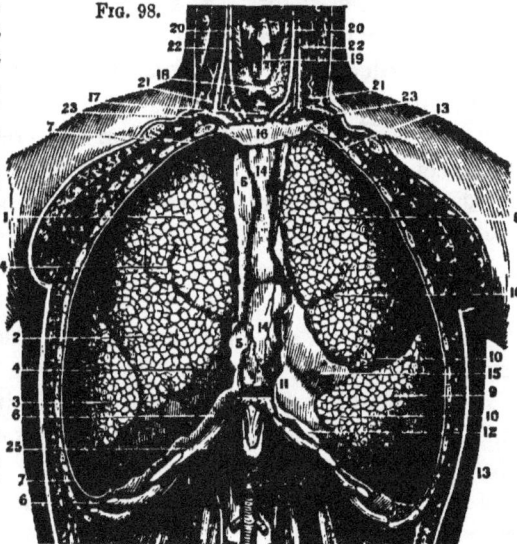

Fig. 98 rep-
resents breast
bone and front
part of ribs,
dissected away
showing the
Lungs, 1, 2, 3,
8, 9, 10, filled
with air, as in
life. 15, posi-
tion of lower
point of Heart,
7, 7, lining of
Chest, corres-
ponding to 7,
7, Fig. 97; 1,
2, 3, the three
lobes of one
side; 8, 9, the
two lobes of
the other.

Describe Figs. 97 and 98. How near to the neck do the lungs appear to be?
Does the windpipe enter top or side of lungs? What keeps them up?

tilages that compose so large a portion of its walls, together with other muscles that assist them, and no position in which the Body can be placed will cause pressure on the lungs, except the windpipe be closed. .

396. THE RESPIRATORY ORGANS INCLUDE as follows:

RESPIRATORY ORGANS = { (Nose Pharynx), Windpipe, Lungs, Trunk-walls, and Diaphragm ;

and all that is necessary in the respiratory process is pure air, active Trunk-walls and Diaphragm, well-formed Lungs, and a good circulation of Blood through them —four distinct topics.

RESPIRATORY TOPICS = { Air. Trunk-walls and Diaphragm. Lungs. Circulation P. or R.

SECTION VIII.

*Digestory Organs.*

397. THE USES OF THE DIGESTORY ORGANS are to supply the Blood with Water and Food, eliminate from and modify the Blood.

398. WATER ENTERS THE BLOOD and exists in it as water, constituting a very large proportion of it.

399. WATER REQUIRES NO PREPARATION, and only needs to be swallowed into a pouch, in the sides of which are numerous blood-vessels to drink it in and mingle it with their contents. (See 8, Fig. 70.)

400. FOOD IS VERY DIFFERENT FROM AIR AND WATER in this respect, that these if pure are always the same, and should vary at different times only in quantity, while Food at different times must not only vary in quantity, but in character very essentially.

401. IT IS EVIDENT that the different organs of the body require that the Blood should be supplied with

___

396. What do —? What are Respiratory topics? 397. What are —? 398. How does —? 399. Why does —? 400. How is —? 401. What —?

different kinds of nutrition : it must also be supplied
with material for warming and cooling itself.

402. AT LEAST THREE DIFFERENT KINDS OF FOOD
must be largely supplied : for nourishing the nervous
organs, for nourishing the muscular organs, and for fuel
(the calorific kind) while cooling, and smaller quantities
of other substances will be constantly or at times wanted.

403. FOOD DIFFERS FROM AIR AND WATER IN THIS
ALSO : that they are found in nature and need no preparation, while food is obtained by labor, needs much preparation, and must be separated from useless or waste
substances incorporated with it.

404. FOOD MAY BE ARRANGED in four classes : *Nutrition*, which is of two very conspicuous varieties, the
nervous and muscular, and several less conspicuous varieties; *Calorific*, which includes much the larger part of
food; *Cooling* and *Waste*, that include unmasticated
seeds and indigestible harmful substances that should
never be eaten, and the fibrous woody parts of vegetables
and fruits, that are useful especially in spring and summer.

$$\text{FOOD} = \begin{cases} \text{Nutritive.} \\ \text{Calorific.} \\ \text{Cooling.} \\ \text{Waste.} \end{cases}$$

405. (THE BODY HAS BEEN CONSTITUTED for activity,
and is kept in the best health when it is properly active.
As it was not anticipated that man would find his food
separated from waste, he has been constituted to separate the useful from the useless substances in his digestory organs; hence it is best that he should do so.)

406. ONE REASON THAT MAN REQUIRES DIGESTORY
ORGANS, and one use of them, is to dissolve out the useful from the waste substance; and as this binds the useful with many fibres closely woven, in some cases the
process will be correspondingly tedious.

---

402. What —?   403. How does —?   404. How —?   Table.   405. How —?
406. What is — ?

407. WHEN WE WISH TO SEPARATE SUBSTANCES intimately united, our first operation is to grind them finely, and then dissolve them in some fluid, if possible.

408. FOR THIS PURPOSE the Digestory Apparatus has been furnished with a Mouth, in which the food can be cut, chopped, bruised, ground, or, best name of all, chewed, at the same time that the Salivary Organs pour into the mouth an abundance of fluid.

409. WHEN SUBSTANCES THAT WE DESIRE TO SEPA-RATE are reduced to a comminuted state and mixed with fluid, we put them into a receptacle where they will be kept warm, and from time to time add such fluids as will favor the process of solution; and as the substances separate we remove them.

410. THE STOMACH is a warm, distensible receptacle, situated as near the mouth as the position of the Lungs and Heart will permit, and on its own account in the very position that is best for it. (See Fig. 70.)

FIG. 99.

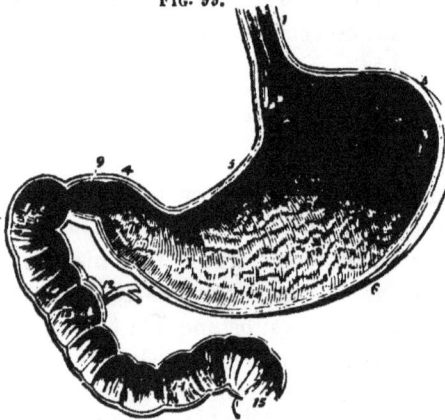

Fig. 99 represents a section of the distended Stomach, and, 10 to 15, Duodenal portion of 2d Stomach; 1, lower portion of Œsophagus; 9, Pylorus, that can close that outlet; 12, duct or tube from the pancreas. The openings of the Gastric Glands are too small to be shown, but they are similar to that of 10, Fig. 87, and 3, Fig. 103.

411. THE SIDES OF THE STOMACH ARE FURNISHED with small organs called Gastric Glands, to pour into the Stomach a suitable supply of proper fluid, by which the further solution and preparation of the food is produced.

407. How do we do —? 408. What furnished —? 409. What do with —? 410. What —? Describe Fig. 99. 411. How —?

412.  THE WASTE FOOD REQUIRES, of course, merely a separation, when it is at liberty to pass on and leave the body.

413.  THE COOLING FOOD is acids, the precise action of which in preventing the production of heat is not known; but they are without doubt merely dissolved, passed into the blood as acids, and there in some way perform their office.

414.  ONLY TWO CLASSES, the nutritive and the calorific, remain to be considered.

415.  IT MUST BE REMEMBERED that the food of animals contains the substances required, not in the condition of chemical elements, but compounded by plants into, or very nearly into, the precise substances needed by the organs.  The question is not merely, does an article or mixture, such as might be made in any chemist's shop, contain the same elements as the Brain or Muscles or Bones, but have they the proper form for serving as food, and also can they be readily dissolved ?

416.  IT IS THE OFFICE OF THE PLANT to compound that the animal may decompound.  The former synthetizes that the latter may analyze.

417.  WHEN ONE ANIMAL EATS ANOTHER, it eats what has before been eaten, and what was a part of a plant at some time.

418.  THE NUTRITIVE PART OF FOOD REQUIRES but little more than to be dissolved out of the substances with which it is combined, and passed into the circulation.

419.  THE CALORIFIC FOOD is of five kinds: 
{ Fat.
Starch.
Gum.
Sugar.
Gelatin.

THE FIVE VARIETIES ARE VERY SIMILAR TO EACH OTHER in chemical composition—will readily burn if thrown into the fire; each, if fed to a pig, will cause him to fatten, and the five will collectively be better than either separately.

420. By GELATIN is meant animal, not vegetable, gelatin. It is merely to be dissolved, when it is ready to pass into the Blood.

421. SUGAR needs only to be dissolved, when it is ready to pass into the Blood, of which it is an element.

422. GUM is a kind of sugar (not spruce gum and the like, which are very unhealthy), and very readily converted into it, or it may pass into the Blood by simply being dissolved.

423. STARCH IS IN THE FORM of grains, too large to pass into the Blood, and, if it could, would obstruct the circulation in the minute tubes.

424. STARCH MUST BE CHANGED INTO SUGAR, and as the chemist can in various ways effect the change, it would not be astonishing if the same could be done in the body.

425. IT WOULD NOT BE LIKELY that the same fluids adapted to dissolve out the Waste substance and leave the Nutritive and Calorific food, would be sufficient to change the Starch.

426. FAT ALSO, THOUGH A COMPONENT OF THE BLOOD, cannot pass into it till an emulsion has been made of it and some other fluid.

427. TWO MORE FLUIDS at least must be supplied to the digestory process, one on account of Starch, and another on account of Fat.

428. IT WOULD BE ADVISABLE NOT to have these supplied till the process of solution and separation had to a degree taken place.

429. ANOTHER RECEPTACLE MUST BE ADDED to the stomach to receive its contents when they have passed through the changes necessary there, and the two fluids must be at once supplied. (17, Fig. 100.)

420. What said of—? 421. Of—? 422. Of—? 423. Of—? 424. Of—? 425. What —? 426. What said of—? 427. Why supply—? 428. What—? 429. Why must—?

Fig. 100, a plan of Di-
gestory apparatus, the head
being divided on the mid-
dle line and turned to the
right, showing the mouth,
1, extending back into the
pharynx, 9, which contin-
ues into 3 the œsophagus,
represented open above, entire at its centre, and open
again below, continuous
with the stomach distend-
ed and the front half re-
moved. 15, is the pyloric
orifice through which the
contents of the Stomach
pass into the Second Sto-
mach, the upper or duode-
nal part of which is re-
presented open; 17 is the
same entire, coiled from
side to side till again opened at 27, just beyond which
it communicates with the
Colon represented open
throughout its entire ex-
tent; 23, duct from Pan-
creas opening at 16; 18,
Gall-bladder; 19, its duct;
20, duct from liver; 21, the
trunk of both ducts open-
ing at 22 (and sometimes
through the same orifice as
23 does). This cut, therefore, illustrates in one view
all the organs of the Digestory Apparatus except the
glands of the Mouth, Sto-
mach and Second Stomach.
The latter two kinds are
too small to be represented
in a cut of this size. 41,
orifice leading to ear; 11,
soft palate; 10, spinal canal; 5, windpipe; 8, lar-
ynx.

FIG. 100.

Describe Fig. 100. What is the position of œsophagus in relation to windpipe and spinal column? How many times across does 2d stomach pass?

430. THIS RECEPTACLE MUST HAVE a great extent of surface in order that the substances may be taken into the Blood as rapidly as they are in a proper condition for it.

431 THE SECOND STOMACH has been made a part of the Digestory Apparatus, and to serve in all the respects mentioned.

432. TO THE UPPER PART OF THE SECOND STOMACH an organ called *Pancreas* has been attached, for the purpose of pouring into it a fluid called Pancreatic Juice, necessary to make an emulsion with the Fat.

FIG. 101.

Fig. 101 represents *s* back surface of Stomach and *l* lower surface of liver, turned up to show *g* Gall-bladder and 6, 5, 4, the head, body and tail of the Pancreas in section to show its duct along its centre opening into duodenum near *d* (see Fig. 100).

FIG. 102.

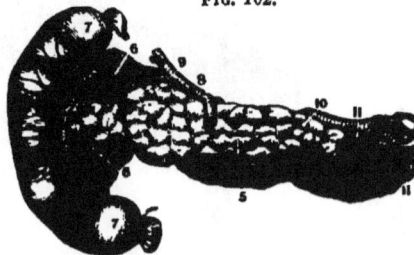

Fig. 102, 6, 5, 4, Pancreas entire; 7, distended portion of 2d Stomach tied at each end.

433. IN THE SIDES OF THE SECOND STOMACH have been placed two kinds of Glands, called Brunners and

Lieberkühns, after their discoverers, which supply a fluid that assists in changing the starch into sugar.

434. To STILL FURTHER ASSIST in the process of separating some of the nutritious substances and fitting them to enter the Blood, the Second Stomach is furnished with another very large Gland, the Liver and its accompanying Gall-bladder, that pour their fluids, Bile and Gall, through a small tube, into the Second Stomach near the orifice of the Pancreatic Duct. (See 20, Fig. 100.)

435. THUS THE SECOND STOMACH, or small Intestinal Canal, is supplied with all required fluids, and being long and with a curious arrangement for increasing the extent of its inner surface, is admirably adapted to keep the food warm and retain it until it is all dissolved and prepared to enter the Blood, or else, worthless, is only fit to be discharged into the Colon.

Fig. 103 shows at 1 how the inner surface of the Second Stomach is increased by folds, as they are called, of the inner surface or mucous membrane of the canal. 4 indicates the sinewy or fibrous structure which is around the tubes, and connects with the serous or outer surface, or of the canal, and also the orifices 3, and the tubes 2 of the tubular or Lieberkühns glands.

FIG. 103.

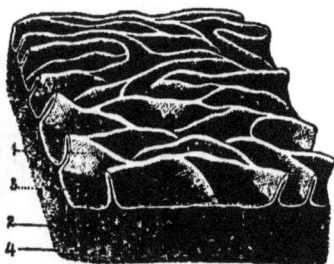

436. THE COLON IS NEEDED as a portable reservoir to receive the waste from digestion and the substances eliminated from the Blood in the various parts of the Digestory Canal and its dependencies.

437. *Inf.*—THE DIGESTORY APPARATUS must be *Modificatory* as well as *Eliminatory*, since the removal of the various fluids mentioned must have an essential influence in modifying the Blood.

438. To SUM UP THE ORGANS NEEDED IN DIGESTION:—There are three pouches or receptacles for the

food: the Mouth, Stomach, and Second Stomach. There are three divisions of the Digestory Canal above the Stomach: the Œsophagus, Pharynx, and Mouth. There are three divisions of the canal below the Œsophagus: Stomach, Second Stomach, and Colon. Six divisions of the central canal. The Second Stomach is subdivided into three divisions: Duodenum, Jejunum, and Ilium. The Colon is subdivided into Vermiform, Appendage, Cœcum, Ascending, Transverse, Descending, Sigmoid flexure, and Rectum; and along the whole canal are situated the organs called Glands, to form fluids from the Blood and pour them into the Canal.

439. THE LACTEALS ARE OFTEN CLASSED, and perhaps most properly, among the Blood-tubes, or that part of them called Lymphatics; for the Lacteals are constituted like the Lymphatics, and perform similar duties in a similar way; yet they are usually reckoned among the Digestory Organs. (See Fig.)

FIG. 104.

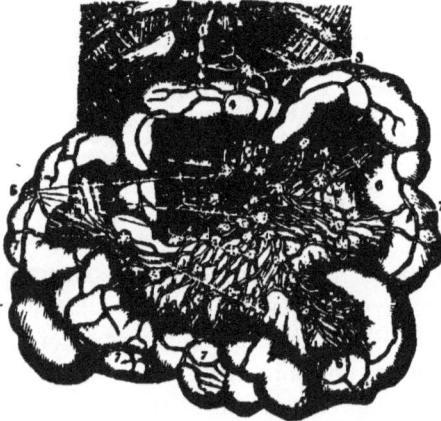

Fig. 104 represents a coil 7 of the Second Stomach turned back and other parts dissected away to show the lacteals, 6, glands, 5, and the main duct, 1, which extends up and connects with the veins in the neck, as shown in fig. To observe how the lacteals commence, see 17, 16, Fig. 87.

440. THE USE OF THE LACTEALS was at one time thought to be to gather all the digested food and lead it into the Blood.

---

Divisions of Digestory canal? Subdivisions —? 439. How —? Describe Fig. 104. 440. What said of —?

441. It is now ascertained that the Lacteals convey but a smaller part of the digested material, cheifly one kind of it, the fat made into an emulsion.

442. The Digestory Organs, including Lacteals, will be as follows:

| | | |
|---|---|---|
| Mouth, | | |
| Salivary Glands, | | 1st Div. |
| Pharynx, | | |
| Œsophagus, | | |
| Stomach, | | 2d Div. |
| Gastric Glands, | | |
| Second Stomach, | 1st Subdiv. | |
| Pancreas, | | |
| Liver, | 2d Subdiv. | 3d Div. |
| Gall-bladder, | | |
| Lieberkühn's Gl'ds, | | |
| Brunner's Glands, | 3d Subdiv. | |
| Peyer's Glands, | | |
| Colon, | | 4th Div. |
| Lacteals, | | 5th Div. |

DIGESTORY ORGANS {

443. Nine Fluids in all are required, produced in as many different kinds of organs, to dissolve out the various useful substances from the food, and prepare them to enter the Blood, as follows:

| | |
|---|---|
| Saliva Parotid, | |
| Saliva Submaxillary, | Stomachic. |
| Gastric Juice, | |
| Pancreatic Juice, | |
| Bile, | Pre-Intestinal. |
| Gall, | |
| Lieberkühn's Glands Juice, | |
| Brunner's Glands Juice, | Intestinal. |
| Mucus, | |

DIGESTORY FLUIDS {

444. *Inf.*—It would naturally be inferred that a process requiring so many fluids must be a very complicated one.

445. It must be remembered that all the fluids are not required for any one element of the food, but that one fluid attacks one element, and another fluid another; thus the matter is very much simplified.

446. *Inf.*—A process requiring so many organs performing so many actions, beyond the reach of ob-

---

441. What —? 442. Write and describe table of —? 443. What —? 444. What
—? 445. What —? 446. What said of —?

servation, it would seem must be shrouded in impene-
trable mystery.

447.  IT EXCITES THE HIGHEST ADMIRATION AND
GRATITUDE to observe the results of the untiring labors
that have been bestowed upon research in this field, assisted by a few fortunate accidents.

448.  THE WONDERFUL PROCESSES OF DIGESTION, of
such immense practical value to man, can now be understood in all their most important and useful aspects by
an ordinary person with moderate application.  There
is no occasion for the student to feel discouraged, as he
will be well satisfied when the process and organs are
treated upon in detail.

## SECTION IX.

### *Eliminatory Organs.*

449.  THERE IS A HARMONY between the constitution of Nature generally and that of the Human Body,
that when perceived cannot fail to excite the liveliest
feelings of admiration and enlist the deepest interest.
This harmony is especially recognized in the study of
the relations of the function of Elimination.

450.  THE LUNGS are important Eliminators of Carbon in the form of carbonic acid; but when hot weather
rarefies the air and elevates its temperature, the action
of the Lungs is correspondingly limited, for less air enters them at each breath, and it is so nearly the temperature of those organs that it is but slightly expanded in
them, and does not therefore distend them to the degree
most highly promoting the action of the air and blood
upon each other, and the blood returns from the Lungs
but partially relieved of carbon.

451.  *Inf.*—IN HOT WEATHER, little heat should be
produced in the Body, so that a feeble action of the
Lungs is desirable at that time.

---

447. What —?  448. What said of —?  449. Where —?  450. What are —?
451. What should be —?

452. The Carbon of the Blood must at all times be removed, and some organ must do it that will not produce heat.

453. The Liver increases its activity in hot weather, and draws from the Blood the carbon produced by the action of the organs, in all of which carbon is a constituent.

454. In hot weather the appetite will not crave calorific food, in which carbon abounds, but will crave fruits, vegetables, acids, waste food, etc., that require a large quantity of Bile and Gall for their perfect digestion, and of course a full degree of activity in the organs supplying them.

Fig. 105.          Fig. 106.

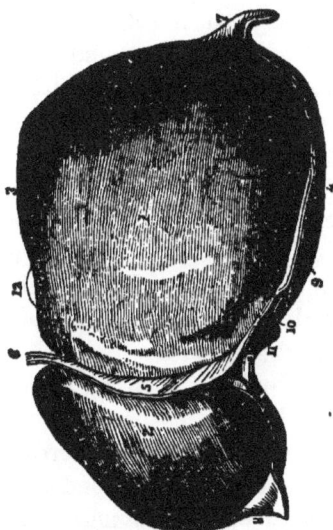 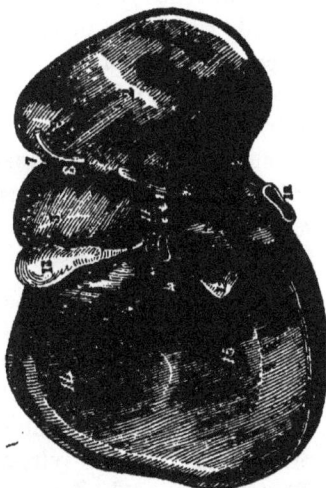

Fig. 105, upper surface of Liver: 1, larger, 2, smaller lobe; 3, front thin edge; 4, back thick part; 10, vein—the upper surface is very convex; 12, gall-bladder.

Fig. 106, lower surface of Liver: 1, right, 2, left lobe; 3, middle lobe; 12, vein, corresponding to 10, previous figure; 13, gall-bladder.

452. What said of — ? 453. What said of — ? 454. What — ? What do vegetables, etc., require ? Describe Fig. 105. Fig. 106.

6

455.  *Illus.*—THE ANIMALS THAT NATURALLY FEED
UPON VEGETABLES (Graminivorous) have a larger supply
of Bile than those that feed upon flesh (Carnivorous).

456.  THE LIVER RECEIVES ITS BLOOD, from which
to form Bile, mostly from the veins of the other digestory
organs, which require, receive, and supply more Blood
when vegetable food is used, than when we use a more
concentrated diet.

457.  THE HARMONIOUS ACTION of the Lungs, Liver,
and other Digestory Organs, of the appetite and the
temperatures of the weather, is very remarkable, but becomes still more so when the relation of muscular
action to the action of those organs is also considered.

458.  THE MUSCLES are very bulky, forming more
than half the weight of the Body; their activity is attended with the production of much heat and with
much material for the production of heat, and with a
correspondingly large demand for food and the digestion
of it.

459.  WHEN THE MUSCLES ARE ACTIVE the Digestory Organs must be correspondingly active, and the
muscular waste that cannot or should not be thrown off
by the Lungs can be thrown off from the Blood by the
Digestory Organs, which are therefore properly Eliminatory Organs, corresponding to the Muscles.

460.  THE HARMONIOUS ACTION OF ALL THESE PARTS,
the balancing (so to speak) of their actions, so that one
increases as another diminishes, or so that several increase
or diminish their activities together, is truly wonderful.
That the action is not accidental will be obvious when
the next case is considered.

461.  THE MIND REQUIRES THE ACTIVITY OF THE
BRAIN as much in hot weather as in cold (and, if proper
food and other treatment are used, is equally allowable),
and its activity must not have any relation to the pro-

duction of heat, and of course not to the Lungs nor to the Digestory Organs as Eliminatory Organs.

462. THE BRAINS WILL REQUIRE especial Eliminatory Organs independent of other activities, and devoted to its one purpose, the activities of which can be controlled solely with reference to Elimination.

463. THE KIDNEYS ARE FURNISHED for the purpose of removing from the Blood a part of the substances thrown into it by the Brain and Nerves, and, as Eliminating Organs, correspond to the Nervous system.

FIG. 107.                                FIG. 108.

Fig. 107, front view of left Kidney: 8, 9, arteries; 10, 11, 12, veins; 7, ureter; 13, suprarenal gland.

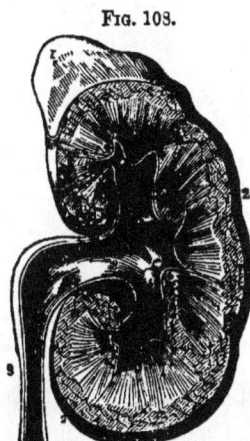

Fig. 108, section of left Kidney: 1, suprarenal capsule or gland; 2, outer skin; 3, part which eliminates; 4, tubes; 7, 8, basin; 9, ureter.

464. THE LARGE BRANCHES OF THE ARTERIES LEADING TO THE KIDNEYS indicate the importance of their office, more Blood being supplied to them, in proportion to their size, than to any other parts except the Brains and Lungs, the latter receiving all of it, as stated.

465. THE KIDNEYS MAY HAVE AN ADDITIONAL OFFICE of Eliminating a part of the material thrown into the Blood by the Bones.

---

462. What —? 463. For what —? Describe Fig. 107. Fig. 108. 464. What do —? 465. What —?

466.  THE KIDNEYS MAY ALSO SERVE the purpose of Eliminating Water when removed for the purpose of diminishing the volume of the circulating fluid, especially as the earthy substances Eliminated by them will require a considerable quantity of fluid to float them away.

467.  WHEN WATER IS ELIMINATED FOR THE PUR-POSE OF COOLING THE BODY it must be exposed to evaporate from an extensive surface, such as is afforded only by the Skin; so that the proper organs must be located in the Skin, where the perspiratory glands are placed.

FIG. 109.

Fig. 109 represents a very highly magnified view of a perpendicular section of a small portion of the Skin. It is divided into three parts : 1, the external layer or cuticle, composed of cells starting up from the basement membrane below, and gradually becoming dry and flattened scales as layer after layer they approach the surface, from which they are worn or drop off. Two curved passages, the outlets of the perspiration, are noticeable, communicating below with four tubes in one case, and two in the other, that are coiled at their lower extremities, and form the perspiratory glands, g. 2 is the papillary portion in which the nerves of touch commence, surrounded by a network of lymphatics and capillaries, supplied through a, an arterial branch. 3 is the true skin composed of sinewy fibres woven around the tubes, nerves, and perspiratory glands. In the lower part the meshes are larger, and filled sometimes with clusters of fatcells, as at e.

468.  ANY VISCID SUBSTANCES that may not with propriety be Eliminated by the Kidneys, the Digestory Canal, nor by its appendages, nor by the Lungs, may be thrown out upon the surface of the Skin.

466. What —?  467. What arrangement necessary —?  Describe Fig. 109.
468. How eliminate —?

469. THE TRUE ELIMINATING ORGANS ARE the *Kidneys* and the *Perspiratory Glands ;* the *Digestory Canal* (and its appendages, the Liver particularly), *Liver*, and *Lungs*, Eliminating in common with other functions.

470. THE ELIMINATIONS may be correspondingly classed: Renal (of the Kidneys), Perspiration, Intestinal, Hepatic (of the Liver), Pulmonary (of the Lung).

$$
\text{ELIMINATORY ORGANS} = \begin{cases} \text{Special,} & \begin{cases} \text{Kidneys.} \\ \text{Perspiratory Glands.} \end{cases} \\ \text{Common,} & \begin{cases} \text{Intestinal Canal.} \\ \text{Liver, etc.} \\ \text{Lungs.} \end{cases} \end{cases}
$$

$$
\text{ELIMINATIONS.} \begin{cases} \text{Renal.} \\ \text{Perspiration.} \\ \text{Intestinal.} \\ \text{Hepatic, or Biliary.} \\ \text{Pulmonary.} \end{cases}
$$

## SECTION X.

### *Modificatory Organs.*

471. EVERY ORGAN OF THE BODY THUS FAR CONSIDERED, it is evident, is constantly modifying the Blood by what it takes from or adds to it.

472. IT WOULD NATURALLY BE SUPPOSED that so many substances thrown into the current of the Blood, by the Lungs from the Air, by the Digestory Canal from the Food, and by the action of all the organs of the Body, together with what would be left by the organs forming fluids for various purposes, would require some modifying influences, to combine and assort the whole in such a way as to render it most serviceable, or to facilitate its exit from the Body.

473. WHAT COULD BE BETTER ADAPTED to the purpose of modifying the Blood than to have an immense number of organs, if they may be so called, so small that they could circulate everywhere with and in the Blood ?

474. THE BLOOD-CELLS, little sacs so small that nearly three thousand of them are found in every drop of Blood, and yet composing not more than one half of it, are one of the abundant means by which the Blood is modified.

FIG. 110.

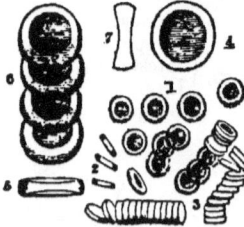

Fig. 110 : 1, 2, 3, a number of Blood-cells, very much magnified. Their shape in various positions is very well shown. They are prone to adhere, as at 3. 4, 5, 6, 7 are the same, still more highly magnified to show the convex, 5, and concave surface, 7, in section, that the same cell will exhibit at different times. Their semi-transparent character is shown at 6.

475. THE BLOOD-CELLS AS THEY FLOAT IN THE BLOOD constantly take from it substances that they yield back to it in a changed and improved condition.

476. THE BLOOD-CELLS ARE NOT SUFFICIENT, though so numerous, to accomplish all the modifying required for the Blood, nor could they be more numerous in it without obstructing its flow.

477. LARGE NUMBERS OF CELLS must be accumulated, forming, with the other necessary parts of a structure, different organs, conveniently located.

478. THE LYMPHATIC GLANDS are small organs, situated in great numbers in the course of the Lymphatic tubes, and chiefly composed of cells that modify the fluid that passes through them, and are also thought to be the producers of the cells that float in the Blood.

FIG. 111.

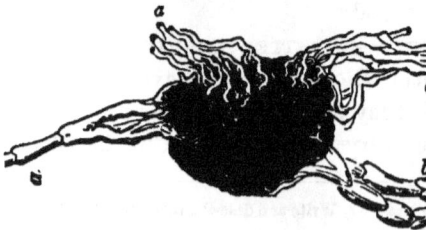

Fig. 111, a section of Lymphatic gland, into which the vessels, a, a, and from which the vessels, b, b, lead. An immense number of cells b line the passages through the gland from a to b.

474. What are —? Describe Fig. 110. 475. What is the effect of —? 476. For what —? 477. How treat —? 478. What are —? Describe Fig. 111.

479.   THE SPLEEN IS a large organ, chiefly composed of cells, and as the Blood circulates among them its contents are modified by the cells, and returned again to the Blood.

FIG. 112.

FIG. 113.

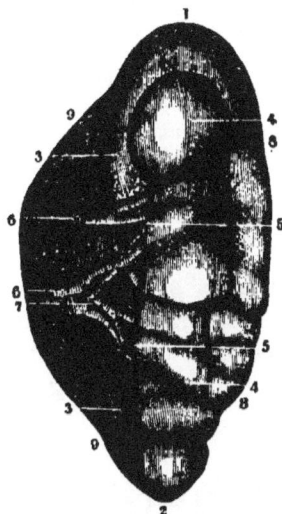
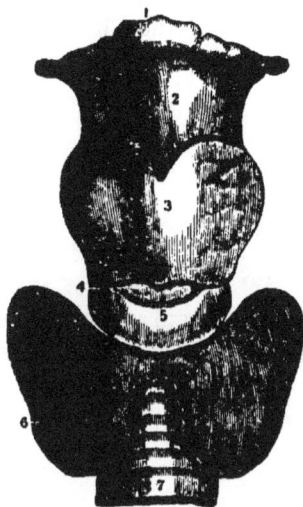

Fig. 112 represents concave surface of Spleen, with which the stomach is in contact; 6, branches of Arteries; 7, Vein.

Fig. 113: 6 represents the Thyroid Gland, like a saddle across the windpipe, 7, just below the Larynx. It is named from the Thyroid Cartilage, 3.

480.   THE THYROID AND THYMUS GLANDS are similarly constructed organs, that, like the others that have no other inlet or outlet except the Blood-tubes, must be modifying organs, though in what peculiar way they modify the Blood is not known.

481.   THE SUPRA-RENAL BODIES are small parts found at the summits of the Kidneys, the use of which has been supposed to be modificatory by some and nervous by others. As they have no outlet nor inlet, except bloodtubes and nerves, we will class them as modificatory.

479. What —? Describe Fig. 112.   480 What is —? Describe Fig. 113.   Can the Thyroid Gland be felt?   481. What are —?

482. In one sense it is not proper to form an Apparatus, by classing the Modifying Organs together, as they do not act connectedly to gain a single result; yet in another sense they may be said to be working together toward a common result. The same remark may be made of the Eliminatory Organs.

483.    Modificatory Organs $=$ $\begin{cases} \text{Spleen.} \\ \text{Blood-cells.} \\ \text{Lymphatic Glands.} \\ \text{Thyroid Gland.} \\ \text{Thymus Gland.} \\ \text{Supra-renal Glands.} \end{cases}$

### CONCLUSION OF CHAPTER V.

484. Thus have all the Organs of the Body been noticed and included under appropriate heads. It will be well to present them all at one view, arranged in connection as Apparatus, and clustered as belonging to their respective centres, as follows:

#### Mentory Organs.

Ganglia + Six kinds of Sensatory Nerves + Organs of Sense $\begin{cases} \text{Ears +} \\ \text{Eyes +} \\ \text{Noses +} \\ \text{Mouths +} \\ \text{Skins +} \\ \text{Muscles +} \end{cases}$

Ganglia

Ganglia

Ganglia + Motory Nerves +

#### Circulatory Organs.

Ganglia + Nerves, (Sympathetic) +

Hearts + Blood-tubes, (Arteries, Veins, Capillaries, Lymphatics) +

#### Sanguificatory Organs.

(Nose + Pharynx) + Larynx + Trachea + Lungs + (Diaphragm + T-w) +

$\begin{cases} \text{(Mouth), Salivary-gl.} + (\text{Pharynx}) + \text{Œsophagus} + \text{Stomach,} + \text{Gastric-} \\ \text{gl., 2d Stomach, (Duodenum, Jejunum, Illium)} + \text{Pancreas} + \text{Liver,} \\ \text{Gall-bl.} + \text{Brunner's, Lieberkühn's, Pey's gl.} + \text{Colon} + \text{Lacteals.} + \end{cases}$

Kidneys + Perspiratory gl. + Hair and Sebaceous gl. +

Spleen + Blood-cells + Thyroid, Thymus, Supra-renal, and Lymphatic gl. +

*(marginal vertical lettering: MOTION ... LIFE ... IS)*

---

485. IT IS NOT SUPPOSED OR PRETENDED that the precise or the entire use or mode of action has been assigned to each organ, but a correct bird's-eye view has been taken of the whole field, and the ground has been laid out in such a way as to be readily and comprehensively studied in detail at the appropriate time.

486. THE GENERAL USES OF ALL THE ORGANS must be such as has been assigned to them, and these will be convenient guides to the student, in present or future studies, toward developing truth and detecting error. The classing of the organs under the appropriate heads of Apparatus enables the student at once to perceive their general relations to each other and to the whole Body, and will enable him easily to remember them, hanging, as it were, in so many clusters, and these also grouped about their centres.

487. THE PARTICULAR MODES OF ACTION of the various organs can only be understood by studying the nature or properties of the various substances of which they are composed, which will be the topic of the next chapter.

It is very desirable to have the student become quite familiar with the general appearance of the organs, their positions and their general uses, before he attempts to master their particular structure and uses. He should endeavor, with his own mind's eye, to see them in the Body itself; should, upon the surface of his own Body, mark out the regions they occupy, and compare them with the representations in the cuts. At this stage of his progress it will be well, therefore, for him to make a very thorough review of the cuts preceding; let him also study those in the atlas appended to the work, which have been thus printed in order that they may be plainer, and that they may be more easily studied in a consecutive manner. Let him also synthetize the whole Body from the point now reached, classing the organs first as Apparatus, then as groups, and then let him point out what organs are double and what single, and thus see in a double manner the correctness of the divisions already made. Then let him review the members, and determine what organs exist in each member, and why they should exist where they do. Let him compare the kinds of organs in the different members, and determine their relative numbers. Also, let him observe and describe the organs that are connected with each centre, and determine their relations to each centre; how, for instance, the mouth serves the commercial capital or centre, and how it serves the political capitol or centre. In particular, let him show the relation of each organ to the Mind.

485. What — ? 486. What are — ? 487. What said — ? How many Sections has Chapter V. ? To what do they correspond?

6*

# CHAPTER VI.

## SYSTEMATIC ANALYSIS OF ORGANS.

### *General Anatomy, Physiology, and Hygiene.*

#### TISSUES, FLUIDS, AND GASES: PROPERTIES.

### SECTION I.

#### *Tissues and their Properties.*

NOTHING affords a more profound source of admiration to the intelligent mind, than simplicity of means for producing important and varied results. Nothing pleases more than to find a result unexpectedly produced by the varied action of a cause previously understood. How astonishing and pleasing to find that all the varieties and shades of color may result from mixing only three, Red, Yellow, and Blue! We look upon the bright red blood almost with veneration, when we learn that the life of the body is in its keeping, and that all the organs are built up, kept in repair and at a proper temperature, by means of that single fluid. How much more of grandeur pervades the heavens when it is discerned that instead of cycles upon cycles, spheres within spheres, the incomprehensible jargon of combined forces of the ancients, one simple law of gravitation pervades all space, and keeps all bodies in their orbits! The causes of organic action will therefore especially delight the mind.

---

What is the subject of Chap. VI.? What is the subject of Section I.? What said of simple causes?

488. The student will conclude, as he glances over the large number of organs and their uses, that they must be composed of a great number of substances.

489. He will be surprised and delighted to learn that only six kinds, modified as circumstances require, with their inherent properties, are necessary.

490. If he carefully inspects all the organs, he will perceive that whatever their forms, textures, attachments, or uses, they are only required to form fluids, or to contract, or to excite, or to be tough and flexible, or to be firm and elastic, or to be rigid—or several of these at the same time.

491. *Illus.*—The Heart is a pouch that requires to be formed of a substance tough and flexible, of one that can contract, of one that can excite contraction, and of one that can form a fluid to keep its surfaces glairy and free from friction—four substances, having four properties.

492. The student cannot name an organ the use of which requires more or other than the six substances with the properties mentioned.

493. Tissue is the name given to the substances of which the organs are composed.

494. Property is the name given to the peculiar characteristic that renders a tissue serviceable in an organ.

495. The names of the Tissues and Properties are as follows :

|  |  |  | PROPERTIES. |
|---|---|---|---|
| TISSUES | Passive | Bony or Osseous, | Rigid. |
|  |  | Gristly or Cartilaginous, | Firm and Elastic. |
|  |  | Sinewy or Fibrous, | Tough and Flexible. |
|  | Active | Nervous, | To Excite. |
|  |  | Muscular, | To Contract. |
|  |  | Secretory, | To Secrete. |

The composition of the organs and the character of the tissue will be, perhaps, made more impressive by the subjoined synopsis of the Tissues composing each organ. The figures refer to the number of varieties of tissue in each organ. It is observable that no organs, except the *Ear, Eye, Nose, Mouth, Pharynx, Larynx,* and *Trunk-walls,* are composed of six kinds of Tissue, and these are compound organs. No other, except the Skeleton, has more than four.

| | | | | | | | | | | |
|---|---|---|---|---|---|---|---|---|---|---|
| GANGLIA | = | — | — | | Sin. | | Sec. | — | | Ner. (W & G) |
| NERVES | = | — | — | | Sin. | | Sec. | — | | Ner. (W) |
| EAR | = | Bony | Gristly | | Sin. | 4 | Sec. | Muscular | 3 | Ner. (W) |
| EYE | = | Bony | Gristly | | Sin. | 4 | Sec. | Muscular | 3 | Ner. (W) |
| NOSE | = | Bony | Gristly | | Sin. | | Sec. | Muscular | 2 | Ner. (W) |
| MOUTH | = | Bony | Gristly | | Sin. | 8 | Sec. | Muscular | 3 | Ner. (W) |
| SKIN | = | — | — | 2 | Sin. | | Sec. | — | | Ner. (W) |
| MUSCLES | = | — | — | | Sin. | | Sec. | Muscular | 2 | Ner. (W) |
| SKELETON | = | Bony | Gristly | 2 | Sin. | | Sec. | — | | Ner. (W) |
| HEARTS | = | — | — | | Sin. | | Sec. | Muscular | | Ner. (W) |
| BLOOD-TUBES | = | — | — | | Sin. | | Sec. | (Muscular) | | Ner. (W) |
| PHARYNX | = | Bony | Gristly | | Sin. | | Sec. | Muscular | | Ner. (W) |
| LARYNX | = | Bony | Gristly | | Sin. | | Sec. | Muscular | | Ner. (W) |
| WINDPIPE | = | — | Gristly | | Sin. | | Sec. | Muscular | | Ner. (W) |
| LUNGS | = | — | Gristly | | Sin. | 8 | Sec. | (Muscular) | | Ner. (W) |
| TRUNK-WALLS | = | Bony | Gristly | 2 | Sin. | 8 | Sec. | Muscular | | Ner. (W) |
| SALIVARY GL. | = | — | — | 2 | Sin. | 8 | Sec. | — | | Ner. (W) |
| ŒSOPHAGUS | = | — | — | | Sin. | | Sec. | Muscular | | Ner. (W) |
| STOMACH | = | — | — | | Sin. | 8 | Sec. | Muscular | | Ner. (W) |
| GASTRIC GL. | = | — | — | | Sin. | 8 | Sec. | — | | Ner. (W) |
| SECOND STOMACH | = | — | — | | Sin. | 2 | Sec. | Muscular | | Ner. (W) |
| LIVER | = | — | — | | Sin. | 8 | Sec. | — | | Ner. (W) |
| GALL-B. | = | — | — | | Sin. | 2 | Sec. | — | | Ner. (W) |
| PANCREAS | = | — | — | | Sin. | 3 | Sec. | — | | Ner. (W) |
| BRUNNER'S GL. | = | — | — | | Sin. | 3 | Sec. | — | | Ner. (W) |
| LIEBERKUHN'S GL. | = | — | — | | Sin. | 3 | Sec. | — | | Ner. (W) |
| PEYER'S GL. | = | — | — | | Sin. | 3 | Sec. | — | | Ner. (W) |
| LACTEALS | = | — | — | | Sin. | 2 | Sec. | — | | Ner. (W) |
| COLON | = | — | — | | Sin. | 2 | Sec. | Muscular | | Ner. (W) |
| KIDNEYS | = | — | — | | Sin. | 2 | Sec. | — | | Ner. (W) |
| PERSPIRATORY GL. | = | — | — | | Sin. | | Sec. | — | | Ner. (W) |
| BLOOD-CELLS | = | — | — | | — | | Sec. | — | | — |
| SPLEEN | = | — | — | | Sin. | | Sec. | — | | Ner. (W) |
| LYMPHATIC GL. | = | — | — | | Sin. | | Sec. | — | | Ner. (W) |
| THYROID GL. | = | — | — | | Sin. | | Sec. | — | | Ner. (W) |
| THYMUS GL. | = | — | — | | Sin. | | Sec. | — | | Ner. (W) |
| RENAL CAPSULE. | = | — | — | | Sin. | | Sec. | — | | Ner. (W) |

496. TEXTURE is the name given to the peculiar form into which Tissues are wrought.

497. *Illus.*—Cotton *batting,* cotton *wadding,* cotton *thread,* cotton *muslin,* etc., are expressions of the tissue cotton wrought in various textures.

498. *Illus.*—Sinewy substance in one texture is *tendon,* in another is *ligament,* in another is skin or *membrane ;* which may be expressed, Sinewy *tendon,* Sinewy *ligament,* Sinewy *membrane.*

499. THE SIMPLICITY OF THE ORGANIC STRUCTURE is observed not only by the modifications of Tissue and by

496. What —?   497. How illustrated?   498. How sinewy textures named?
499. What said of — ?

its differences of texture in different cases, but by the differences in its quantity, and in the form and other peculiarities of Organs; which must be noticed when each organ is considered in detail.

500. *Illus.*—THE HEART AND STOMACH ARE COMPOSED of four of the same kinds of Tissue. Both organs require a framework of sinewy tissue formed into pouches, but the size and form of the organs must differ. The sinewy tissue must in both cases be wrought so as to leave meshes, but in a different order, to receive muscular tissue that must be differently arranged to produce different kinds of motion in the two cases. Nervous tissue must be inserted in each according to the muscular arrangement. The outer surface of each must be formed of the same variety of secretory (the serous) tissue to produce the same variety of fluid for the purpose of preventing friction; but while the Heart should be lined with the same variety of secretory (serous) tissue, the Stomach must be lined with a variety that will secrete a more viscid fluid, and the Stomach must also be furnished a variety to secrete the gastric juice. They are therefore very much alike, and yet very distinctly different. The Heart and Diaphragm are just alike in tissural constitution, and only differ in size, form, and position; the Diaphragm being the Heart spread out, and the Heart being the Diaphragm in the form of a pouch.

501. SYSTEM is the name given to all of any kind of Tissue in a living thing *arranged* as it naturally exists.

502. *Inf.*—There are as many systems in the Body as there are Tissues.

$$\text{THE HUMAN SYSTEM} = \text{the} \left\{ \begin{array}{l} \text{Bony} \\ \text{Gristly} \\ \text{Sinewy} \\ \text{Nervous} \\ \text{Muscular} \\ \text{Secretory} \end{array} \right\} \text{Systems.}$$

503. SYSTEM IS ALSO USED, though not with perfect propriety, for naming the whole of any kind of parts of similar structure.

---

500. Illustrate how —? The heart and diaphragm. 501. What is —? 502 Write and explain table of —. 503. How —?

**504.** *Illus.*—The Arteries or Veins of the Body taken together are called the Arterial System, the Veinous System, etc.

**505.** FROM TISSUES, either of two modes of synthesis of the Body may be chosen.

**506.** TISSUES MAY BE FORMED into systems, and these united into what will most properly be called the Human System; or,

**507.** TISSUES MAY BE WROUGHT into Organs, these arranged as Apparatus, etc., and the Body produced.

### SECTION II.

#### *Fluids and their Properties.*

**508.** FLUIDS are not, strictly speaking, essential parts of any organ; yet as they are essential to the action of each, and, in common with Tissues, are composed of organic and chemical Elements, derived from air, water, and food, they may be properly classed under the analysis of organs.

**509.** THE NECESSITY FOR FLUIDS is evidently of three classes: 1st, All substances that are to be moved from one place to another, must be in a fluid condition. 2d, The surface of the Body must be protected by fluids. 3d, The Digestory processes, as we have seen, require fluids for dissolving food.

**510.** FLUIDS, THEREFORE, MUST BE of three classes, *General, Surface,* and *Digestory,* of which the character of the last has been shown.

**511.** THE SURFACES of the Body are of three kinds: 1st, that of the external skin; 2d, those of the lining of the air and food canals; and, 3d, those where the organs only are in contact with each other.

**512.** THE SURFACE OF THE EXTERNAL SKIN RE-

QUIRES an oily fluid that shall prevent too great evaporation from the skin (not evaporation of the perspiration which flows out of its tubes on to the surface of the oil), and protect it from other external influences.

513. THE OILY FLUID IS of three varieties, the common, the ear-wax, and the viscid fluid of the Meibomian glands that keeps the tear-fluid from pouring over the lids.

FIG. 114.

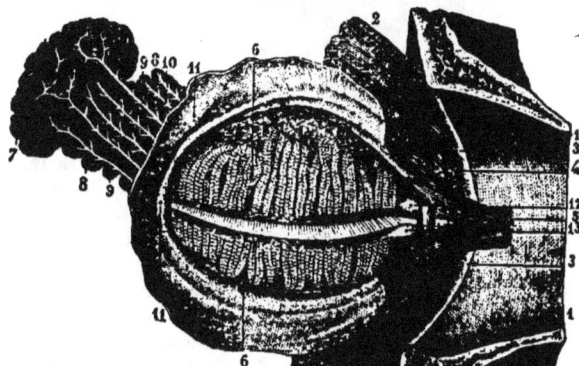

Fig. 114 represents the left eye-lids, cut through as far from their opening as possible, and the Lachrymal Gland drawn from its place, and with the lids turned toward the nose to show the inner surface of the lids, in which the Meibomian Glands, 6, are seen, opening at the edges, between which the lashes are seen. 14, numerous openings of the ducts, 9, 10, from the Gland, 7, 8. 12, 13 are the minute openings at the inner corner of the lids, through which the tear-fluid passes to the nose.

514. THE SURFACE OF THE AIR AND FOOD PASSAGES requires a slimy but not oily fluid, called mucus, that will not too readily evaporate, and will facilitate the passage of the air and food; and as the evaporation in the nose and mouth is greater than elsewhere, they should both be still further moistened.

515. THE LINING OF THE NOSE is kept moist by the mucus, and the tear-fluid that passes into the nose after serving its purpose in the eye.

513. How class —?  Describe Fig. 114.  514. What do—?  515. What said of—?
What is the quantity of the tear-fluid?  Is it always the same?

FIG. 115.

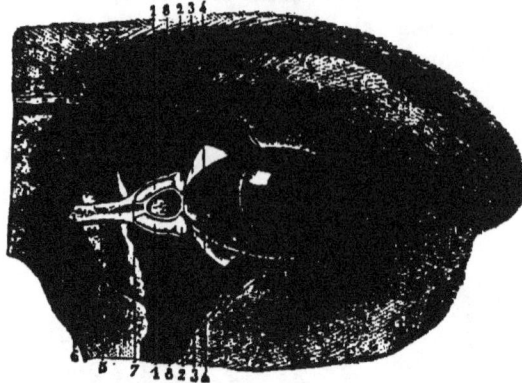

FIG. 116.

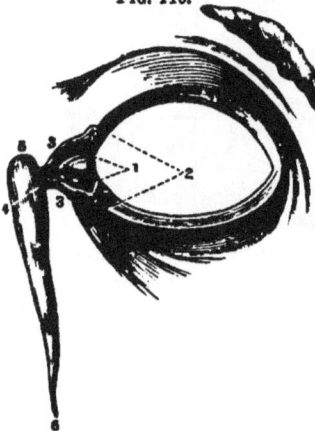

Fig. 115, 2, the same as 12, Fig. 114; 1, ducts leading the tear-fluid from 2 to 5, covered by 7.

Fig. 116, 5, same as in 115, extending into the nose at 6. 1, the same as 2, and 3 same as 1, of 115. 4, where 3 joins 5. (The three preceding cuts beautifully illustrate the tear apparatus, for so it may be called.) The tear-fluid, formed in the gland, Fig. 114, passes down to the eye, and over it, as shown by 115, to the openings 2, from which it glides through the ducts down into the nose. (See Plate 4.)

516. THE LINING OF THE MOUTH is kept moist by the saliva, in addition to the mucus, which is supplied to the back part of the mouth in large quantities by the Amygdaloid glands (tonsils), to facilitate the swallowing of food.

517. THE SURFACES OF THE ORGANS IN CONTACT WITH EACH OTHER REQUIRE a very watery fluid, called serum or serous fluid, simply to prevent adhesion and friction.

Describe Figs. 115, 116. What are remarkable characteristics of tear-fluid? 516. What said of —? 517. What said of —?

518.  The Serum is of three varieties: 1st, the common; 2d, the synovial, that lubricates the joints or bursæ, is a little more viscid, and in some of the joints has an appropriate arrangement for supplying it in large quantities; and 3d, the tear-fluid, that has a small portion of salt to make it beautifully transparent, and is supplied by an organ for the purpose.

519.  The General fluids are of three kinds, Flesh-juice, Lymph, and Blood.

520.  Flesh-juice is required by all the organs to keep them in a proper condition of pliability, etc.

521.  Lymph, or white Blood, is a slightly-colored watery fluid, containing an immense number of white cells, from which it is supposed the red cells of the Blood are formed.

522.  Lymph exists in all parts of the Body except the Brain and Nerves.  Its rationale is not understood. The Lacteal fluid is considered as one form of Lymph.

523.  Blood, the common necessity and common result of all the fluids and all the organs, is everywhere found: its water serves as a solvent and a vehicle to whatever is thrown into it; its living cells work as they circulate; its character is constantly changing, yet ever the same, owing to the wonderful provision that the action of one part of the Body is ever balancing the action of another; and the velocity of its current is such that its entire resources are every few moments present to every minute portion of the Body.  It equalizes the heat of the Body by swiftly coursing through every part, gathering heat from one and distributing it to another; and when the whole Body is too warm, it pours out its watery portion upon the surface of the skin to be freely evaporated, and thus carry off the surplus heat; or, if the Body is too cool, it yields a portion of its water to the kidneys.

---

518. How class —?  519. How class —?  520. What said of —?  521. What is —? 522. Where does —?  523. What is —?

**524.  Summing up :**

$$
\text{FLUIDS}
\begin{cases}
\left.\begin{array}{l}
\text{Blood,} \\
\text{Lymph,} \quad 2 \text{ var.} \\
\text{Flesh-juice,}
\end{array}\right\} \text{General.} \\[4pt]
\left.\begin{array}{l}
\text{Serum,} \quad 3 \text{ var.} \\
\text{Mucus,} \\
\text{Oil,} \quad\quad 3 \text{ var.}
\end{array}\right\} \text{Surface.} \\[4pt]
\left.\begin{array}{l}
\text{Saliva,} \quad 2 \text{ var.} \\
\text{Gastric-juice,} \\
\text{Pancreatic-juice,} \\
\text{Bile,} \\
\text{Gall,} \\
\text{Brunner's gland-juice,} \\
\text{Lieberkuhn's gland-juice,} \\
\text{Peyer's gland-juice,}
\end{array}\right\} \text{Digestory.}
\end{cases}
$$

### SECTION III.

#### *Gases.*

**525.  It is not known** how many kinds of gases are uniformly present in the Body, nor is their use or mode of action well understood.

**526.  Carbonic Acid and Oxygen Gas,** in larger or smaller quantities, can always be found in the Blood, and are essential to its efficiency; the others found at times are probably accidental.

**527.  Thus Analysis finds** the three forms of matter, Tissues, Liquids, and Gases, essential to the composition and action of Organs.

**528.  It is evident** that in passing from the study of the Organs and their uses to that of Tissues, Fluids, Gases, and their Properties, the character of the study has changed from that of special parts to that of general parts, or those common to many organs.  Hence,

**529.**  The study of parts above Tissues is called Special, and the study of all below Organs is called General Physiology, Anatomy, and Hygiene.  For the former the unassisted eye is sufficient, for the latter the microscope is necessary.

---

524. Table of —?  525 What —?  526. Where — found?  527. What does —?
528. What —?  529. Define special and general.

530. It must not be supposed that passing from the study of Organs to that of Tissues is a step downward, in the sense of inferiority of importance.

531. The study of Tissues is more important than that of Organs, for evidently their properties wholly depend upon those of their Tissues.

532. Each Tissue in an Organ acts independently, and an organ exhibits not the combined but the collective properties of its Tissues; therefore,

533. The perfection of the structure of the whole Body depends upon the perfection of its Tissues, Fluids, and Gases. But,

534. Tissues do not exist as such in the Blood; only their elements exist in it.

535. It is a question whether any of the other Fluids exist as such in the Blood. Probably some of them do; others certainly do not.

536. Each Tissue has the property, under proper influences, of forming itself from elements furnished by the Blood, and drawn from it by the Tissue.

537. The formation of perfect Tissue requires three conditions: 1st, Perfect Tissue to form itself; 2d, Perfect influences; and, 3d, Perfect elements furnished by the Blood. Default in either will produce a deranged organ.

538. The Tissues, or the influences acting upon them, in one organ may be perfect, and those in another imperfect; while if the elements furnished are imperfect, all the corresponding Tissues of the whole Body must be imperfect.

539. *Inf.*—The Body must not be studied as composed of, nor the Tissues as composing, Systems; but the Tissues must be studied as parts of Organs, that their liabilities to derangement may be understood.

540. *Inf.*—To understand Tissues, their Elements, and how they are produced, must be understood.

# CHAPTER VII.

## Systematic Analysis of Tissues and Fluids.

### ORGANIC AND CHEMICAL ELEMENTS.

**541.** Hitherto the several characteristics of parts have merely exhibited the sum of the characteristics of their subdivisions; guided by this indication, the Analysis has been obvious, and easily made.

**542.** The Analysis of Tissues will exhibit entirely new features; and the student will perceive that in entering upon the study of General Physiology, Anatomy, and Hygiene, he has entered upon a new field of inquiry, and must adopt new methods of investigation.

**543.** He will be well rewarded, as he will reach the very heart and marrow of the practical phase of the subject, which will also present to his mind the most exquisite, delicious, seedful thoughts.

**544.** The characteristics of Tissues are not the sum of those exhibited by their components.

**545.** The properties of Tissues depend upon their own nature; and the only reason to be given why it is so, is, that it has been so ordained.

**546.** *Illus.*—The reason why muscular tissue contracts is, that it is its nature to contract, and cartilage or gristle is elastic because it is its nature to be so, and not because any of its components alone have any measure of it, for they do not.

**547.** The components of Tissues cannot therefore be indicated by their properties, but must be

---

541. What has been —? 542. What —? 543. Why —? 544. What —? 545. On what —? 546. What is —? 547. What said since —?

sought in some other way, which we will proceed to find.

548. THE TISSUES MUST VARY IN QUALITY in different cases; therefore they must be compounds—for if simple substances, they would be always the same.

549. THE TISSUES CANNOT BE DEFINITE COMPOUNDS, for if they were they would be uniform, which they are not.

550. TISSUES MUST THEREFORE BE TRUE COMPOUNDS, no part of the characteristics of which are possessed by their components alone, but which, nevertheless, by their different proportions, affect the quality of the Tissue.

551. THIS CHARACTERISTIC OF TISSUE, viz., being a true but yet indefinite compound, shows that it is endowed with life, for it is only in the domain of life that such a thing can be; indeed, it is a distinguishing characteristic of life to produce indefinite compounds, while chemistry always produces definite compounds.

552. THE THREE ACTIVE TISSUES, AT LEAST, exhibit power when they act. Every contraction of a muscle not only, but every nervous action, and every secretion, exhibits or represents power; but,

553. THE EXHIBITION OF POWER IS ALWAYS a destructive process, is always not only attended with, but produced by, decomposition of substance.

554. EACH ACTION OF TISSUE IS THEREFORE ATTENDED BY OR DEPENDENT UPON decomposition of a corresponding amount of substance.

555. *Illus.*—Cattle are sometimes chased, or "run," as it is termed, by dogs, just before being killed, for the purpose of making their meat tender. It will not keep long in such a case, as it has to so great a degree been made to undergo decomposition already, and is not worth as much for food on this account.

556. THE TISSUE AS IT DECOMPOSES must be corre-

spondingly reproduced, in order to preserve its perfect
state for action.

557. ONE OF THE MOST REMARKABLE AND INTEREST-
ING PROPERTIES OF TISSUE is its power of self-growth,
or of reproducing itself.

558. TIME is one of the elements required for perfect reproduction; for if the Tissue is decomposed before
sufficient time has been allowed for its perfect formation,
the result is incomplete and pernicious.

559. IT MIGHT AT FIRST THOUGHT BE SUPPOSED that
as the decomposing Tissue necessarily contains all the
components of Tissue, it would only be necessary for the
Tissue to re-form itself from them.

560. IT MUST BE CONSIDERED, however, that if the
Tissue could reproduce itself from its decomposed substances, just as much power would be required for recomposition as had been obtained by decomposition; so
that nothing would be gained.

561. IT WOULD BE A GREAT FALLACY to suppose that
power can be exhibited without it has first been obtained
from some source, and without the destruction of a corresponding amount of substance.

562. THE TISSUES, THEREFORE, MUST BE SUPPLIED
with and form themselves from components that are
themselves indefinite compounds, possessing (or, so to
speak, storehouses of) the power exhibited when the Tissues decompose.

563. THE SELF-PRODUCING PROPERTY OF TISSUES IS
LIMITED to forming themselves from components previously compounded and previously prepared.

564. SINCE THE POWER TO BE EXHIBITED BY THE
TISSUES cannot be produced in the Body without an equal
expenditure of power, it must be condensed or stored in
the components of Tissue by the action of plants, from
which all food is directly or indirectly obtained.

---

557. What is —? 558. What said of —? 559. What —? 560. What —?
561. What —? 562. With what —? 563. How —? 564. What effect —?

565. Plants thrust their roots into the ground and their branches and leaves into the air, and, basking in the rays of the sun, gather from the air, water, and earth the various elements that, under the influence and power of the sun's rays during a long period of time, they combine into indefinite compounds, possessed of the power that has thus grown up in them.

566. As coal slumbering in the Earth contains heat embodied, during ages past, in the growing plants of which coal is the remains, so do the nutritious parts of plants embody power derived from the sun during their production, which power will come out again when their decomposition takes place. The coal may lie for a day or for ages—the heat locked in it will appear whenever and wherever it returns to the form it had. So the grain of wheat may remain a month or a year unchanged, yet contains a latent power, the source of which was the sun, that will be exhibited whenever decomposition takes place.

567. The influence of the sun's rays, acting through the mechanism of plants, stores up power in their nutritious parts. This power is not lost when they are eaten, but follows the substances in which it exists till they have become Tissues, and decompose, when it again exhibits itself.

568. The substances furnished by plants doubtless require some further preparation before they are quite ready to become components of the Tissues; and it has been seen that they are subjected to the action of secretions, and it may be to secretory processes in the course of the Blood-circuit.

569. The Blood-cells are merely very simple plants floating in the Blood; the process of secretion, which is the only office they can perform, being a vegetative or plant-like process.

---

565. What said —? 566. What said of —? 567. What said of —? 568. What said of —? 569. What are —?

570. SECRETION is at the same time both a productive and a destructive process. The substance produced is obtained by the expenditure of power, and that, as said before, is produced by destruction or decomposition of substance.

571. BY SECRETION the power that was in one form will appear in another. It is a transformation of power. By secretion, power is, so to speak, wrought up and condensed into higher capabilities, and made to serve more exalted purposes.

572. ALL THE TISSUES ARE THEREFORE SUBSIDIARY to two, muscular and nervous; and therefore the whole Body is constructed with a view to the production of those two Tissues and the appropriate exhibition of their properties.

573. MUSCULAR CONTRACTION AND NERVOUS EXCITATION are the two grand properties that distinguish animal from vegetable life, and to them it is subservient.

574. VEGETABLE LIFE REACHES up into animal life, is the basis of it, and blends with it. The former may exist without the latter, but the latter cannot without the former.

575. *Inf.*—Food should be selected with reference to its adaptability to nourish the two Tissues, as it is not supposable that two Tissues so different can be equally well nourished by the same kind of aliment.

576. *Inf.*—Not only must the right elementary substances be used for food, but they must also be properly compounded before they are eaten.

577. IT CANNOT TRUTHFULLY BE SAID that we understand all the steps in the production of Tissue or the precise number of them. They are numerous, and substances assume various forms from the time they are first drawn in by the plant till they appear as perfect tissue; much time and power are also consumed in the produc-

tion; but the general plan and idea have been correctly exhibited.

578. THE EVOLUTION OF NERVOUS SUBSTANCE, especially the gray or active part, is the most time-taking, elaborate, and expensive of all the tissues.

579. PLANTS PRODUCE THE COMPONENTS OF NERVOUS TISSUE in comparatively very small quantities, and if the nervous tissue is very active they must be selected with care.

580. THE COMPONENTS OF NERVOUS TISSUE abound more in certain kinds of meat-diet than they do in plants, as would naturally be the case.

581. THE ELABORATE PROCESS OF PERFECTING THE NERVOUS TISSUE, combined with the necessity for its constant operation during waking hours, makes it necessary that it should have periods of complete repose, as in sleep, during which no process requiring its action should take place.

582. SINCE THE PROCESS OF SECRETION IS VEGETATIVE, it can go on during sleep, at a moderate degree, without the influence of the nervous tissue; and as the Heart and the organs concerned in breathing rest three fourths of the time, the process of producing heat and circulating Blood may take place without the necessity of long periods of complete inactivity in their ganglia.

583. PROXIMATE OR ORGANIC PRINCIPLES, OR ORGANIC ELEMENTS, are the names given to all the components produced in plants or in the Body adapted to its nutrition.

584. WHEN THE ORGANIC ELEMENTS ARE ANALYZED by the chemist, they are found to be composed of at least thirteen, usually sixteen, and sometimes, but probably accidentally, nineteen simple or chemical Elements.

585. SIMPLE OR CHEMICAL ELEMENTS are so called because they are the simplest substances known, having

never been analyzed; therefore the Analysis of the Body can be carried no farther.

586. CHEMICAL COMPOUNDS are produced by the combination of Elements. The character of one does not indicate that of the other.

587. *Illus.*—Water, a liquid, is a chemical compound of Oxygen and Hydrogen gases.

588. PLANTS USE some of the Elements in their simple, but mostly in a compound, state.

589. ANIMALS AND MAN USE one element, oxygen, in the air breathed; one chemical compound, water, as a drink—another, salt; also Vegetable, Organic, or Proximate Principles as food.

590. IT IS EVIDENT, from the method of compounding the Tissues, that water and air are used, not to form the Tissues, but to produce heat, or as vehicles for equalizing temperatures, and bringing away substances decomposing in the Tissues.

591. IN THE SYNTHESIS OF MAN from *Chemical Elements*, the first step will be to *Chemical Compounds;* the second, to *Proximate Elements;* the third, to *Tissues;* the fourth, to *Organs;* the fifth, to *Apparatus;* the sixth, to *Groups;* the seventh, to *Members;* the eighth, to the *Body;* the ninth, the MIND, being added, to MAN.

592. IT SEEMS TO BE PROVED that the Analysis and Synthesis of Man thus exhibited is correct, from the fact that there are general terms in common use corresponding to the divisions, and no more; as follows:

| Human Constitution | corresponds to | Elements. |
|---|---|---|
| "    System | " | " Tissues. |
| "    Organism | " | " Organs. |
| "    Apparatus | " | " Apparatus. |
| "    Mechanism | " | " Groups. |
| "    Body | " | " Members. |
| Man | " | " Body and Mind. |

---

586. What are —? 587. What is —? 588. What do —? 589. What do —? 590. What —? 591. What steps —? 592. What —? Write table.

# CHAPTER VIII.

## SYSTEMATIC SYNOPSIS OF PRACTICAL SUGGESTIONS.

### *Golden Hint - Words.*

593. A REVIEW OF THE PRECEDING CHAPTERS WILL EXHIBIT the idea that all the practical suggestions pertaining to the welfare of Man, as related to his Body, may and should be arranged under four heads, and three divisions under each.

594. THE FIRST HEAD INCLUDES all those suggestions having reference to the direct relations of Mind and Body, which are briefly summed up in the three words, *Educate, Exercise, Arrange.*

595. THE SECOND HEAD INCLUDES all those suggestions having reference to promoting the circulation of Blood, also briefly summed up in the three words, *Rub, Clean, Clothe.*

596. THE THIRD HEAD INCLUDES all those suggestions having reference to the substances received into the Body, summed up briefly in three words, *Air, Water, Food.*

597. THE FOURTH HEAD INCLUDES all those suggestions of a miscellaneous character pertaining to the Body generally, expressed concisely in three words, *Sleep, Repose, Habits.*

598. THUS DO TWELVE HINT-WORDS EXPRESS the

---

593. What will — ? 594. What does — ? 595. What does — ? 596. What does — ? 597. What does — ? 598. What — ?

topics of a complete Hygiene, arranged under four
heads, as follows:

| Educate; | Rub; | Air; | Sleep; |
|---|---|---|---|
| Exercise; | Clean; | Water; | Repose; |
| Arrange; | Clothe; | Food; | Habits. |

## Personal Attractiveness.

599. A REVIEW OF THE PRECEDING ANALYSIS WILL
PROVE that attractiveness must be dependent on one or
all of six things: 1st, the *Natural Constitution ;* 2d, upon
the *Health ;* 3d, upon acting favorably through the
*Senses ;* 4th, upon exhibiting proper *Emotions*, a good
disposition; 5th, upon *Intellectual Culture ;* and, 6th,
upon graceful *Motions.*

600. IN REGARD TO ALL, except the first, knowledge
and culture are essential to a desirable effect. The first,
if good, may be a great blessing; if bad, it may be partly
corrected or nearly balanced by assiduous application in
regard to the other five. The twelve hint-words will
always be as serviceable in regard to personal appearance as they are in respect to Health.

601. IT WILL BE OBSERVED, that the points to be
considered correspond to the different kinds of Apparatus, Motory, Intellectory, Emotory, Sensatory, and to
those having direct relation to the making and circulating of the Blood, all of which are, in a measure, dependent on their inherent constitution, and in part on care.

602. THUS DOES A GENERAL ANALYSIS OF MAN
DEMONSTRATE that the study of Hygiene is one of the
most fruitful, as well as germful, that can occupy the
attention of the Mind; and the further prosecution of
the subject under the head of Synthesis will, it is hoped,
prove still more interesting, since it will exhibit a large
number of practical details.

# INTRODUCTION

## TO DETAILED SYNTHESIS.

———◆◆◆———

SOME persons delight to do, say, and write that which is novel, merely for the sake of novelty; while others dare not leave the beaten track, either distrusting their own judgment or fearing criticism. Again, some persons are attracted by any novelty, merely because it is novel; while others, for the same reason, condemn it without examination.

In this work, however, everything has been written with sole regard to the good of the pupil; and it is anxiously desired that teachers should examine and use it in the same spirit, and without bias or prejudice.

Neither are they presented with any untried novelty; the success which has attended the course pursued in this work may assure them that in their hands it also will produce all the results they desire.

The preceding remarks have been made because, though the succeeding Part of this work, being Synthetic, is more like the plan usually pursued than the Analytic method, and the inferential style of the First Part, yet there are in the method and style of the Synthetic part conspicuous, distinguishing features that cannot fail to be noticed as novel; and the question may be asked, Why not adopt some of the usual methods?

They are not discarded for the sake of something new or peculiar, but because they are not sufficiently systematic and compact, and because scholars, by their diligent use, do not acquire that thorough understanding of these most practical of all subjects that is desirable and possible. For when there has been success, the teacher, rather than the book, has given it, since success has not been uniform, but exceptional.

It has been suggested that, in some cases, it may be advisable to commence recitations with the Synthetic part: some of its features are due to its having been prepared with reference to this suggestion. In such cases, however, it will be best to have the class spend a few of the first recitation hours in reading the Analytic part—nor will it be amiss to have the same thing done with the Second Part before commencing recitations in it; for some parts of the Body are so interwoven and so intimately related, that it is a great assistance to a pupil to first gain a general familiarity with all the parts before he studies their details. In fact, the general knowledge gained by reading the work, at least the Synthetic part, under the eye and questioning of a teacher, especially if reviewed in the same manner, will be all that is desirable for practical purposes, and none the less profitable because pleasantly obtained.

It will also be sometimes convenient, in the Synthetic part, to have the capitalized words necessary to fill out the questions, and which have hitherto commenced a paragraph, located in other portions of it. It is not denied that this method of asking questions is sometimes very simple, sometimes leading, and that answers to them are always very easily given; but it is remembered that teachers often have but little time to devote to any one lesson, either to devise questions or make explanations; and it is also noticeable, that by this plan questions may be very numerous without occupying too much space, thus developing in recitation all the practical facts of the lesson, which should be the sole aim in pursuing this study. Teachers may be fairly promised that if they will ask the successive questions and require the corresponding answers, they will, by the time the work is recited, develop in the minds of their pupils a thorough and in every way practical knowledge of the subject; a knowledge not merely of what is in the book, but of what is in the Body, as well as the why and wherefore of its construction; a germinal knowledge that will not vanish when the language through which it was received is forgotten; a knowledge ideaful, that will be a leaven, disciplining their minds as well as informing them.

# SYSTEMATIC

## HUMAN

# PHYSIOLOGY, ANATOMY, AND HYGIENE.

## PART II.

### DETAILED SYNTHESIS.

———•◦•———

## CHAPTER I.

### GENERAL PHYSIOLOGY, ANATOMY, AND HYGIENE:

#### PROPERTIES.

##### SYSTEMATIC SYNTHESIS OF ELEMENTS.

###### SECTION I.

*Chemical Elements.*

1. CHEMICAL ELEMENTS is the name given to sub-stances that cannot be or have not been analyzed, and are therefore supposed to be simple.

2. EACH SIMPLE ELEMENT IS ENDOWED with pecu-liar characteristics, inherent, natural, ordained, and un-varying.

3. *Illus.*—GOLD, IRON, when pure, are Elements uniformly the same. OXYGEN AND NITROGEN, which nearly constitute the air we breathe, are Elements. How different the characteristics of the four !

———————————•———————————

What is the topic of Part II. ? Of Chapter I. ? Of Section I. ? 1. What —? 2. How —? 3. What are —?

4. ALL MATERIAL THINGS CAN BE REDUCED by Analysis to the Chemical Elements, for it is by that very process that the existence of most of them is known.

5. ALL MATERIAL THINGS ARE CONSTITUTED of Chemical Elements, either by the process of mixture or compounding. The two processes and their results are very different, and very noteworthy by the Physiological Student.

6. A MIXTURE OF THE CHEMICAL ELEMENTS EXHIBITS merely the collective properties of the separate Elements. Ingredients is the proper name of the constituents of a mixture.

7. *Illus.*—THE AIR is a mixture, chiefly of the two Elements, Oxygen (about one fifth) and Nitrogen (nearly four fifths). If the proportion of the ingredients be varied, the characteristics of the air will be correspondingly varied, but not changed.

8. A COMPOUND OF CHEMICAL ELEMENTS IS such a union of them as exhibits to our view entirely new characteristics, never dreamed of till seen. Indeed, the chemist in his laboratory witnesses more wonderful realities, more unlooked-for results, more admirable creations, than the most extravagant fancy has ever imagined it was in the province of enchantment to produce. The constituents of a compound are appropriately named components.

9. *Illus.*—WATER IS a chemical compound constituted of two gases, Oxygen and Hydrogen, the latter the lightest known substance, and on that account used to elevate balloons. What a wonderful difference between these two Elements and their Compound! If the proportions of the components of a chemical compound are varied, the characteristics will be changed, and it will be a new compound.

10. SOME THINGS ARE partly mixtures and partly compounds.

11. *Illus.*—WATER, AS USUALLY FOUND, is a mixture of the compound Water proper and Air, the proportion of Oxygen being considerably larger than in the atmosphere, enabling fish to live in the mixture as they could not in the compound alone.

---

4. How —? 5. How —? 6. What does —? 7. What is —? 8. What —? 9. What —? 10. What —? 11. What —?

12. ALL THE CHARACTERISTICS MANIFESTED BY SUBSTANCES are not dependent alone on those of the Elements.

13. THERE ARE A NUMBER OF WHAT ARE CALLED INFLUENCES CONSTANTLY EXERTED UPON THE ELEMENTS AND THEIR COMPOUNDS, that modify their characteristics, causing them to unite or decompose, and exhibit a variety of powers or forces that, uninfluenced, they could not do.

14. IT IS QUITE AS IMPORTANT TO TAKE NOTICE OF THESE INFLUENCES as it is to consider the nature of the Elements, for without the *influences* the Elements would be of no avail.

15. THE SUN POURS DOWN three kinds of influences, the *Heat, Light,* and *Chemical rays,* each of which is powerful, at different times, in modifying the action of the Elements.

16. *Illus.*—SUGAR IS a compound of three simple elements, Oxygen, Hydrogen, and Carbon; but they will never unite so as to form sugar, unless influenced by the sun's rays.

17. ELECTRICITY OR MAGNETISM EXERTS oftentimes a powerful influence on the combinations of the Elements.

18. *Illus.*—THE PROPER PROPORTIONS OF OXYGEN AND HYDROGEN to form Water may be mixed in a vessel, yet they will not of themselves unite; but if a stream of electricity is sent through them, they unite instantly and become Water.

19. VEGETABLES EXHIBIT a secretory influence, causing their growth; and ANIMALS a nervous influence, modifying the Elements, and causing them to exhibit characteristics not wholly their own.

20. There are, therefore, THREE CLASSES OF INFLUENCES, the *celestial,* the *terrestrial,* and the *vital,* that modify the characteristics of the Elements as exhibited in their compounds.

12. What said of —? 13. What effect of —? 14. Why —? 15. What does —? 16. What —? 17. What does —? 18. What said —? 19. What do —? 20. What —?

7*

21. The characteristics of any compound will depend therefore upòn the Elements of which it is composed, as modified by the influences exerted upon it.

22. In brief, therefore, it may be said that every compound is compounded of *Elements* and *Influences.*

23. The Human Body consequently must be constituted of *Elements* and *Influences;* and

24. The characteristics of each part of the Body must be dependent on the *Elements* of which it is constituted and the *Influences* exerted upon them.

25. *Inf.*—The issues of life and death, of health and sickness, must depend upon a correct understanding of the Elements and their modifying influences.

26. The number and kind of Elements constituting the Body can only be ascertained by chemical analysis.

27. The number and kind of Influences essential for constituting the Body can only be determined by carefully noticing the circumstances in which the compounding of the Elements takes place, and noting the modifying influences at work.

28. The chemical Analysis of the Body determines the fact that there are uniformly thirteen of the same elements present in it; usually three more are found, but they are not essential, and are in quantities so minute as not to be worth mentioning in this work.

29. Two things are evident: 1st, The thirteen Elements, properly influenced, possess all the endowments necessary to constitute the Body; and, 2d, It can possess only those characteristics that the endowments of the thirteen Elements, properly influenced, can confer: its sphere is therefore limited within certain bounds.

30. It must, however, be considered that the same Elements uniting chemically in different propor-

tions, or in the same proportion but in a different order, exhibit entirely different characteristics; and therefore,

31. *Illus.*—*A*, *B*, *C*, THREE DIFFERENT ELEMENTS, may unite together directly; or, *A* and *B* may first unite, and the compound then unite with *C*; or, *A* and *C* may first unite, and that compound with *B*; or *C* and *B* may first unite, and then with *A*: thus from the same proportions of the same three Elements, four compounds, with very different characteristics, may be supposed to be produced. Vary the proportions of either, and four more might result. So that the thirteen Elements are surely enough to allow all the compounds required for exhibiting all the diversities of characteristics needed in the Body. Indeed,

32. IT WILL BE FOUND, UPON ANALYSIS OF THE DIF-FERENT PARTS OF THE BODY, that several of the thirteen Elements are in very small quantities, and exist in only a few parts, and that, in fact, the active parts of the Body are constituted of the very small number of six Elements.

33. THE THIRTEEN ELEMENTS ARE NAMED Oxygen, Hydrogen, Carbon, Nitrogen, Phosphorus, Sulphur, Calcium, Magnesium, Silicon, Potassium, Sodium, Chlorine, and Iron.

34. IRON exists in the Blood-cells, and may have some subsidiary office in the preparation of the higher tissues, or serve a purpose as a carrier of Oxygen.

35. CHLORINE is a gas, but is not thus found in Nature. It is one of the components of common Salt, and is therefore a very common article, and in that form is found in most of the liquids of the Body. It is also a very important component of Gastric juice.

36. SODIUM is a silvery-colored, very light metal; not found thus in Nature. It is the other component of Salt, the technical name of which is Chloride of Sodium, and in that form exists in most of the liquids of the Body.

37. POTASSIUM is a bluish gray metal; not found simple in Nature: combined with Chlorine, it exists as a

component of Chloride of Potassium in small quantities in most of the liquids of the Body.

38.   SILICON in very small quantities is a component of hair.

39.   MAGNESIUM is a white metal, that does not exist in Nature pure, but is found in large quantities as a component of Carbonate of Magnesia and of Magnesian Limestone.   It exists in small quantities in the Bones and some other parts of the Body.

40.   CALCIUM is a metal, the basis of lime, in which form, combined with carbonic acid, it is a component of the Bones, and in small quantities of other parts.

41.   SULPHUR is found in small quantities in all the liquids and tissues of the Body, except fat.   Its special office is not known, but it is evidently a necessity to Human life in all its forms.

42.   PHOSPHORUS, a now well-known waxy substance, used in making friction matches, not found pure in Nature, is an element of nearly all the Tissues.   It exists, combined with lime (phosphate of lime), in large quantities in the bones, giving hardness to them.   It also exists in the Brain and Nerves, and especially abounds in the gray part of the Brain.   *It is one of the most important Elements, since the student must supply food containing it abundantly, if he would use his Brain actively without exhaustion.*

43.   NITROGEN is a gas constituting nearly four fifths of the Atmosphere, and is an important Element of all the Tissues and liquids in the Body, except fat.   It is not however taken into the Blood from the atmosphere, but must come in as a component of food.

44.   CARBON is a solid, well-known substance, under the form of coal or diamond.   In the Body it is always combined, being a component of Bones in connection with lime (carbonate of lime), assisting with

phosphate of lime to give them hardness; combined with lime, potash, and soda, it exists in several of the fluids; it is a component of all the tissues; it and Oxygen combined are a gas, called carbonic acid gas, found in the Blood and expired at every breath.

45.   HYDROGEN is a very light gas, not existing pure in Nature, but very abundant as a component of water. It is an element of all the Tissues and liquids in the Body, and is obtained in both food and drink.

46.   OXYGEN is a gas, constituting nearly one fifth of the Atmosphere, and eight ninths of the weight of water. It also unites with Carbon, forming carbonic acid gas, which in its turn unites with lime, potash, and soda.    It is also an element in each of the tissues and fluids.    It therefore constitutes a large part of the weight of the Body, much more than any other element, or than all of them.    It is obtained in the air breathed, drank in the form of water, and eaten in the food.

47.   THE THIRTEEN ELEMENTS MAY BE REVIEWED as follows, under several heads:

| | | | | |
|---|---|---|---|---|
| Oxygen, | Nitrogen, | Calcium, | Potassium, | |
| Hydrogen, | Phosphorus, | Magnesium, | Sodium, | Iron. |
| Carbon, | Sulphur, | Silicon, | Chlorine, | |

48.   THE FIRST GROUP OF THREE ELEMENTS are components of all the tissues and liquids of the Body, including the fat.    This might at first view be thought to argue their superior importance.

49.   THE SECOND GROUP OF THREE ELEMENTS are components of all the tissues and liquids, except fat and the sinewy tissue.    They are very conspicuous in the Brain, and are obtainable only from food, not from air or water.    They need the previous preparation that food undergoes.    Sinewy Tissue is comparatively inert; so is FAT: its valuable power consists in producing heat.

---

45. Describe —.   46. Describe —.   47. How —?   48. What said —?   49. What said of —?

50. THE SECOND GROUP OF THREE ELEMENTS IS MUCH THE MOST IMPORTANT IN THE BODY, since the manifestation of the nervous and muscular powers is dependent on them.

51. THE ELEMENTS OF THE THIRD GROUP are chiefly useful in giving hardness to parts; they serve admirably in the framework.

52. THE ELEMENTS OF THE FOURTH GROUP are chiefly useful in forming the fluids of the Body.

53. IRON, though an essential Element, does not, strictly speaking, constitute any part of any tissue or fluid, being part of the contents of the Blood-cells, ect.

54. ANOTHER IMPORTANT FACT IS SUGGESTED by a review of the third and fourth groups. Their compounds in the Body are merely chemical compounds, requiring to produce them no higher order of influence than that which is chemical or terrestrial.

55. THE ELEMENTS OF THE FIRST TWO GROUPS, HOWEVER, enter into compounds that cannot be produced except under vital influences.

56. THE REALLY ACTIVE VITAL PROPERTIES OR CHARACTERISTICS are dependent therefore upon not more than the small number of the six Elements of the first two groups, combined and acting under appropriate influences.

## SECTION II.

*Proximate or Organic Elements or Principles.*

57. THE CHEMICAL ELEMENTS MAY COMBINE, by virtue of their own nature, or under the influence of the sun's rays, or that of electricity; but they will produce merely chemical compounds, which are very different from, very far short of, vital or organic compounds.

---

50. What said —? 51. What said —? 52. What said —? 53. What said —? 54. What —? 55. What said —? 56. What said of —? 57. How —?

58. THE RIGHT KIND AND PROPORTION of the chemical Elements must be selected, and under appropriate influences they must combine and form chemical compounds; but that is not enough.

59. THE CHEMICAL COMPOUNDS MUST BE SUBJECT to a vital or secretory process, that will, under other proper influences, produce a different kind and higher order of compound than chemistry can produce.

60. *Illus.*—SUGAR IS COMPOSED of the Elements of the first group, and requires the influence of the sun's rays for its production; but it also requires the influence of secretion. Those Elements never will combine to produce sugar, unless in addition to all other influences they receive that of a living substance, and a living substance peculiarly and nicely constituted.

61. *Illus.*—A SMALL BUD is taken from a sweet-apple tree, and inoculated in the limb of a sour tree; both continue to grow, they receive the same upward-moving sap, they hang in the same air, and each seems to receive the same Elements; but each possesses its own peculiar power, one producing sweet, and the other acid, fruit.

62. LIVING SUBSTANCES THEREFORE POSSESS a secretory or vital power of compounding; and if the right elements, in a proper condition, come under that influence, and other influences assist the process, peculiar and valuable compounds are sure to be produced.

63. VARIOUS SUCH COMPOUNDS ARE PRODUCED BY PLANTS AND CALLED *Organic* or *Proximate Elements*, or *Principles*, because it is from them that the animal tissues and fluids are chiefly formed.

64. ORGANIC DIFFER FROM CHEMICAL COMPOUNDS in several respects, but in one very important particular.

65. CHEMICAL COMPOUNDS ARE always definite, that is, not only composed of the same Elements, but in the same proportions, and the slightest change of proportions changes the whole character; so that the Chemical compound must be made right, or not made at all.

66. THE ORGANIC ELEMENT IS ALWAYS COMPOSED

of the same elements, but the proportions have usually a wide limit, and the quality of the compound may range correspondingly.

67.  *Illus.*—IF AN EGG OR A CHICKEN WERE A CHEMICAL COMPOUND, they would always be the same thing; but if a hen be fed upon corn, the yolks of her eggs will be a bright yellow, and have a good flavor. If she be fed upon flour, the yolks will be pale and insipid. If fed upon meat, especially such as worms, fish, grasshoppers, &c., the yolks will have a very deep color, and very fine, rich flavor. In either case a chicken can be hatched, but it cannot be possible that the three will be alike.

68.  ORGANIC ELEMENTS OF THE SAME KIND WILL THEREFORE VARY very much in quality; they may be called by the same name, but not have the same virtues.

69.  THE CHARACTER OF THE COMPOUNDS PRODUCED BY PLANTS will be determined by the soil in which they grow, by the air that surrounds them and the water they receive, by the influences derived from the sun, and by the character of the plant itself.

70.  A VERY IMPORTANT CHOICE IS THEREFORE TO BE MADE in the selection of food, for it is not to be supposed that a good Body can be composed from poor material.

*Remark.*—The herbage of some pastures is much better than that of others, as is proved by pasturing the same cows in both; for in one case they yield excellent butter, and indifferently good in the other. Different cows also produce different qualities of butter and milk when fed in the same pasture; but however good the cows, they cannot produce good milk if fed upon poor pasturage. They will also produce cheese differing in quality according to their food. The wool of sheep also will be affected in quality as well as quantity by the food they eat.

71.  IT MIGHT BE EXPECTED, as is the case, that the constitutions, and of course characteristics, of people, would be greatly affected by all the influences that affect the plants upon which they live, such as soil, air, &c.

*Remark.*—Wheat differs in its character in different lots; so do other grains and parts of vegetables; so do meats: and it must not be inferred, because articles of food are called by the same name, that they have the same value. All articles of vegetable or animal composition will vary so much in different cases, that where the preservation or restoration of health is of great importance, and very difficult to accomplish, the greatest care must be taken to select the best qualities as well as kinds of food.

72. THE ORGANIC PRINCIPLES COMPOUNDED BY PLANTS, for the use of Man, are of two classes; those adapted merely to produce heat (or fatten), and those adapted to nourish his Body.

73. THE ORGANIC PRINCIPLES ADAPTED BY THE PLANTS FOR CALORIFIC PURPOSES are Starch, Sugar, Gum, Fat, composed of the Elements of the first group, in different proportions. These, compounded under the influence of the sun's rays, store up or condense, so to speak, his heat, until such time as they are burned, when they give out merely what they have received.

*Remark.*—The Calorific Organic Principles may be, and in some cases are, separated from other parts of plants in which they grow, and used by themselves, or mixed with other articles; in other cases they are eaten in a combined state, as they exist in the plant, when they will not be distinguished by the ordinary observer. For instance: the sap of cane is expressed, boiled down, and sugar obtained; starch is also separated from the other parts of potatoes, corn, wheat, etc., and used; or, the grain and potatoes are cooked and eaten in the form of bread, etc., and their starch is eaten with their other components.

74. PLANTS ALSO COMPOUND, principally in their seeds, the Organic Principles necessary to serve the purposes of nutrition to tissues abounding in Nitrogen, Phosphorus, Sulphur, Oxygen, Hydrogen, and Carbon. These nutritive Organic Principles also are compounded under the powerful influence of the sun's rays, which are stored in them until the time occurs for their power to be exhibited. These are able to receive and store up more than the sun's heat; that can be done by the first group of Elements: the second group can store up more, viz., the chemical or actinic power; and the entire nutritive Principle, composed of both the first and second groups,

72. What said of —? 73. What said —? How compounded? How are calorific elements used? 74. What do —? How compounded?

will be able to exhibit heat in addition to other power.
It therefore may be expected, as is the fact, that when
the nutritive Principles are decomposed in the animal,
they will, in addition to the peculiar power they exert,
either produce heat or supply a compound that can by
further decomposition exhibit heat.

> *Remark.*—THE NUTRITIVE PRINCIPLES MUST EXIST in other parts be-
> sides the seeds, since graminivorous animals, eating only the herbage,
> supply all their tissues with nutriment; but they eat a very large quan-
> tity of food.  Granivorous animals eat less to gain the same nutriment,
> carnivorous still less.

75.  UNFORTUNATELY WE DO NOT UNDERSTAND the
precise nature of the sun's influence upon vegetable construction; therefore it cannot be spoken of very definitely, and only a brief and unsatisfactory statement of
leading facts can be given.

76.  THE NUTRITIVE ELEMENTS MUST BE COMBINED
under the influence of the sun's rays, since the plant
cannot alone do the work; nor can it, even if assisted by
chemical influences, combine the Elements into Organic
Principles without the sun's aid : therefore the power of
the sun may be said to be represented in the Principles
that are indeed compounds of Chemical Elements and
chemical, solar, vegetable, and sometimes animal, influences.

77.  THE VITAL INFLUENCES OF A PLANT ARE always
constructive, growing, or productive of growth, while
chemical influences are either constructive or destructive, according to circumstances; and, on the other hand,
strictly speaking, all animal processes are destructive.
Secretion is truly vegetable in its character, though in
the animal modified by nervous influences in some cases.
Thus, as chemical influences reach up and coact in the
plant, so do chemical and vegetable reach up and coact
in the animal.

78. THE PLANT STORES UP that the animal may destroy, as it must destroy itself in order to act. Chemical influences come in on the one hand to assist the plant in producing or compounding, and on the other to assist the animal in destroying or decomposing. The action of the sun's influence is also compounding in the plant; it promotes the vegetative in the animal as well as in the plant. Hence the exposure of the Body to the influences of the sun is essential to the enjoyment of health, and children in particular should be caused to play in the full light of the sun, since they are growing, and need all growing influences.

79. THE ANIMALS, HOWEVER, COMPLETE THE PROCESS begun by the plant, and perfect their tissues from the Principles furnished, to a degree beyond what the plant can do, for it can secrete neither muscular nor nervous tissue. The vegetative process in the animal can be, if it is not always, influenced by the nervous tissue; for nothing is more evident than that the eye and mouth water under mental influences, and that the quality of the liquids is always affected by the existing state of the mind.

80. WHILE ON THE ONE HAND ANIMALS cannot begin to compound Elements so low in the scale as plants can, PLANTS ON THE OTHER HAND cannot compound to so high a degree as animals; yet the plants cannot compound from so low a point as the simple elements. Even the plants must be supplied with food, that is, prepared or compounded Elements.

81. THE PROCESS OF COMPOUNDING THE ELEMENTS MAY BE CLASSED in three stages: the chemical, the vegetable, the animal. The former two we have considered in this chapter; the third will be most appropriately treated in the next. chapter, in connection with Tissues, to which we will now advance.

---

78. Why does —? What said of sun's influence? 79. Do —? What said of mental influence? 80. What said of —? 81. How —?

# CHAPTER II.

### SYSTEMATIC SYNTHESIS OF PRINCIPLES INTO TISSUES.

#### PROPERTIES.

## *Introductory Remarks.*

82. IT IS VERY EASY TO DETERMINE the number and characteristics of the thirteen Chemical Elements in the Body, and of the Tissues into which the Elements are compounded.

83. IT IS ALSO EASY TO CLASSIFY the influences under which the compounding takes place: they are chemical, vegetable, and animal, of which the first can operate alone, and the first and second together; or the second in succession to the first, and the first, second, and third together; or the third in succession to the first and second together; or the third in succession to the second, it having acted in succession to the first; so that the first must always act before or when the second does, and the first and second before or when the third does.

84. IT IS EQUALLY EASY TO CLASSIFY the compounds of Elements in the Body, since they are, 1st, merely chemical, and, with the exception of *Water* and *Carbonic Acid*, contain some of the Elements of the third and fourth groups, and are useful merely in the passive hard tissues and in the liquids; 2d, the Calorific or Oleaginous, or non-Nitrogenous compounds of the Elements of the last group, united under chemical and vegetable influences; 3d, the albuminous, nutritious, nitrogenous compounds of the first group of Elements, and two or all

---

82. What —? 83. What —? How can they operate? 84. What —? What are they? Name the thirteen elements in groups.

of the second, united under chemical and vegetable influences, and brought to the highest condition by animal influences. In addition to these compounds, the simple Element, oxygen, of the first group, and iron, may be mentioned as necessary in the Body.

85. SUMMARY OF CHEMICAL COMPOUNDS AND PROXIMATE PRINCIPLES IN THE BODY:

1st, CHEMICAL
- Hydrochloric Acid.
- Silica.
- Carbonate of Lime, Soda, and Potassa.
- Phosphate of Lime, Magnesia, Soda, and Potassa.
- Chloride of Sodium and Potassium.
- Water.

2d, CALORIFIC
- Gum.
- Sugar.
- Fat.

3d, ALBUMINOUS
- Albumen.
- Albuminose.
- Fibrin.
- Casein.

4th, EXCRETORY
- Carbonic Acid.
- Urea.
- Bilin verdin.
- Cholesterine.
- Lactic acid.
- Creatin, &c.

5th, SIMPLE ELEMENTS used in the Body
- Oxygen.
- Iron.

86. IT IS VERY DIFFICULT, on the other hand, to determine by what *precise steps* the thirteen simple Elements are compounded into the six Tissues.

87. AN EGG THROWS a flood of light on this intricate subject, since the contents of the shell are compounded into all the tissues.

88. THE MEAT OF AN EGG IS COMPOSED of one form of Albumen, called egg-albumen, compounded of the Elements of the first and second groups; a portion of calorific fat, compounded of the Elements of the second group; and the chemical compounds, including elements of the third and fourth groups, and combined more or less intimately with the albumen and fatty matter.

89. IT IS THEREFORE CERTAIN of what all the tissues can be composed, and through what form the Elements

pass in one stage of their progress, as the experiment of forming tissues from the meat of the egg daily takes place, all the tissues being formed from it, and exactly all of it being used.

90.   NOW IT FOLLOWS that all the Elements of the tissues must exist at one time, in their course of development, in the form of albumen, since it is not conceivable that the tissues can be formed in several different ways.

91.   AGAIN, MILK IS a substance by which, in many instances, all the food of the creature is supplied, and that liquid is composed of albuminous substances, calorific fat and sugar, and chemical compounds: it is, in fact, the meat of an egg diluted with water, and enriched by sugar and more fat.

92.   IF THE PARTS OF PLANTS USED AS NUTRITIOUS FOOD ARE EXAMINED, they will also be found composed in part of albuminous substances, called vegetable *albumen*, *fibrin* (gluten), and *casein* (cheese or curd).

93.   ONE OF THE FIRST OPERATIONS OF DIGESTION is to change the vegetable albumen, fibrin, and casein, and also the animal albumen, fibrin, and casein, into albuminose; a form of albumen very similar in all respects, and in the number and kind of its Elements precisely like the egg-albumen.

94.   IT APPEARS, therefore, that *Albumen* is a compound or station-point, from which an observer may look back toward the simple Elements, and forward toward the Tissues.

95.   THE PRECISE STEPS FROM THE SIMPLE ELEMENTS TO ALBUMEN, or from it to the Tissues, are not well understood.

96.   BUT THE ALBUMEN AND OTHER CONTENTS OF THE EGG-SHELL do not of their own accord assume the form of the tissues of the chicken; the contents must, in the

---

90. What —? 91. What —? 92. What —? 93. What is —? 94. What appears? 95. What said of —? 96. What said of —?

first place, be kept at a temperature of about 100°, and, in the second place, be acted on by a minute portion of animal tissue already appropriately compounded.

97. THE CHEMICAL INFLUENCE OF THE HEAT WITH THE ANIMAL INFLUENCE OF THE TISSUE will soon begin to work like leaven, and first one variety of tissue and then another will start into existence, each also growing by additions to itself, till all the varieties of Tissue are formed in proper proportions, and all the contents of the shell are changed into tissues, and a perfect chicken appears.

98. IF THE CHARACTER OF THE MINUTE PORTION OF TISSUE IN THE EGG BE EXAMINED, it will afford much instruction upon the important subject of the constitution of the tissues.

99. IF AN EGG-SHELL BE CAREFULLY CUT OPEN, the yolk will be seen suspended, so that a very easily discerned spot upon its surface is always uppermost.

100. IF THIS SPOT IS EXAMINED WITH A MICROSCOPE, a minute cell or sac will be found, formed of an albuminoid substance as thin as that of the thinnest soap-bubble, and called homogeneous, viz., alike in its character throughout.

101. THE CELL CONTAINS a liquid substance, in the midst of which float minute, irregular collections of matter called granules; of what, or how, they are composed, is not certain.

102. THE SUBSTANCE OF THE WALLS OF THE CELL IS of such a nature, that materials can pass into and out from the cell very freely, the substance-wall of the cell having very little tenacity, and seeming to serve its purpose by surrounding its contents rather than by enclosing them.

103. AT ONE OR MORE POINTS IN THE WALLS OF SUCH CELLS there is a peculiar construction called its

nucleus, from which the granules appear to take their rise in some cases, and in others from which new cells spring up.

104. IN SOME INSTANCES THE GRANULES DEVELOP into new cells, and in some they become arranged in the form of fibres.

105. THUS THE EARLIEST APPEARANCES IN THE FOR-MATION OF TISSUES ARE the formation of granules, fibres, and membrane, in a homogeneous Fluid containing the proximate principles essential to the constitution of the Tissues.

PRIMARY PHYSICAL ELEMENTS $\left\{\begin{array}{l}\text{Homogeneous Liquid.}\\ \text{Homogeneous Granules.}\\ \text{Homogeneous Fibres.}\\ \text{Homogeneous Membrane.}\end{array}\right.$

106. THE IMMEDIATE CAUSE OF THESE PRIMARY FORMATIONS IS the influence of the previously formed membrane constituting the walls of the cell, without which the homogeneous fluid can never be developed into tissue.

107. ALL LIVING THINGS, WHETHER VEGETABLE OR ANIMAL, take their rise from such a cell: its components in that form and condition have the power of communicating to other similar material the power they themselves possess, with such modifications as the exigency requires.

108. THE CELLS HAVE TWO POWERS: 1st, that of forming themselves from the appropriate materials, and 2d, that of modifying their contents, each of which is called the power of secreting, though, as they differ in their character, they should be called by different names; and in this work the former will be called *nourishing*, and the latter *secreting*.

109. IT IS ALSO REMARKABLE that the parent-cells will bestow upon their offspring the power of secreting

substances different from those the parent-cells themselves could secrete.

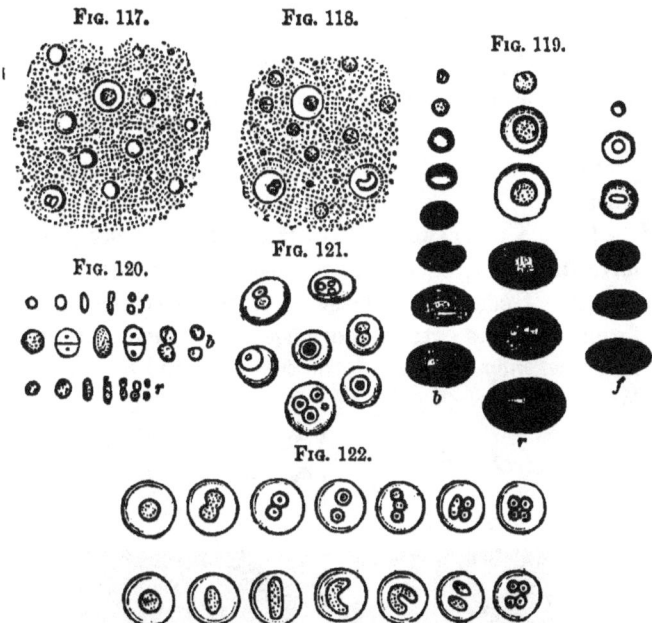

FIG. 117.  FIG. 118.  FIG. 119.

FIG. 120.  FIG. 121.

FIG. 122.

Fig. 117 represents a portion of the fluid chyle, in which are exhibited minute granules and larger globules, some of them appearing like, and doubtless being, cells.

Fig. 118, the same, after the addition of acetic acid.

Fig. 119 represents, *f*, the developing blood-cells of a Fish (Haddock); *r*, those of a Reptile (Frog); *b*, those of a Bird (Turkey).

Fig. 120, the same as the previous, after the addition of acetic acid.

Fig. 121, cells from spleen. In some the nucleus is seen dividing, in others it is divided, and again subdivided. The process is still better seen in

Fig. 122, representing the white blood-cells undergoing changes, and their nuclei getting ready to form new cells. Still further representations of cells are given on the next pages.

110. IT IS NOT KNOWN upon what modification the power of secreting the different contents of different cells depends; their walls seem to be alike, but are different in different cases.

Describe Fig. 117. Fig. 118. Fig. 119. Fig. 120. Fig. 121. Fig. 122. 110. What —?

8

## SECTION I.

## *Secretory Tissue.*

111. SECRETORY TISSUE MAY THEREFORE BE DEFINED as a homogeneous albuminoid substance or membrane, compounded of the Elements of the first and second groups, under the influence of secretory tissue previously existing.

112. VEGETABLE AND ANIMAL SECRETORY TISSUE have the same general appearance and constitution.

113. SECRETORY TISSUE EXHIBITS three general forms: 1st, the extended or sheet-like, when it is called *basement membrane;* 2d, the *tubular;* and 3d, the *cellular.*

SECRETORY TISSUE { Basement Membrane. Tubular. Cellular.

FIG. 123.

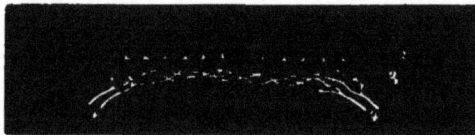

Fig. 123 is a beautiful plan of cells, 1, resting on a basement membrane, 2, beneath which is the fibrous layer, 3, containing the blood-tubes, 4, The spot in the centre of the side of the cell is its nucleus. (For tubular form of membrane, see *i*, Fig. 137.)

114. THE CELLULAR FORM OF SECRETORY TISSUE EXHIBITS numerous varieties, each of which changes its form more or less, according to circumstances, such as pressure, etc. Whether the form of the cells has anything to do with the peculiar character of the secretion formed by them, is not known; it probably has, since each different substance is secreted by a cell of a peculiar form.

115. THE MOST CONSPICUOUS CELLS ARE, 1st, the

---

111 How —? 112. What said —? 113. What does —? Write table. Describe Fig. 123. 114. What does —? 115. What —?

Blood and Lymph cells (the red and white Blood-cells), Fig. 110; 2d, the cells of the external skin (the cuticle), called the cuticular cells, Fig. 109; 3d, the cells of the surface of the air and food passages, secreting mucus, hence called mucus cells, also epithelial cells, Figs. 88, 77; some of them are furnished with cilia, and are called ciliated cells; 4th, the cells of all the surfaces of all the organs moving against each other, secreting serum, hence called serous cells, also called pavement or tessellated cells. Others of minor importance will be noticed in connection with the organs to which they belong.

FIG. 124.    FIG. 125.    FIG. 126.

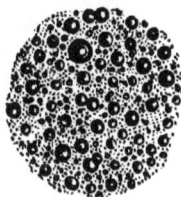    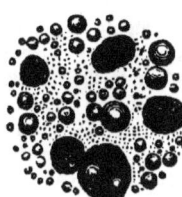    

FIG. 128.    FIG. 129.

FIG. 127.

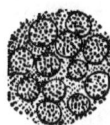    

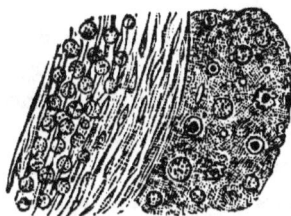

FIG. 130.

FIG. 132.

FIG. 131.

    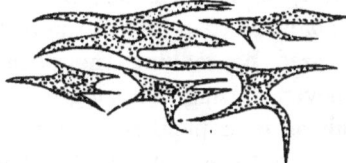

Describe Fig. 124, 125, 126. Describe Fig. 127. Describe the Figures from 128 to 132.

FIG. 133.      FIG. 134.

FIG. 135.

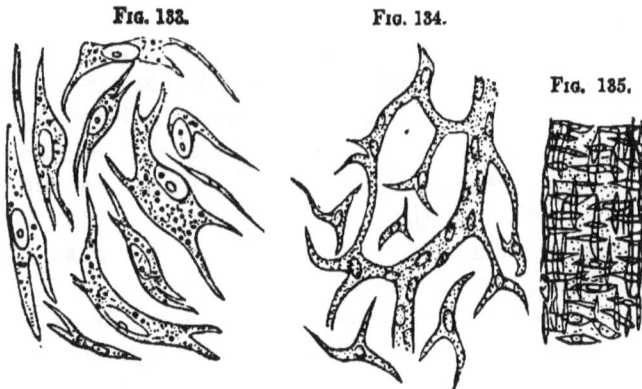

Figs. 124 and 125 represent highly magnified portions of milk; their granules, globules, and cells, being very apparent. Fig. 126, Blood-cells differing in appearance according to their position; there are two white ones. Fig. 127, portion of chyle of dog, showing granules, globules, and cells, some with fibrous appearance. Figs. 128 to 135, differently formed cells, dividing in 132, extending into fibres in 131, uniting to form capillaries in 134, or arranged, as in 135, to form an arterial tube. (See p. 171.)

116. It is EVIDENT that the amount of substance that can be secreted will depend upon the amount of secretory tissue concerned.

117. WHEN A LARGE AMOUNT OF SUBSTANCE MUST BE SECRETED, it will be necessary to correspondingly increase the amount of secretory tissue; and if the space allowable is not sufficient to spread out the membrane, it must be formed into tubes or into cells, by which a greater amount can be packed in a given cubical space.

118. IF THE CELLULAR FORM RESTING ON AN EX-TENDED SURFACE IS NOT SUFFICIENT, as it is in case of the general, cuticular, mucus, and serous surfaces, the amount of secretory tissue in a given space can be in-creased by making the surface on which the cells rest uneven, raising it in ridges or sinking it in grooves, or raising it in papillæ or burrowing it into the form of tubes with branches and pouches on the sides of them,

the whole of these covered or lined with appropriate cells. Fig. 88.

FIG. 136.

Fig. 136 is a rude plan of the same parts as in 123. *a*, Basement membrane; *b*, cells; *c*, fibrous under-structure; all as in an extended membrane, without indentations or eminences.

FIG. 137.

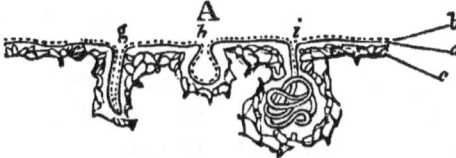

Fig. 137 represents the same, forming a short tube at *g*, a pouch, crypta or follicle at *h*, and a coiled long tube at *i*.

FIG. 138.

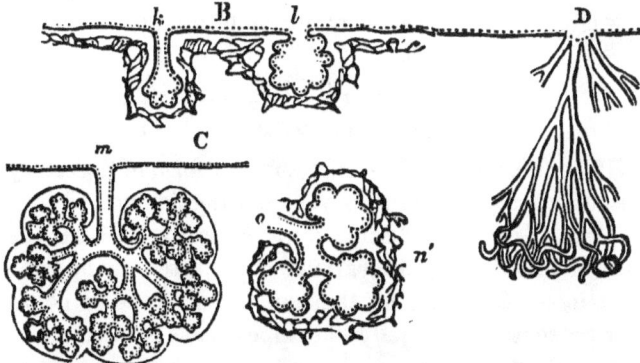

Fig. 138 represents, by a rude plan, different modes of increasing surface, *k* and *l* being simple follicles, with sides a little corrugated. At *o* is the outlet to three like *l*, while *m* is the common outlet of many, or of a racemose gland. At *D* there is a common outlet of many tubes, or of a tubular gland. *l* is, in fact, a simple gland.

119.   GLAND is the name given to any means taken for increasing the number of cells in a given space.

Describe Fig. 88 in reference to surface.   Describe Fig. 136.   Fig. 137.   Fig. 138. 119. What is a — ?

Glands are called simple or compound, according as they are simple tubes, or branched (racemose) with clusters of little pouches (cryptæ or follicles) about them.

120. THE CELLS OF ALL THE SURFACES and of the Glands proper have beneath them a layer of basement membrane, from the nuclei of which the cells take their rise.

121. THE MODE OF CELLULAR GROWTH AND ACTION is this: the cells take their origin from the nuclei of the basement membrane, and enlarge to maturity, at the same time filling themselves with their appropriate secretion; when mature, they are loosened from the nuclear spot, and if at the surface, they break or dissolve away, yielding their contents to fulfil their purpose; if not at the surface, they are crowded up by those growing beneath them, and either dissolve, as in some cases, or, drying, fall off as scales, as in case of the skin.

122. THUS SUCCESSIVE CROPS OF CELLS ARE GROWING in contact with the basement membrane, and successively wasting away in the very act of fulfilling their office.

### SECTION II.

#### *Muscular Tissue.*

123. MUSCULAR TISSUE IS another form of Secretory Tissue, having peculiarities very distinguishing.

124. MUSCULAR TISSUE IS IN THE FORM OF cells, the walls of which are to appearance constituted of the same albuminoid Elements as those of other kinds of cells; their contents have a jelly-like appearance. (See Fig. 78.)

*Remark.*—Muscular Tissue is sometimes described as composed of tubes, as if the ends of the cells had broken down, allowing all in one filament to form one continuous tube.

125. THE MUSCULAR CELLS ARE PECULIAR in this, that they have no relation to basement membrane for their production.

---

120. What beneath —? 121. What —? 122. What said —? 123. What —? 124. What —? 125. How —?

126.  THE MUSCLE-CELLS ARE ESPECIALLY DIFFERENT FROM THOSE OF THE SECRETORY TISSUE proper in this, that the substance secreted in them is not of use in the form in which it is secreted, but only when it is decomposed and destroyed as muscular tissue.

127.  MUSCULAR TISSUE IS PASSIVE while it remains muscular tissue, and as soon as it becomes active it changes, for it is the change that produces the power of activity or contraction.

128.  SUCH TISSUE IS NOT produced by the plant, nor can it be; it is the province of the animal to exhibit the power developed by the plant, and to decompose what the plant has compounded.

129.  THE CONTENTS OF THE MUSCLE-CELL exhibit all the six elements of the first two groups; but whether the walls of the cell can secrete their contents from the albumen of the blood directly, or whether it is necessary that the albumen should undergo some process of preparation, is not known; but the latter is probably true.

130.  THERE ARE TWO FORMS OF MUSCLE-CELLS, that of the striated and that of the non-striated. In the former the cells are bead-shaped, and arranged end to end (see Fig. 78); in the latter case the cells taper at each end, and are sometimes found single and sometimes clustered without regularity.

131.  THE USE OF THE MUSCLE-CELLS IS to contract. How the effect is accomplished is not known, nor is the *rationale* of the act even conjectured. It is certain that each contraction is attended with a corresponding amount of decomposition of muscular secretion, to be renewed only by a corresponding amount of substance eaten.

132.  IT IS ALSO CERTAIN that there is no other power so economically used as the power of the muscle-cells.

133.  IF THE SOURCE OF THIS POWER IS TRACED BACK

126. How —?   127. When —?   128 What said —?   129 What said —?
130. What —?  Describe Fig. 78.   131. What —?  132. What—?  133. What said—?

it will be found to be drawn from the rays of the sun
that were condensed by the plant in its albuminized
products.

## SECTION III.

### *Nervous Tissue.*

134. NERVOUS TISSUE IS of two kinds, the gray,
cellular, ganglionic, central or active, and the white,
fibrous, tubular, or passive parts.

135. THE GRAY NERVOUS is one form of secretory,
but differs from it in the same general respects as the
muscle-cells.

136. THE GRAY TISSUE DIFFERS by not having any
relations to a basement membrane, and by secreting a
substance that is not of use until it is decomposed, and
because its use is destructive, or the result of destruction.

137. THE GRAY TISSUE IS COMPOSED of cells, with
walls constituted like all cells, filled with a jelly-like
granular substance, and also surrounded by a similar
substance, in which the cells seem to be very thickly
imbedded.

FIG. 139.

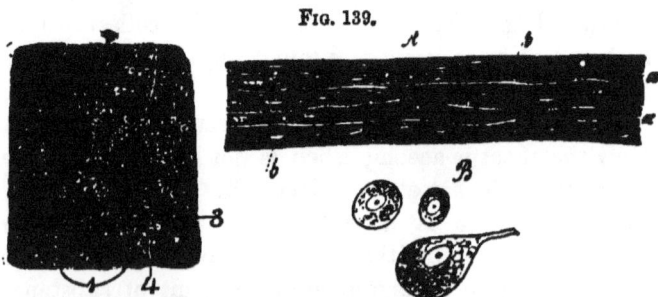

Fig. 139 represents nervous tissue: 1, two large cells continuous with
nerve filaments; 8, undeveloped cells with nuclei; 2 and 4, a chain of cells
connected by filament; B, nerve-cells much enlarged, showing nucleus, its
spot, the granules, and one extending point or pole in case of the largest
cell; a a, filaments of white nervous tissue.

134. Of what kinds —? 135. What —? 136. How does —? 137. How —?
Describe Fig. 139.

138. THE CELLS SEEM TO GROW from the granules, and as they reach maturity portions of the walls elongate and become continuous with the filaments of white tissue constituting the nerves or commissures.

139. THE CONTENTS OF THE CELLS, as well as the substance in which they are imbedded, is constituted of the six Elements of the first two groups, the proportion of phosphorus being larger than in any other tissue.

140. IF THE NERVOUS MATERIALS ARE TRACED BACK, it will be found that the powers that Nervous Tissue can exhibit are condensed by the plant under the influence of the sun's rays, in the albuminoid products of the plant, and that the power of the nervous cells is a representative of so much power as the plants have stored, and no more.

141. ALL POWER THAT CAN BE MANIFESTED BY THE NERVOUS CELLS IS exhausted when their substance is decomposed, which must be replaced from food eaten, containing the appropriate Elements properly prepared.

142. IT IS NOT PROBABLE that the nerve-cells can secrete their important contents directly from the albumen of the Blood, but it is probable that it is wrought up to a certain degree before the nervous Tissue performs upon its Elements the finishing secretion that endows them with the highest and most wonderful power that matter is ever capable of exhibiting, and enables them to fulfil duties most intensely important to human welfare.

143. WHETHER THE SUBSTANCE ABOUT THE CELLS is affected by them or not, is not known; the probabilities favor the idea that it is.

144. THE WHITE NERVOUS TISSUE does not exhibit the cellular arrangement (though originally constituted from cells), nor is it necessary that it should, as its office is passive so far as it is understood, having merely to communicate the influences that originate at either extremity of the fibres that it constitutes.

138. How do —?   139. What said of —?   140. What —?   141. What said —?
142. What —?   143. What said of — ?   144. What said —?

8*

## SECTION IV.

### *Sinewy Tissues.*

145. SINEWY TISSUE EXISTS in the form of minute threads or fibres, and hence is usually called fibrous tissue; but as other tissues assume a fibrous structure, and as the character of sinews which are composed of this tissue is well known, the above name is much the best.

146. SINEWY TISSUE IS CONSIDERED under two heads, that of the white or inelastic, and the yellow or elastic.

147. WHITE SINEWY TISSUE gathers its fibres into small bundles, or into bands more or less extended, the fibres of which adhere so closely that it is difficult to separate them. They are exceedingly strong, flexible, and inelastic, and therefore well adapted for use in different parts of the Body, which is, in fact, largely composed of them, arranged in various forms or textures.

FIG. 140.

FIG. 141.

Fig. 140 represents a portion of Sinewy Tissue with its fibres pulled apart; Fig. 141, the same, surrounding spaces from which the natural contents have been dissolved. In both figures the tissue is very much magnified.

145. How does —?   146. How —?   147. What said —?   Describe Fig. 140. Fig. 141.

148. YELLOW SINEWY TISSUE is also very strong, but it is likewise elastic; its fibres do not lie parallel to each other, but are branching, curled, and irregular. It exists by itself in only a few places in the Body, but in many it is mixed with the white.

FIG. 142.　　　　　　　FIG. 143.　　　　　FIG. 142.

 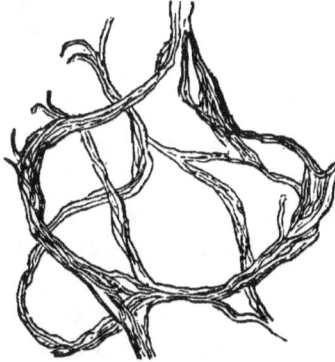 

Fig. 142 represents two portions of highly magnified Yellow Tissue, differing in size of fibres.

Fig. 143 represents the same, around the air-cells of the Lungs. By its action the air is expelled at each breath, and at each inhalation it must be overcome by the internal pressure of the air. The perfection of breathing, therefore, is chiefly dependent on this tissue.

149. THE TWO TISSUES SEPARATE OR TOGETHER ARE WROUGHT into many different forms or textures—of *Ligaments, Tendons, Membranes, Fascia, Areolar Texture*, etc.

150. IN LIGAMENTS THE TISSUE IS ARRANGED in cords, bands, or caps, extending between' or around parts that are fastened together, as in the case of the bones. The fibres lie nearly parallel, or are braided together, as may be required.

151. TENDONS ARE LIKE ligaments, except that at one or both extremities they are attached to muscular fibres, which makes a peculiar arrangement of the fibres of the tendon necessary where they unite with the mus-

---

148. What—? Describe Fig. 142. Describe Fig. 143. 149. How are—? 150. How is—? 151. What—?

cular part. The cords on the back of the hands are examples.

FIG. 144.

Fig. 144 represents both Ligaments and Tendons, the former binding the latter in their places. 2, 3, 4, tendons; 5, 6, muscles; 1, finger-bone. The tendons can be noticed extending into and blending with the muscular part. A muscle is said to be composed of its muscular part and its tendon; an expression that seems contradictory, but it is not so in intention.

152. MEMBRANES are of two classes, the dense and open. The only difference consists in the distance of the fibres apart.

153. IN MEMBRANES THE FIBRES ARE INTERWOVEN in every direction, leaving larger or smaller meshes between them to be occupied with other parts. In some instances the meshes are very few and small, in other cases they are large and numerous.

FIG. 145.

Fig. 145 represents one form of membrane woven from White tissue in narrow bands, and Yellow Tissue in fibres. A few fat-cells in a row and in a cluster are shown. The meshes of different membranes vary very much in form and size.

154. THE SKIN is composed of an external layer of cellular tissue called cuticle, beneath which it is chiefly composed of Sinewy Tissue, the meshes of which are

very fine near the cellular surface, but which grow larger and more numerous lower down.

155. *Illus.*—The fine surface of a kid glove and the openness of the inner side is an excellent example of the structure of such a membrane. The membrane at the surface of any bone can be peeled off and show the character of a dense membrane.

156. FASCIA ARE merely dense membranes of a sheath-like form, performing special purposes.

157. AREOLAR OR CONNECTIVE TEXTURE (also called tissue and cellular) is merely Sinewy Tissue in the form of bands of small or considerable width, intersecting each other in every direction so as to form spaces of very small or considerable size if they should be distended. These spaces are usually merely moistened with serum, but sometimes it collects, producing one form of dropsy. They can be distended with air, when it will be found that the air will creep from one space to another till all parts of the body will be distended.

FIG. 146.

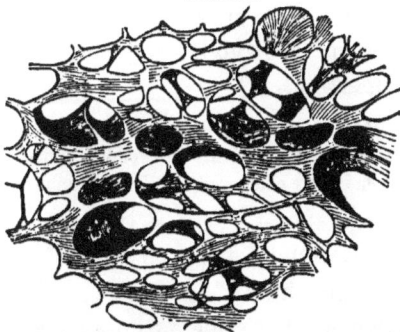

Fig. 146 represents a small portion of one form of Areolar texture, with its spaces distended and communicating. They evidently vary in size, and may be as small as a mustard-seed, or as large as an egg. As usually found, the sides of these spaces touch.

158. *Illus.*—BUTCHERS sometimes introduce one end of a pipe into a piece of meat, and blow air through the pipe into the areolar texture of the meat, and give it an appearance of being fat.

159. *Illus.*—In Paris, a child with an enormous head was exhibited to excite compassion and obtain charities. The parents, being questioned by the authorities, confessed that they had tied a bandage around the child's head to prevent the air from passing below it, and made an

opening through the skin above, introduced the end of a pipe, and forced in air from time to time, gradually distending the skin to a wonderful degree.

*Remark.*—Sometimes the chest is injured in such a manner that a communication is established between a lung and the areolar texture, in part constituting the walls of the chest; at each respiration, a little air will be forced into the areolar texture, until, in some cases, that of the whole body will be distended, producing a very grotesque appearance. As soon as the passage of the air from the lung is checked, that which is in the areolar texture will be absorbed by the Blood, and the distension will diminish as rapidly as it was produced, without causing any unpleasant consequences to the injured person, thus showing the inert or passive character of the areolar or connective texture.

160. It thus appears that the Areolar texture loosely connects and binds together all parts of the Body, and exists from head to foot; and it will be found that as we go through the skin its meshes increase till they become those of the Areolar texture.

161. The Sinewy Tissue also exists as a part of the two tissues next to be mentioned.

162. The sinewy textures are composed wholly of white or wholly of yellow, or of both, in varying proportions.

163. *Illus.*—At the knuckles the skin is composed mostly and at the elbow almost wholly of Yellow, while on the front of the arm the skin is chiefly composed of White, Tissue.

164. It is evident that a large amount of Sinewy Tissue will be required in the Body; and in fact more than one half of its entire weight is composed of this tissue, so that some authors have described the Body as being composed of Sinewy Tissue wrought with a multitude of meshes of various forms, in which the other parts were packed; and indeed, if all the other tissues could be dissolved out, the form and size of every organ would yet remain fully represented by the Sinewy Tissue, so much is it the framework of every part.

160. What —? 161. Where does —? 162. How —? 163. How knuckles and elbow composed? 164. What —?

165. THE SINEWY TISSUE IS EVIDENTLY a passive tissue, and does not require the elementary constitution of the active tissues.

166. THE ANALYSIS OF SINEWY TISSUES SHOWS that they are entirely wanting in phosphorus, and have but a very little sulphur and an inferior proportion of nitrogen.

167. THE WHITE AND THE YELLOW DIFFER in the arrangement of their elements, since from the former Gelatine can be obtained, but not from the latter; there is also a difference in the proportions of their elements.

168. WHEN THE WHITE TISSUE IS BOILED awhile in hot water it is dissolved, and upon cooling forms a jelly called gelatine, or gluten technically, or glue in commerce.

169. THIS SUBSTANCE, GELATINE, *is never found in the blood*, though it is eaten as food, and is the component of the larger part of the Body.

170. THE FACT MENTIONED SHOWS that Gelatine must be formed from the Blood by the Tissue in the very act of its formation; that the Gelatine decomposes when it goes back from the tissue into the Blood; and that the Gelatine eaten must serve some other purpose than nourishing any tissue, for it has the Elements of no other but the sinewy; therefore it must be useful only in producing heat, to which purpose also the decomposing sinewy Tissue can be applied.

171. THE SINEWY TISSUE IS CONSTITUTED IN PART of Nitrogen, and therefore it must be obtained from some part of the food that contains Nitrogen; and as in the chicken it must have been obtained from the albumen, so it must be at all times; but through what changes the albumen passes, to free it of phosphorus and a part of its sulphur and nitrogen, is not known. It is supposed that the fibrin of the Blood is one stage in the process of forming sinewy tissue. It may also

---

165. What —? 166. What said of —? 167. How do —? 168. What effect —? 169. What said —? 170. What does —? 171. How —?

be formed, as some think, from the decomposing active tissues that were formed from albumen.

172.  **SINEWY TISSUE IS PRODUCED** by cells extending into the form of threads, and attaching themselves together either at their ends or side by side.

173.  **BUT THE QUESTION MAY BE ASKED,** What makes these fibres assume their appropriate positions in so many different textures ?

174.  **THE ANSWER IS,** that the influence of the tissues upon each other causes them to assume those special arrangements of form that adapt them to their purpose, so that, as a minute portion of tissue can determine the compounding of elements or principles, so does a larger amount of tissue or several tissues determine the arrangement of the parts of the tissue in a texture.

*Remark.*—Sinewy tissue, being a dense and passive mass when in the form of Ligament and tendon, and Blood-vessels not being permissible in those textures, is very slow in restoration when sprained or strained. Time, patience, and rest, are the chief curative means. The thousand nostrums and external applications used are useless, or worse than that. If the part is too warm, cool it, if too cool, warm it, and promote circulation by rubbing, if not painful. After a time, if the part is stiff, but not painful, induce action by forcibly moving it, and gradually increase the extent of motion. It must be expected that more time will be required to restore parts badly sprained than to unite broken bones.

## SECTION V.

### *Gristly or Cartilaginous Tissue.*

175.  **GRISTLE OR CARTILAGE IS** a somewhat firm, flexible, and elastic tissue, of which the frame of the ear and the tip of the nose are good examples.

176.  **CARTILAGINOUS TISSUE IS DIVIDED** into Cartilage and Fibro-Cartilage, or Gristle and Sinewy-Gristle.

177.  **GRISTLE PROPER IS COMPOSED** of a dense gristly substance called hyaline or chondrine or cartilagine,

which is very much like Gelatine, in which are imbedded a great number of cells containing a small amount of earthy matter, like that of the Bones.

FIG. 147.

Fig. 147 represents a portion of cartilage or gristle with nucleated cells imbedded in it. Some of them have two nuclei, and appear to be dividing to form new cells. The fibro-cartilage will be represented, and cartilage further illustrated, in connection with Bony Tissue. (See Fig. 41.)

178. FIBRO-CARTILAGE OR SINEWY-GRISTLE is composed of a varying proportion of fibres and gristle, all the way from a very few fibres up to nearly all fibres and a very few cells.

179. THE GRISTLY TISSUE IS passive in its offices, and does not require the activity-giving Elements of the second group; it is therefore composed, like the Sinewy Tissue, of inactive Elements, with a small portion of earthy matter combined.

*Remark.*—As Cartilage or Gristly tissue is liable to compression for a considerable length of time, it would not be proper to have Blood-tubes extend through it. Therefore its nourishment, like that of ligaments and tendons, must come into it from Blood-tubes near its surfaces. The passage of nutritious substances into and of waste substances from the inner portions of the cartilages must be slow, though not as slow as in the case of ligaments; and of course their restoration from injury will be correspondingly tedious.

### SECTION VI.

#### *Bony or Osseous Tissue.*

180. IF A BONE IS EXPOSED TO A CONSIDERABLE HEAT for some time, it becomes very brittle and crum-

Describe Fig. 147. 178. Describe —. 179. What said of —? How nourished? 180. What —?

bles easily, showing that the substance that gave tenacity has been burned out.

181. IF A BONE IS EXPOSED TO THE ACTION OF DILUTED NITRIC OR MURIATIC ACID for a time, an earthy substance is taken from it, and a flexible gristly substance of the form and size of the original bone is left.

FIG. 148.

Fig. 148 represents the animal or gristly portion of a bone tied in a knot.

182. THESE EXPERIMENTS SHOW that bones are constituted of two classes of substance, one called the earthy, and the other the animal part—it is about one third the whole.

183. THE ANIMAL PART IS ALSO CALLED gristle, but not properly, for it is found to be composed of gelatine, like white sinewy tissue; but it has not a fibrous arrangement. Another name is Osteine.

184. BONES THEN ARE COMPOSED of gelatine, with a large but variable proportion of earthy substance, phosphate of lime and magnesia, and carbonate of lime, intimately combined throughout, so that both the gelatine and the earths have the entire size and form of the bone.

185. BONE IS FORMED by having cartilage or fibrous membrane, first constituted in the position which the bone is to occupy; then the cartilaginous matter is exchanged for the gelatinous, or the fibrous arrangement of it is changed, and the earthy matter is deposited in increased

proportions from the first till the close of life, producing great brittleness of the bones in advanced years.

FIG. 149.

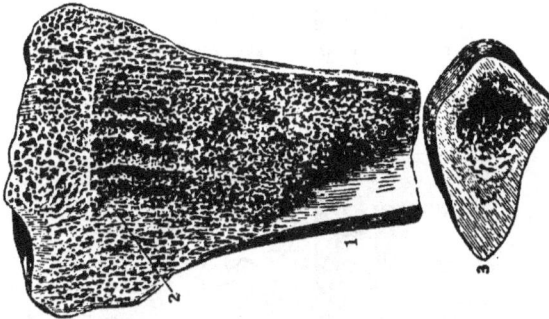

Fig. 149 represents, at 3, a small portion taken from the centre of a bone, having a thick, dense exterior, and a central canal; at 1 the dense part growing thinner, enclosing 2, a mass of curiously wrought bone, looking like very fine sponge, and called cancellated. The appearance can be seen in almost any bone.

FIG. 150.                FIG. 151.

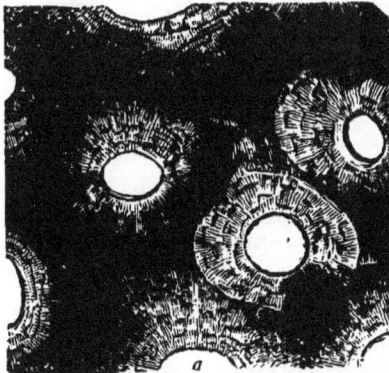

Fig. 150 represents a magnified portion of bone, and the arrangement of matter around the canals, $a, b$, forming hollow rods; $c$, tissue packed between the cylinders. The canals are channels for the location of blood-tubes, nerves, etc., for nourishing the bone.

Fig. 151, $a$, represents, in black, one canal, $H, c$, and very much enlarged lacunæ, arranged around it; $b$, an irregular cluster between cylinders. The branches from the lacunæ are seen to communicate in some cases. Thus nourishment has the opportunity of reaching every part of a bone.

Describe Fig. 149. Describe Fig. 150. Describe Fig. 151. When will bones most readily be broken?

186. BONY TISSUE IS ALSO CONSTRUCTED in very dense laminæ, as at the surface of the bones, or in a cancellated manner, as at the ends of all and in the centres of many, or with large central canals, as in the long bones—in all of which spaces marrow is deposited.

FIG. 152.                    FIG. 153.

 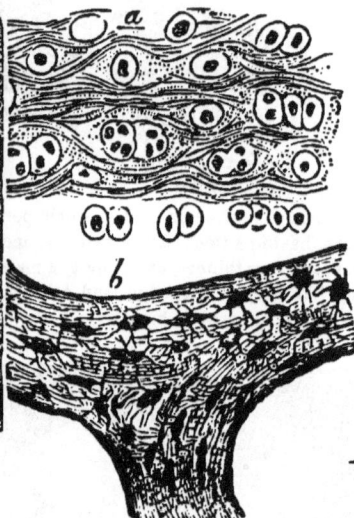

Fig. 152 represents a portion of bone beneath a portion of cartilage, attached and forming the surface of a joint. The cells in it are, near the bone, perpendicular to it, and near the surface parallel to it. The bone appears in laminæ, with numerous lacunæ and several Haversian canals, 7, in their midst.

Fig. 153 represents a portion of fibro-cartilage; a, more highly magnified, and bone, b, the configuration of the laminæ being indicated, and the lacunæ, in black, irregular spots, with their canaliculi radiating from them, being very distinctly shown. These are cavities.

187. THE BONY TISSUE is also deposited around minute canals, called Haversian, after their first describer, and around small cavities called lacunæ, that are exceedingly numerous; so that though Bony Tissue may seem to be very dense, it will be found to be excavated

186. How — ? Describe Fig. 152. Describe Fig. 153. 197. — how deposited? Is the bony tissue really dense?

in every direction, and made very light without materially diminishing its strength.

188.   ON THE OTHER HAND IT MAY BE SAID that Bony Tissue is constituted as a texture of hollow bony rods packed together in different directions, such as will give to each part of a Bone its greatest strength, which will be clearly shown in the description of the different Bones.

*Remark.*—Bony Tissue has a multitude of Blood-tubes in every part, and is, of all the passive tissues, most readily restored.

## SECTION VII.

### *Comparison of Tissues.*

189.   A COMPARISON OF ALL THE TISSUES SHOWS that there are three active and three passive.

190.   A COMPARISON OF THE ACTIVE TISSUES SHOWS that they are permanent varieties of the cellular, and are all albuminous, yet different from each other, and the substances within the cells are called *Nervine, Musculine, Celline.* The latter is of many varieties.

191.   A COMPARISON OF THE PASSIVE TISSUES SHOWS that they may be considered as varieties of the sinewy tissue with additions; that they are gelatinous, with or without the combination of earthy matters, and called *Osteine, Chondrine* or *Cartilagine,* and *Gelatine.*

192.   THE BODY THEREFORE IS COMPOSED of three *gelatinous* and three *albuminous* tissues; and the organic substances of which these tissues are composed, as they exist in the living body ready for action, have the six distinctive names that indicate clearly the correctness of the tissural classification.

|  |  |
|---|---|
| GELATINOUS . . . . . . | Gelatine.<br>Chondrine.<br>Osteine. |
| ALBUMINOUS . . . . . . | Celline.<br>Musculine.<br>Nervine. |

---

188. What —?  Do Blood-tubes exist here?  189. —shows what?  190. — shows what?  191. — shows what?  192. How —?  Table.

193.  ALL THE ACTIVE TISSUES ARE CONSTITUTED in part of phosphorus; none of the passive tissues are, except as a component of their earthy constituents.

194.  *Query.*—MAY NOT THE ACTIVE TISSUES, in performing their action, so far decompose that their phosphorus and a part of their nitrogen and sulphur will be no longer of value in the Body, while the balance of the nitrogen and sulphur, together with their oxygen, hydrogen, and carbon, may, after some modification, be used to constitute the gelatine of the passive tissues?

195.  *Remark.*—THE SUBSTANCES ELIMINATED BY THE KIDNEYS after unusual activity of the brain, would tend to sustain the suggestion just made.  At such a time the eliminations of the kidneys will abound in ammonia composed in part of nitrogen, and in phosphates.

196.  AN ENTIRE REVIEW OF THE NECESSITIES OF THE BODY, as exhibited by the constitution of the Tissues and their demands, shows that there is a requirement for three classes of substances in the Blood: 1st, those from which the Albuminoid or active tissues can be formed; 2d, those from which the Gelatinous or passive tissues can be formed; and 3d, those which are Calorific.

197.  OF THE THREE CLASSES OF SUBSTANCES, the plant furnishes only two, the 1st and 3d, and therefore the 2d must come from one or both of the other two.  But as gelatine contains more elements than exist in the calorific substances, it cannot be formed from them, and therefore must be directly or indirectly produced from the albuminoid class.

*Inf.*—This view suggests another argument in favor of the idea that the gelatinous tissues are constituted from the decomposing active tissues, with the addition of such chemical compounds as are necessary.

*Remark.*—When Bones are broken, or the other gelatinous tissues are to be repaired, it is not improbable that they may be constituted directly from the albuminous food, of which *much at such times is required, and should be eaten.*  Such tissues, at such times, cannot be rapidly repaired from a slop diet, and unless there is serious objection, a hearty diet and plenty of it should be allowed; soft-cooked eggs being one of the best articles, and meats being preferable to other diet.

*Remark.*—One of the most important lessons to be learned from a review of all the tissues is, that *Time* is one of the essential elements in their production, and it will be found that the more *Time* the plant has for perfecting its compounds the better, and that a perfect work cannot be done by the animal tissues without *Time* is allowed for their composition. The *Brain*, in particular, requires *time* for its perfection, and if it is made active too long, or with too brief intervals of repose, its cells do not become mature and perfect, and nervous excitement, irritability, and other disorders will be manifested or ensue. The constant activity of city life requires long periods of repose. Students should sleep much and well; and for recuperation of any part, repose, rest, and quiet, are necessary.

## SECTION VIII.

### *Compound Tissues.*

198. ADIPOSE TISSUE IS usually treated as if it were a distinct Tissue, but it is merely cells, the walls of which

FIG. 154.    FIG. 155.

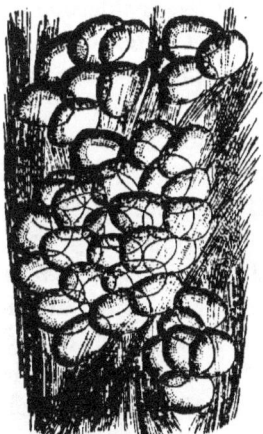 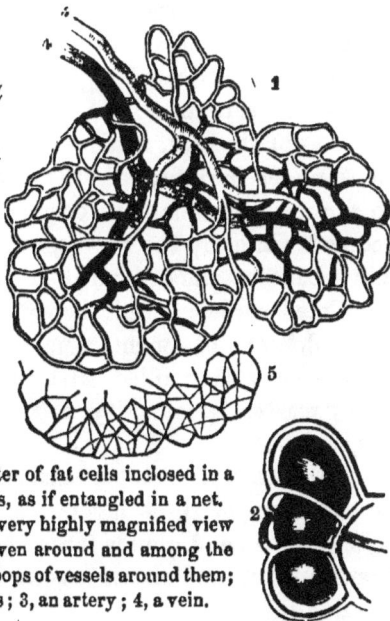

Fig. 154 represents a cluster of fat cells inclosed in a small number of sinewy fibres, as if entangled in a net.

Fig. 155 represents, at 1, a very highly magnified view of capillary blood-vessels woven around and among the fat cells; 2 is three cells with loops of vessels around them; 5, a compressed cluster of cells; 3, an artery; 4, a vein.

What said of time? Why does repair of brain require time? 198. What —? Describe Fig. 154. Describe Fig. 155.

are albuminous, filled with fat, and kept in their position by sinewy fibres. There may be one or many cells together, and they can increase or diminish as any other cells do.

*Remark.*—In milk, the fat or butter is contained in albuminoid cells, that must be broken by churning, in order that the butter may be gathered; and the more perfectly this is done, and the more completely the inclosing cell is removed from the butter, and is left pure, the better it will keep. If rancid, it can be greatly improved, if not made sweet, by thoroughly washing out the decomposing cells. This can be done best in sweet milk. (See Fig. 124.)

199. There are also THREE COMPOUND MEMBRANES of very considerable importance, to be described before proceeding to the consideration of organs; they are called the *skin*, and *mucous* and *serous membranes*, and correspond to the three general surfaces of the body; the external, the air and food, and the internal surfaces, furnished with the three kinds of cells, the epidermic or cuticular, the epithelial or mucous, and the tesselated or serous, and lubricated by the three surface-fluids, oil, mucus, and serum.

200. THE SKIN, as already stated, IS COMPOSED of a thick under part formed of sinewy fibres, woven so as to form meshes that are occupied by the Blood-tubes, red and white, by the nerves, and by perspiratory, hair, and sebaceous glands, and their outlets. The uneven or papillated surface of this layer is covered by the exquisitely delicate basement membrane from which spring up the cells that, filled with watery fluid and a little horny matter, become dry, flattened scales as they reach the surface.

201. THE MUCOUS MEMBRANE IS correspondingly CONSTITUTED of three layers, the sinewy having Blood-tubes, nerves, and glands in its meshes, a basement membrane upon it, and one or several layers of cells forming the free surface.

202. THE SEROUS MEMBRANE IS like the preceding, except that it is thinner, and has but a single layer of thin cells forming its surface, pouring out a watery fluid called serum. It is called by many different names in different places, though always one thing; for example, *Pleura*, at the surface of the Lungs, *Peritoneum*, in the Abdomen, *Dura-Mater*, when lining the Skull, &c.

203. THE LATTER TWO COMPOUND MEMBRANES enter into the composition of a large number of organs; and though they are compound parts, they are usually considered and spoken of as elements of organs. As they are common to many, it is proper that they should be briefly described under the head of General Physiology.

## SECTION IX.

### *Liquids.*

204. THE CHIEF PART OF ALL THE LIQUIDS IN THE BODY IS FORMED of Water, which, in fact, is the liquid of the Body, having more or less of various substances dissolved in it.

205. WATER SERVES FOUR PURPOSES IN THE BODY: 1st, as a vehicle it transports the substances dissolved in it, and also heat, from place to place; 2d, it moistens the soft solids of the Body, in fact, makes them soft and pliable—for when it is dried away they become hard; 3d, by evaporating from the Skin and Lungs it removes a corresponding amount of heat; 4th, when removed from the circulation by the Kidneys, the quantity of Blood is correspondingly diminished, passes round the quicker, produces more heat in a given time, circulates it faster, &c.

206. WATER IS merely a chemical compound; serves simply mechanical purposes in the Body, and is introduced and used by physical means alone.

---

202. What —? What its special names? 203. What said —? 204. Of what —? 205. How does —? 206. What —?

9

207. THE NECESSITY FOR WATER IN INCREASED QUANTITY IS SIGNIFIED by thirst, which very properly exists when water passes away by the action of the Skin to promote the loss of heat, but does not exist when Water is removed by the Kidneys to increase heat.

208. BLOOD IS a very complex fluid, composed largely of Water, in which, as a vehicle, digested food, oxygen from the air, and effete material from the tissues are dissolved.

209. BLOOD HAS THEREFORE a double character: it is composed of useful and useless material.

210. BLOOD OBTAINS FROM FOOD calorific material in the form of fat and sugar, and nutritious material in the form of liquid albuminose, with which is united the necessary chemical compounds to supply the passive tissues.

211. BLOOD OBTAINS FROM DECOMPOSING TISSUE a variety of compounds that are useless in the particular tissue from which they are thrown off, yet may in part be of use in some of lower grade, or may be burned, producing heat, or else must be eliminated.

212. THE BLOOD IS THUS THE COMMON SOURCE of materials of all the tissues and all the liquids, and is the common receptacle of materials discarded by them; yet none of the substances composing the tissues exist in the Blood in the same condition as in the tissues, nor do most of the fluids. Some of the fluids are merely drawn from the Blood.

213. IN ADDITION, the Blood is composed in part of an immense, almost incredible, number of Blood-cells. In health there are nearly three thousand in every drop. Less than seventeen hundred are never found, and sometimes twice that number: increase or diminution by any great number indicates disease.

---

207. How —? 208. What —? 209. What —? 210. What does —? 211. What does —? 212. Of what —? 213. What said —?

214. THE BLOOD-CELLS CONSTANTLY DISSOLVE, and new ones take their place in rapid succession. It is computed that not less than twenty millions of them die every moment and yield their contents to the Blood, and as many come into it in the same time.

215. THE BLOOD-CELLS ARE OF TWO KINDS, red and white, the latter about one tenth in number, and the source of the former by means of appropriate changes.

216. THE RED BLOOD-CELLS give color to the Blood, and by their change in form affect its shades.

217. THE FORM OF THE RED CELLS IS AFFECTED by the reception and discharge of carbonic acid, which is received in the different parts of the Body from the decomposing tissues, and darkens the color of the Blood; it is given off in the lungs, when the form of the cells again changes and their color brightens.

218. THE CELLS CONTAIN a semi-liquid substance of an albuminous character, not definitely understood, called Globuline, and a red coloring matter called Hæmatine, in which the iron of the Blood is found; and as it is found in combination with other coloring matters, and only with such in the Body, it is thought to be of prime consequence in that respect.

219. THE CELLS DOUBTLESS ARE floating plants that prepare the constituents of the Blood for their higher position in the tissues.

220. IF THE CELLS ARE STRAINED OUT OF THE BLOOD, the remaining portion, if allowed to stand, will separate into a fibrous mass called fibrin, or the clot, and a liquid portion called the serum of the Blood.

221. IF BLOOD BE DRAWN AND STIRRED WITH RODS, fibres of the same substance will cling to them.

222. IT IS A QUESTION how or where this fibrin of the Blood is formed and what end it serves. It has nearly the same constitution as albumen. Most argue that it is

---

214. How do —?  215. Of what —?  216. What the effect —?  217. How is —?
218. What do —?  219. What —?  220. What —?  221. What —?  222. What —?

albumen changing into tissue; but it must be allowed that its *rationale* is not understood.

223. THE SERUM OF THE BLOOD CONTAINS its albumen, calorific compounds, chemical compounds, and the waste material of the tissues.

224. PHYSICALLY, BLOOD MAY BE DIVIDED into cells, fibrin, and serum. (Fat may also be mentioned.)

225. LYMPH IS a watery fluid, slightly tinged, resembling the serum of the Blood in many respects, though not of course so complex. It is sometimes called the white Blood. It contains cells like the white ones of the red Blood, and of various sizes, and which, as they develop, become genuine red cells.

226. FLESH-JUICE IS a watery fluid like serum.

227. SERUM IS a fluid composed of water and a very small portion of albumen, to produce glairiness. The percentage of albumen varies in different cases, being largest in the joints.

228. MUCUS IS about one half water and the other half albumen, producing a viscid liquid of a very slimy and glairy character.

229. FAT IS, in man, properly speaking, an oil, as it is liquid at the temperatures of the Body. It forms parts of the Blood, and is deposited in cells in various parts of the Body. It is a component of the Brain, and is poured out on the surface of the Skin to protect it from harm.

230. FAT IS OBTAINED from the food directly, also from the starch eaten, and perhaps also from sugar. It abounds in meat, and exists also in many vegetables and in the seeds of grain, being especially abundant in corn.

231. *Remark.*—THE OTHER FLUIDS OF THE BODY ARE ADAPTED to special purposes, and can be better studied in connection with the particular organs to the use of which they are related. Several of the preceding also must again come under observation in describing the organs with which their varieties are associated.

---

## SECTION X.

### *Gases.*

232. THE BLOOD HAS A CAPACITY of taking into itself a certain amount of gas—of, so to speak, dissolving it. Now it does not drink all the air, but makes a choice, taking the oxygen and leaving the nitrogen. This it does in the lungs and in the stomach, when air is swallowed in the saliva, as it is especially while speaking.

233. THE OXYGEN OF THE BLOOD MAY BE OF USE in two ways: it may combine with some of the constituents of the Blood in its circuit and produce heat, or it may seize upon some of the decomposing elements in the tissues and assist in bringing them into the Blood. It is certainly of great use, and should be plentifully received: the capacity of the Blood to receive it should be constantly satisfied.

234. CARBONIC ACID IS a gas that is constantly forming in the different parts of the Body, both by the decomposing tissues and the calorific food. Any surplus quantity must be speedily removed, for it is very deadly, and at times accumulates very fast.

235. OTHER GASES ARE PRODUCED by disease, indigestion, etc., but their number is not definite nor uniform, nor is their quantity.

### *Practical Review.*

236. A REVIEW OF WHAT HAS BEEN SAID upon Tissues, Liquids, and Gases, and their constituents, will convince any one that they are topics of the utmost importance in reference to the Laws of Hygiene.

237. IN THE FIRST PLACE, IT HAS BEEN DEMONSTRATED that certain substances must be selected for food, since the Body is composed of only a few.

---

232. For what —? 233. How —? 234. What —? 235. How —? 236. The effect —? 237. What —?

238. In the second place, it has been shown that the active Tissues undergo changes by use, as do the liquids and gases, and therefore the food taken must correspond to the activity; and if the food cannot be had or cannot be digested or otherwise prepared properly, the activity of the tissues must be limited by the facts existing.

239. In the third place, it has been proved that food adapted to produce heat is different from that adapted to nourish the tissues, and therefore should be eaten abundantly when the Body is to be warmed, as in cold weather, and sparingly in warm weather.

*Remark.*—It must be evident, from what has been said, that keeping the Body properly clad and sheltered will prevent the escape of heat, and save the necessity for a corresponding amount of food. Proper clothing is therefore an economy. It will be equally advantageous to apply the same deductions to the care of animals. Horses blanketed, and cattle kept in apartments artificially warmed during cold weather, are kept at less cost, and do better. A good digestion of a plentiful supply of food is healthy; therefore, the body should not be kept so warm as not to need nor demand a good supply of food.

240. In the fourth place, it has also been made to appear rational that the effete substances of the tissues are in part burned or can be in the Body, so that activity of the tissues is directly or indirectly a source of heat.

241. In the fifth place, it has been argued that the Elements constituting food vary in proportions in the same kinds, which are therefore of very various qualities, and should be selected with care.

*Remark.*—It will not be unworthy of consideration, as the previous chapters have proved, that the food necessary for furnishing the Blood with its requirements may be obtained from different sources at very different rates of expense. For example: starch, sugar, and fat are calorific, and there cannot be any appreciable difference in the amount of heat that similar weights of each will produce. Starch is much the cheapest, as it is more readily produced; but it requires additional preparation in the body, and food must be used to supply the power required in the preparation, so that sugar will be nearly if not quite as cheap at double the price. Butter will produce no more heat than any fat. Eggs are among the very cheapest articles of nutrition. Fowls, being warmly clad, are easily supported, and a great deal of what they eat is costless in summer; hence their eggs are producible at less cost than any other kind of meat. Different kinds of breadstuffs are produced at varying costs, and are of different values. Beans and cabbages are, among these classes of food, probably the cheapest, and if properly prepared are exceedingly wholesome. The brown part (the middlings, not the bran) of wheat is really the best of it, and can be bought at half the price of flour. Note in particular that the BRAINS of animals being the most highly wrought of any meat-food, are the best for man, especially for students and for "nervous" people. Brains, also, if rightly cooked, are very delicious, and of course must be easily digested. Millions of dollars' worth of this best of food are yearly thrown away in this country. Verily, ignorance is costly.

242.  IN THE SIXTH PLACE, IT HAS BEEN CLEARLY EXHIBITED that plants compound our food with different degrees of perfection under different circumstances, and that similar different effects will be produced by animal action upon the components of tissues.

*Remark.*—THIS FACT SHOULD BE PARTICULARLY WEIGHED in selecting meats, for it is not to be supposed that good tissues can be composed in the human body from the meat of imperfect animals, as many are that are brought into the market (that cannot be considered cheap at any price, unless it is good.) The process of corning is particularly injurious to meats, since it takes out a large part of their phosphorus. It is a fit method of preserving only fats.

243.  IN THE SEVENTH PLACE, IT HAS BEEN MADE EVIDENT that drink should be taken to satisfy thirst, and that it should be water.

244.  IN THE EIGHTH PLACE, IT HAS BEEN FOUND NECESSARY to health to supply the Blood with an abundance of oxygen, only to be obtained from pure air; that this want is constant, by night as well as by day, for an accumulation of carbonic acid in the blood is a poison of the most deadly kind.

---

What said of eggs? Fowls? Other substances?   242. What —?   *Remark.*
When —?   243. What —?   244. What —?

# CHAPTER III.

SYSTEMATIC SYNTHESIS OF TISSUES INTO ORGANS.

*Introductory.*

245. TISSUES ARE CONSTITUTED by the process of compounding Elements and Proximate Principles.

246. ORGANS ARE CONSTRUCTED by weaving together Tissues.

247. TISSUES EXHIBIT characteristics entirely different from those that could be shown by their components separately, so that neither suggests the other, and there seems to be a new creation.

248. *Remark.*—IT IS VERY DIFFICULT FOR THE MIND TO PERCEIVE that the characteristics of a compound, as well as those of a mixture, are wholly owing to the properties of their Elements; and because neither by itself can exhibit any part of the characteristic of the compound, the MIND CANNOT EASILY APPRECIATE THE FACT, that to exhibit new characteristics in combination with other Elements is one of the properties of every Element. From habits of thought and expression the MIND IS APT TO JUDGE that where entirely new characteristics are observed, there must be some new substance, or some new creative power; and thus a practical idea of great value is not realized.

249. *Illus.*—THE PROPERTY OF EXHIBITING THE CHARACTERISTICS OF WATER belongs to Oxygen in combination with Hydrogen, and to them alone—to no other Elements either separately or conjointly; and the fact

that conjointly they are water is one of their characteristics, but not one that either can exhibit alone.

250. *Illus.*—THE SIX ELEMENTS OF THE FIRST TWO GROUPS CONJOINTLY ARE nervous tissue, if combined in proper proportions and under proper circumstances; and that such is their property should be their most distinguishing, as it is their most exalted, characteristic.

251. ORGANS EXHIBIT only those characteristics that can be discerned in the tissues of which they are constructed; so that if the tissues are given, it is easy to determine what an organ constructed from them can do; or, if what the organ can do is given, it is easy to say of what tissues it is constructed.

252. TISSUES ARE CONSTITUTED to possess useful properties.

253. ORGANS ARE CONSTRUCTED to make a practical use of the properties of Tissues.

254. TISSUES VARY in quality, according to the proportions of their constituents and the circumstances under which they were constituted.

255. THE PERFECTION WITH WHICH AN ORGAN WILL FULFIL ITS USES will depend upon the character of the Tissues of the Body, since organs can be no better nor worse than the Tissues of which they are constituted.

256. *Remark.*—ONE KIND OF TISSUE THROUGHOUT THE BODY may be good or bad, and the rest be the opposite; or a Tissue in one organ may be good or bad, and be the opposite in other organs, either for a brief time or during a person's life, owing to accidental circumstances in one case, and being inherent in the other.

257. AN ORGAN MAY ITSELF BE COMPOSED OF PERFECT TISSUES AND NOT RECEIVE PROPER INFLUENCES from other organs, and thus its perfect action will be interfered with. When this is the case, the organ is said to be functionally deranged; when its own tissues are imperfect, it is said to be organically deranged.

258. IN THE STUDY OF ORGANS THE PRIMARY OBJECT WILL BE to ascertain of what Tissues and how they are

250. What —? 251. What do —? 252. How —? 253. How —? 254. How do —? 255. What said of —? 256. What said —? 257. What of —? 258. What —?

9*

constructed; how they are brought into action, whether singly, successively, or combinedly; and the result of such action.

259. It WILL ALSO BE A SUBJECT OF INQUIRY whether an organ acts singly, successively, or combinedly, and what relations it has with other organs.

260. THE GENERAL APPEARANCE OF THE ORGANS must also be described, their color, size, form, surface, and position; which last point indicates the organs to be first described, since there is but one class of organs the positions of which can be independently described, and to which also the position of all the rest can be conveniently referred.

## SECTION I.

### *Skeleton.*

261. THE SKELETON IS REQUIRED, as a framework from head to foot, to be strong as a support and protection, yet light, that it may be easily carried, and constructed with many joints, that it may be flexible. Its surface must be sufficiently extended to allow all parts a proper attachment to it, yet not so great as to be unwieldy or unsightly.

262. ALL THE DESIRED RESULTS ARE MOST PERFECTLY ATTAINED by the appropriate use of four tissues: the Bony, to construct the hard part; the Gristly, to give elasticity and a perfect finish to the joints, and to form the entire frame of some parts; the Sinewy, to bind all parts snugly together, and to sheathe the bones and cartilages; and the Secretory, to secrete a fluid (Synovial) to prevent friction at the joints.

SKELETON IS CONSTRUCTED OF
$$\begin{cases} \text{Bones.} \\ \text{Cartilages.} \\ \text{Ligaments.} \\ \text{Synovial Membranes.} \end{cases}$$

---

259. What —? 260. What said —? Subject of Section IL 261. Why —? 262. How —? Table.

263. THE SKELETON IS DIVISIBLE on the middle line of the Body into two similar halves, or two wholes, as a person prefers to speak of it, most of the bones in each being complete and undoubtedly double, as the ribs, arm-bones, etc., and others being seen as double only when observed with a philosophical eye, as the vertebræ.

264. THE BONES ARE CLASSED as long, tabular or broad, irregular, and short; examples of which will be recognized by an observation of the illustrations in the plates.

265. THE BONES ARE CONSTRUCTED with very uneven surfaces, and with many projecting points, called *processes*, to allow surface for the attachment of tendons and ligaments, and also to give the muscles a leverage power.

266. THE HARDNESS OF THE BONES not only increases with the age of a person, but it differs in different bones at the same age, and indeed in different parts of the same bone.

267. THE GRISTLE OR CARTILAGE differs in its form, thickness, and elasticity, at different joints. In some parts of the body it gradually changes to bone, as in the case of the lower extremity of the breast-bone or sternum, and the cartilages of the ribs; while in other cases it never ossifies, as changing to bone is termed.

268. *Remark.*—It is easy to determine if a fowl is young, by observing if the point of the breast-bone yields.

269. THE GRISTLE THAT COVERS THE ENDS OF BONES IN FORMING JOINTS is thicker in the centre than at the sides of the ball, and thicker at the sides than at the centre of the socket.

270. THE FORMS OF THE JOINTS are various. In some cases, as in the skull, the bones are locked together immovably; in other cases, as those of the thigh and shoulder ball-and-socket joint, there is a very extensive

motion; at the elbow and knee are examples of hinge-joint; that of the wrist is a compound hinge, like the ankle, while in the lower arm and neck there are rotary joints. Other joints allow a still more limited motion, to which attention will be drawn.

271. THE LIGAMENTS that bind the bones together are of three kinds: the flat or strap, the round, and the capsular. The first is the most common. They are all arranged so as to permit desirable motion, but to check all other. (See Fig. 2, Pl. 17 and Pl. 18.)

272. *Remark.*—The action of muscles also tends to bind the organs together at the joints. The pressure of the air is of considerable effect, for if all the ligaments of a hip-joint should be cut, it would require a weight of forty pounds to draw the bone from its socket.

273. THE JOINTS ARE LINED WITH serous membrane, that continually secretes the glairy, anti-friction synovia or synovial fluid. In early life this membrane covers the entire surface of the cartilages of a joint; but in later years the cartilages themselves appear to touch, and the synovial membrane is found lining only the sides of the joint. Sometimes the membrane is laid in folds and fringes in the sides of the joint, as in the knee, so that it may form a larger quantity of fluid. (Fig. 6, Pl. 17.)

274. IN SOME OF THE JOINTS THERE IS an extra or friction cartilage, to promote freedom of motion in the joint. (See Fig. 162.)

FIG. 156.

Fig. 156 represents the Periosteum peeled up from a part of the bone upon and to which it grew.

275. THE BONES AND CARTILAGES ARE COVERED with a thin, dense, finely-textured membrane of sinewy tissue, called periosteum on the bones, and perichondrium

on the cartilages; on the head it is called pericranium. It follows the nutritious canals leading into the bone, and lines the central and medullary canal and the areolæ of the spongy portion, under the name of endosteum.

276. THE CENTRAL CAVITIES OF BONES, both their canals and the areolæ of the spongy portion, are, in health, filled with marrow, the central canal with a soft fat, and the spongy part with a reddish, fatty substance. Both are oil deposited in nucleated cells.

277. THE MARROW OF THE BONES SERVES to deaden the effect of jars, and prevent them from reaching the brain, and is a stock of calorific matter that can be taken up to produce heat when a supply cannot be had from without, or, from feebleness of the organs, cannot be prepared from food.

### Special Bones of the Body.

FIG. 157.

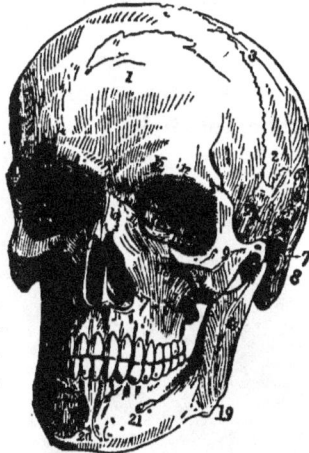

Fig. 157 represents a front and side view of a skull. The forms of the cranium and the facium differ very much in different cases. This is an advanced skull, since the suture or joint, like that at 3, that in early life is exhibited up the centre of the forehead, has been obliterated by the right and left FRONTAL BONES becoming consolidated. A similar suture is sometimes seen at maturity in the middle of the upper jaw. The lower jaws become consolidated at 20, very early. The skull is in early life literally divisible into two.

278. THE SKULL is composed of the CRANIUM, including all above and back of the eye-sockets, and the FACIUM, added to the lower part of the Cranium.

---

276. What said of —? 277. What does —? What is the effect of water in a tumbler on its vibrations? Describe Fig. 157. 278. How divide —?

279. THE CRANIUM, or brain-case, is arched in all directions in which it is exposed to danger, as that form is the best for resisting the effects of blows. Where it is covered by the facium it is flat and very irregular.

280. THE THICKNESS OF THE WALLS OF THE CRANIUM varies in different cases and in different parts of the same cranium, from the sixteenth of an inch up to an inch, averaging from an eighth to a half. Those parts are the thickest that are most likely to receive blows, being the back and lower parts.

FIG. 158.

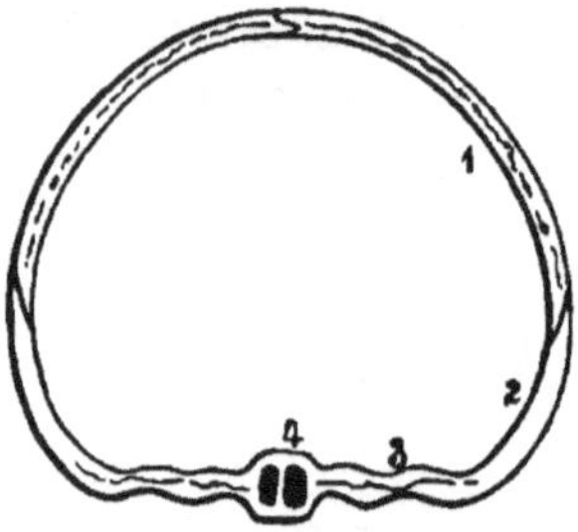


Fig. 158 represents a cross section of the Cranium, just in front of its middle. 1, the PARIETAL or side bone, jointed with its opposite, above, and below bevelled within the rising wing, 2, of the SPHENOID; 3, its level portion; 4, its cells. The proportionate thickness of the Cranium to its cavity is truthful; the setting of the arch above within the buttresses, 2, is also excellent. The central dark line in the walls above and below is the diplœ.

281. THE CRANIUM IS COMPOSED of three TABLES or plates, an outer or fibrous, an inner or vitreous, and a middle, the diplœ.

FIG. 159.

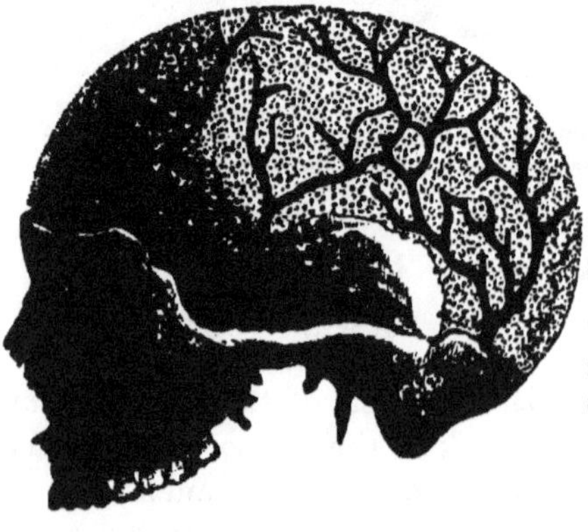

Fig. 159 represents the external table removed and the diplœ brought into view. It has the same spongiform structure as is found in most bones, with large canals for the location of tubes, as illustrated by the figure.

---

279. What form has —? 280. What said of —? Describe Fig. 158. 281. How —? Describe Fig. 159.

282. THE OUTER TABLE IS tough, and its edges very irregular, and intermatched or even dovetailed.

283. THE INNER TABLE is more brittle, and its edges abut evenly by a harmonious joint.

284. THE CONSTRUCTION OF THE CRANIUM IN THREE TABLES of such different characteristics prevents blows upon the head from jarring the brain, as the tables will not vibrate in harmony, and hence they deaden or scatter the effects.

285. THE CRANIUM IS COVERED with a thin, dense, sinewy membrane, called PERICRANIUM. It adheres closely to the bones, but outwardly its fibres become continuous with the areolar texture of the head.

286. THE CRANIUM IS LINED with a thicker, dense, sinewy membrane, called DURA-MATER. It adheres closely to the bones, but on its inner surface, toward the Brain, it is finished with a layer of basement membrane, covered with a layer of serous cells, that keep their surface constantly moistened with serum, and prevent any friction as the brain moves against them.

287. THUS THE CRANIUM IS SEEN TO BE a dome-shaped roof thrown over the Brain; and if the latter were removed the Cranium would remain an arched vault of wonderful beauty, and having a perfect finish, smooth and glairy, against which the delicate brain might rest in safety.

Fig. 160 represents a section of the head and neck upon the middle line. 2, 3, 4, are the skull and its three tables. The membrane covering the Cranium can be distinctly seen below 4, as it extends down joining the lining at the edge of the hole in the base of the skull, from which it extends up, lining the cranium, and down, lining the spinal canal of the spinal column or back bone, within which, 51, the spinal cord, cut off above, is seen, not quite filling the canal. On the opposite side of the canal its lining can again be traced and followed up within the Cranium till it reaches around to the commencing point.

Where the lining of the Cranium reaches its central line, near where the section is made, it leaves the cranium and extends down quite to the

FIG. 160.

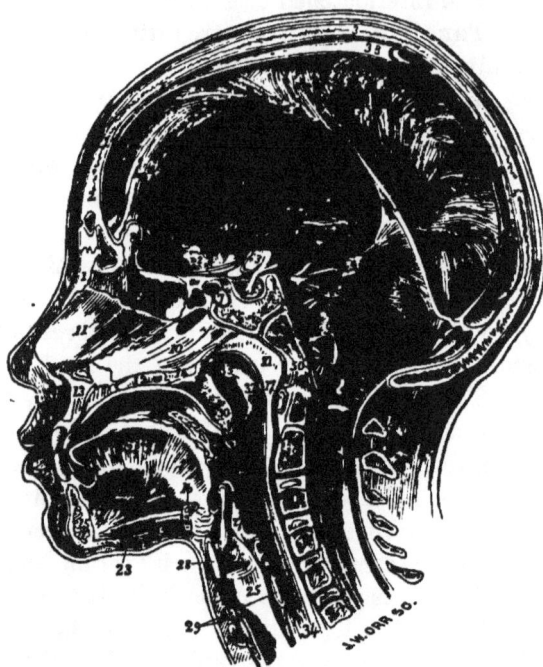

bottom of the front part, and two thirds down the back part, and down
the centre to the arched line shown, under which connections between the
right and left brains stretch across.

Thus the Cranium is divided into its right and left cavity. Into the
right one the eye looks and sees 39 and 35, on the inner surface of the right
wall of the Cranium. From the form of the partition, 36, it is called the
FALX or sickle. From its lower edge it extends back again up to the cranium, joining it near where it left it, but, spreading a little, it leaves a
triangular space, 38, that properly lined becomes a vein, called a sinus.

From a point against 4 the lining also extends forward, arching upward,
and forms a shelf called TENTORIUM, to the upper surface of which the
back part of the Falx is attached. The Tentorium supports the back part
of each large Brain, and covers and protects from pressure the small
Brains. The Falx supports the upper brain when the head rests upon
either side. The Cranium is thus divided into three cavities, neither completely closed, the two upper ones corresponding to and containing the
right and left cerebri, and the lower one the cerebellum.

---

Describe Fig. 160. What relation has the back of the nose to the spinal column? What the mouth?

288. The Cranium is usually said to be composed of eight bones, in which two bones only are considered as double, and the rest are looked upon as single. The parietal and temporal are usually so distinct at mature years that they could not be counted otherwise than double; but during growth, and sometimes in mature years, the suture up the middle of the frontal bone is distinct, and at an earlier period all the bones are distinctly double.

Fig. 161.

Fig. 161 represents a side view of the skull, with the bones slightly separated. 1, Frontal; 2, Parietal; 3, Occipital; 4, Temporal; one not numbered, in front of 4, Sphenoid; one faintly represented below 8, Ethmoid; 5, Nasal; 6, Molar; 7, Superior Maxillary; 8, Unguiform; 9, Inferior Maxillary. Between 3 and 4 is seen a small bone, several of which are apt to be found in the course of those sutures, or seam-joints. They are named Ossa-triquetra.

289. The best classification of the bones of the Cranium, being most philosophical, is that which makes twelve of them, six right and six left; for some of the sutures are obliterated early, others later, and

sometimes all are, leaving the cranium one bone. For description of each bone, see Plates and their descriptions.

290. THE CRANIUM HAS NO OPENINGS above; those it has are all below, and in a very small space, for the transmission of Nerves and Blood-tubes. (See Plate 11, Figs. 5, 6.)

291. THE SPHENOID, ETHMOID, AND (in males) FRONTAL BONES have cavities in them called cells (see Fig. 148, and Fig. 4, Pl. 11), the precise use of which is not known, except it may be to enlarge the bones without increasing their substance.

292. THE FACIAL BONES ARE very irregularly shaped, and attached to the lower front portion of the Cranium, forming various cavities for the lodgment of the organs of sense, and those required in mastication.

293. THE FACIAL BONES CAN BE BEST DESCRIBED in connection with the organs for which they are constructed, except as they are described in connection with the Plates.

294. IT IS TO BE OBSERVED IN PARTICULAR, by means of the preceding cut, that the lower portion of (7) the upper jaw-bone, that extends up by the side of the nostrils hangs down some distance below the bottom of the Cranium, and of course the cavity of the mouth is still lower.

295. Hence, if the skull is placed upon the spinal column, the eye could look through the cavity of the nose against the upper part of the column, or if the fingers should be passed up behind the facium and in front of the spinal column, and turned forward, they would enter the passages for the nose. The same IDEA IS ILLUSTRATED BY FIG. 160.

296. THE OBJECT OF TAKING PARTICULAR NOTICE OF THE NASAL BONES, and the manner in which they hang from the Cranium, is to understand the relation of the

passages through the nose and mouth to the upper part
of the throat or pharynx, that extends up in front of the
spinal column, and between it and the facium, to the
under surface of the cranium.

297. IT IS ALSO WORTHY OF NOTICE HERE that a part
(Petrous) of the TEMPORAL bone, in which the internal
ear is wrought, is connected by an arm with the facium
(Fig. 1, Pl. 9), and, properly speaking, is to be counted
as a part of it.

298. THE PETROUS IS FORMED distinct from the
temporal bone, and afterwards, in man, consolidates with
it, though it does not in all animals; and as it contains organs of sense, and is constructed with reference
to them, it is to be classed with the facium, constructed
for similar reasons.

299. THE EIGHT BONES OF THE EAR MAY BE DE-
SCRIBED in connection with that organ.

300. THE SKULL NEEDS BUT TWO MOVABLE JOINTS,
similar, since they unite the lower jaw to its sockets.

FIG. 162.

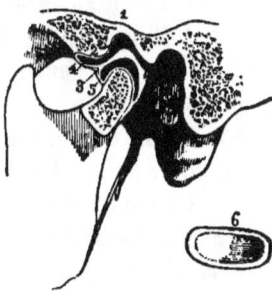

Fig. 162 represents the joint of the
lower jaw, that is very curious on account of the friction cartilage 6, introduced in the joint at 3, there being a
lubricating space above and below it.
This is required on account of the great
and frequent pressure produced here in
eating.

301. *Remark.*—SOMETIMES in gaping this bone slips forward, and
the jaw cannot be raised. Press the back part of the jaw down and
backward with sufficient force to carry it into place.

302. *Inf.*—Very little CARTILAGE OR gristle, very little SINEWY TIS-
SUE in the form of ligaments, IS REQUIRED BY the construction of THE
SKULL.

303. THE CRANIUM IS FORMED by the deposit of earthy matter in a thick membrane, that at the same time loses its fibrous character.

304. THE PROCESS OF CHANGING THE MEMBRANE into bone is not completed at birth; hence the soft spot, fontanelle, on the top of the head, that gradually closes by the continued process.

305. THE PROCESS OF DEPOSITING THE EARTHY MATTER commences at several points, called points of ossification, in each (the right and left) bone, and proceeds till the edges of the bone are reached; and in the outer table, points of an irregular shape mutually extend across the line and interlock, as represented by the sutures.

### Vertebral Column.

306. THE VERTEBRAL COLUMN IS a strong, elastic column of bones called vertebræ, and cartilages, from their position between the vertebræ called the intervertebral substances, the whole bound together in addition by numerous ligaments (see Fig. 6, Pl. 16).

307. THE TWO UPPER VERTEBRÆ (indeed, they can hardly be called vertebræ) are worthy of a distinct description.

308. THE UPPER VERTEBRA IS CALLED *Atlas* because, as the god sustained the world, so does this bone sustain the head upon its two shoulders, to which it is so jointed as to allow the nodding motion of the head.

309. THE SECOND VERTEBRA IS NAMED *dentatus*, from the tooth-like process or pivot that extends up from it through the central hole in the atlas above, a ligament passing across from side to side of the atlas behind the pivot, and thus separating it from the canal proper in which the spinal cord is located. A ligament also extends from the upper point of the pivot to the

---

303. How —? 304. What said of —? 305. What is —? 306. What is —? 307. What said of —? 308. What —? 309. What —?

skull, binding both atlas and dentatus to the skull, yet allowing sufficient motion to each (see Fig. 5, 4, Pl. 16).

310.   THE USE OF THE DENTATUS IS to allow a rotary motion to the head, as in negation.

311.   THE TWENTY-TWO LOWER VERTEBRÆ are very similar to each other in all general respects, increasing in size and thickness from top to bottom (see Pl. 12, 13). They are composed of body, arches, and seven processes.

312.   THE BODIES OF THE VERTEBRÆ are cylindrical blocks of bone, slightly flattened, or even concave, on the back, not very dense at the surface, and spongiform within.

313.   INTERVERTEBRAL SUBSTANCES situated between the vertebræ, as represented in the figures, are fibro-cartilages, having but few cells, the fibres woven very densely at and near the margin, but not as closely toward and in the centre, the minute meshes being filled with serous moisture and chondrine.

FIG. 163.

Fig. 163 represents a section of one and part of another vertebra with the intervening cartilage, of which 1 represents the fibres curving outward, 2, curving inward, and 3, the more yielding centre, containing cartilage cells.   It is, therefore, both ligament and cartilage.

314.   THE ELASTICITY OF THE INTERVERTEBRAL SUBSTANCES is their interesting and useful characteristic. They possess a much greater degree of firmness than India rubber, and considerable pressure is required to

make them yield or to put them on the stretch. They have a double elasticity, and equally tend to restore themselves whether compressed or stretched, to lift the Body up when bent forward, and to pull it up when bent back, or to do both when it is bent to either side. They, of course, grow tenaciously to the vertebræ above and below them.

315. THE BODIES OF THE VERTEBRÆ AND THEIR CARTILAGES·FORM a tapering column of alternate thick discs of resisting bone and thin discs of yielding elastic cartilage, that, viewed in front or behind, should appear straight, but to a side view should present the several curves Fig. 1, Pl. 12, which are owing to the varying thicknesses of the back and front part of the bodies and cartilages, and are necessary for giving proper form to the cavities of the Trunk, carrying the head without jar and flexing the column with ease.

316. THE ARCHES OF THE VERTEBRÆ are thin portions of bone extending from the sides of the back part of the bodies and arching around so as to enclose a space, hole, or foramen, about as large as a thumb, so that in the column there is a spinal canal back of the bodies; in this the spinal cord is located.

317. A NOTCH EXISTS in each side of the arch close by the body (see preceding Fig.). Through this notch the corresponding nerve extends.

318. THE ARTICULATING PROCESSES are small portions of bone extending up and down from the arches just back of the notches. The upper processes being adapted to the lower ones of the vertebra above, a joint is formed, that, in the preceding Fig., is represented as surrounded by a capsular ligament, *l*, binding them together, yet allowing motion. It is lined by synovial membrane, and a thin cartilage covers the surfaces of the processes. It is called the spinal pivot-joint.

FIG. 164.    FIG. 165.

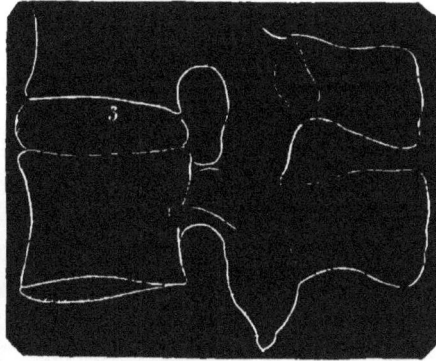

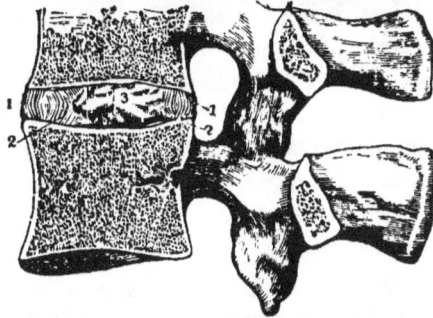

FIG. 166.

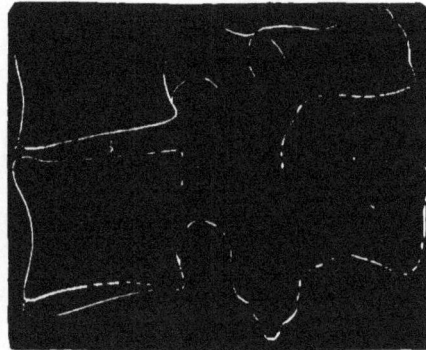

Fig. 164 represents the spinal column, curved in the region of the chest, as the result of pressure upon its cartilages. The line, *a*, shows the point to which the column would reach if its cartilages were natural. Fig. 165 represents, 3, the cartilage extended; Fig. 166, the same, 3 compressed; the spinous processes, *s*, being made to approach in the former case, and to separate in the latter.

319. THE PIVOT-JOINTS OF THE SPINAL COLUMN are the pivots upon which the vertebræ turn; so that when the column is bent forward, the whole thickness of the intervertebral substance must be compressed (see preceding Fig.), and when the column is bent backward, the whole substance is stretched, though in each case the front part is a little more affected than the back part.

320. Thus THE VERTEBRÆ OF THE SPINAL COLUMN are a series of levers with an elastic prop under their front end; and when all these props yield, the front ends of the levers approach and tend to form a circle; the length of the column, if measured through the pivot-joints, being the same in each position, but the apparent height of the person becoming so much less as the thickness of the props is lessened.

321. CONTINUED PRESSURE UPON THE INTERVERTE-BRAL SUBSTANCES tends to make them permanently thinner, and repose after action tends to restore them. Alternate pressure and repose promotes their highest degree of perfection.

322. *Illus.*—At night a person is not quite as tall as in the morning. By a long horseback-ride the height will be lessened for a time. A French physiologist says a son lost an inch during one night's dancing. A professor informs the author that, during a hunting excursion from 8 till 4, he lost three quarters of an inch, that was not regained till the second morning after. Usually one quarter to one half inch is lost during the day.

323. *Inf.*—THE ERECTNESS OF FORM MUST USUALLY BE DEPENDENT upon the condition of the intervertebral substances.

324. THE CAUSES OF CONTINUED DEFORMING PRES-SURE upon the intervertebral substances are, holding one position too long, and tight clothing. The latter is the chief cause.

325. *Inf.*—AN ERECT POSTURE too long continued is not as bad as any other position.

---

319. What are —? 320. What said of —as levers? 321. What said of —? 322. Illustrate. 323. On what —? 324. What are —? 325. What said —?

326. To prevent deforming pressure upon the intervertebral substances, let the clothing be free and easy; occasionally change the position of the body, and frequently take repose in a reclining posture, if the health is delicate.

327. *Remark.*—When a child, more particularly a youth, is growing rapidly, especially if of light complexion, the cartilages do not as rapidly become firm, nor ought they, for several reasons that will be given under the various heads to which they belong, to sustain the spinal column erect under those circumstances: they will readily yield, and the column appear curved forward about the shoulders. If it does not curve laterally no matter (it seldom will unless badly treated); as soon as the child is older and the cartilage hardened, erectness will be regained, if the clothing is not allowed to constrict nor any one position to fatigue him. He should also have much time for reclining repose. If, on the other hand, the clothing is tight, and apparatus worn or injunctions laid down to maintain one position, the cartilages will be sure to yield upon the sides, and unfortunate lateral curvatures will result, that will be with great difficulty, if at all, corrected. Nature, *Nature*, Nature, give NATURE play, and she will reward the deference with litheness of motion, graceful elasticity, and the thousand blessings she is in the habit of bestowing upon her followers.

*Inf.*—The seats upon which scholars sit should be easy, with backs sloping at a proper angle, while children should also have frequent recesses, or stand or move about, nor be required to maintain the same position for an hour at a time.

328. The Spinous processes are portions of bone extending from the back point of the arches, being short, inclining downward a little, in case of the upper seven (very small in case of the upper two), and allowing considerable backward motion, longer and inclining downward very much in case of the next twelve and allowing but slight backward motion, and in the lowest five extending straight back, short and stout, and allowing extended backward motion; therefore,

329. The Vertebræ are subdivided into *Cervical* (neck), *Dorsal* (back), that correspond to the ribs, and of course chest, and the *Lumbar* (loin).

826. How —? 827. What said —? What will ensure curvature? What said of seats? 828. What are —? 829. How —?

10

330. THE LATERAL OR TRANSVERSE PROCESSES are short portions of bone that extend from each side of the back part of the arches of the vertebræ, for the attachment of ligaments, tendons, and especially in the dorsal region for the attachment of the ribs. (Fig. 7, Pl. 16.)

331. THE SACRUM, so called because by some of the ancients considered sacred and offered in sacrifices, is a part of the spinal column in one sense, and in another is a part upon which the spinal column may be said to rest.

332. THE SACRUM IS COMPOSED of five vertebræ, consolidated into one triangular bone, wedged both downward and backward between the hip-bones. (See Fig. 2, Pl. 14.)

333. THE SURFACE OF THE SACRUM is very even, presenting a gentle curve on the inside, and very uneven and irregular on the backside, affording attachment to numerous ligaments and tendons.

334. THE CENTRE OF THE SACRUM IS OCCUPIED by the lower portion of the spinal cord, and has several pairs of holes in its front and back surfaces for the position of nerves.

335. THE SACRUM IS FASTENED TO AND SEPARATED FROM THE LOWEST LUMBAR VERTEBRÆ by a decidedly wedge-shaped intervertebral substance, which makes the twenty-third in the spinal column.

336. THE COXCYX IS in one sense the termination of the Sacrum, being composed of four small bones, sometimes ossified into one with the Sacrum, but usually having one or more limited joints. (See Fig. 1,2, Pl. 13.)

337. THE LIGAMENTS OF THE SPINAL COLUMN are few. There are twenty-four pairs of yellow, sinewy, and of course elastic ligaments extending between the upper and lower edges of the back part of the arches of the vertebræ. A broad ligament extends from the skull

down the front surface of the bodies of the vertebræ to the very extremity of the coxcyx (see Fig. 6, Pl. 16), and another is found on the back face of the bodies, that is, within the canal. Several others exist, of minor importance.

$$
\text{SPINAL COLUMN DIVISIONS} = \begin{cases} \text{Cervical} & = 7 \text{ Vertebræ} = \begin{cases} 2 \\ 5 \end{cases} \\ \text{Dorsal} & = 12 \quad " \\ \text{Lumbar} & = 5 \quad " \\ \text{Sacral} & = 5 \quad " \quad \text{consolidated.} \\ \text{Coxcygeal} & = 4 \quad " \quad \text{rudimentary.} \end{cases}
$$

$$
\text{SPINAL COLUMN composed of} \begin{cases} \text{Bones} & = \text{Vertebræ} & = \text{Bony } T. \\ \text{Cartilages} & = \begin{cases} \text{Intervertebral} \\ \text{Articular} \end{cases} = \text{Gristly } T. \\ \text{Ligaments} & = \begin{cases} \text{White} \\ \text{Yellow} \end{cases} = \text{Sinewy } T. \\ \text{Synovial Membrane} & = \begin{cases} \text{Secretory } T. \\ \text{Sinewy } T. \end{cases} \end{cases}
$$

338. Thus THE SPINAL COLUMN FINISHED stands before us one of the most admirable pieces of mechanism the mind can contemplate, exceedingly strong yet beautifully elastic by means of its curvatures and cushions, securely protecting, at the same time, in its flexible canal the vital spinal cord, and affording, though so neat and compact, ample surface for the attachment of all the ligaments and tendons that are adapted to produce or limit its motions. Resting upon it, the head is carried without a jar; and hanging upon it, all the delicate vital organs enclosed in the walls of the trunk are carried with equal safety.

339. THE RIBS, 12 in number (sometimes 10, sometimes 14), SERVE the purpose of levers in controlling motions of the spinal column, and as a framework to enclose and protect the organs of the Chest.

340. THE RIBS ARE CONSTRUCTED with very irregular forms, and singularly adapted by their peculiarities to be strong, while at the same time they are very light.

341. THE RIBS ARE ATTACHED by a head to a socket, formed in the back part of the sides of the bodies of each

Describe Fig. 5, Pl. 16. Write table of Div. Write 2d table. 338. What said —? 339. What do —? 340. How —? 341. How —?

of two adjoining vertebræ, and in the substance between them.

342. From their union with the vertebræ THE RIBS CURVE back and down, forming a joint with the transverse process of the lower vertebræ (see Figs. 7 and 8, Pl. 16), after which they begin to make a circuit, sweeping down and around, as shown by Pls. 12 and 13, forming the frame of the chest, and standing out like levers for muscles to seize upon and control the movements of the spinal column and all its dependent parts.

343. THE LENGTH OF THE RIBS increases from the first to the eighth, and from it to the last they diminish.

344. A CURIOUS GROOVE is found just within the lower edge of the ribs, within which is lodged the costal artery, that is thus admirably protected from injury, and also kept warm.

345. THE FRONT ENDS OF THE RIBS are continuous with their cartilages, which increase in length from the first to the seventh, and then diminish.

346. THE INCLINATION OF THE RIB (costal) CARTILAGES is down in case of the first, horizontal in case of the second, and upward in case of the rest.

347. THE JOINTS OF THE COSTAL CARTILAGES are, in case of the first pair, immovable, they growing directly to the breast-bone (Sternum); the next six pairs have movable joints, and are bound by ligaments to the sternum (see Fig. 2, Pl. 18); the next three pairs of cartilages are pointed and joined by continuation with those above; while the lower two are but tips to their ribs, that are hence called free or floating ribs, since only one end is fastened.

348. THE STERNUM or breast-bone is in early life composed of two bones and a cartilaginous point. The former are soon ossified into one, and the point usually follows in the same course by the prime of life.

---

342. How do —? 343. What said —? 344. Where —? 345. How —? 346. What said —? 347. What said —? 348. What said —?

349. THE STERNUM INCLINES outward at its lower extremity when natural.

350. A GENERAL VIEW OF THE FRAME OF THE CHEST SHOWS (see Fig. 4, Pl. 13), that its back part is composed of the dorsal portion of the Spinal Column, and a portion of the ribs that curve backward before they extend along the sides of the chest, which they alone form, in front it being formed of the ribs, their cartilages, and the sternum.

351. THE CHEST IN FORM IS a double cone, the largest circumference being about the middle of it. It is flattened in front and also behind; is shortest in front and longest on the sides.

352. THE CHEST IS PARTIALLY DIVIDED into right and left by the body of the spinal column, the front surface of which is situated nearly one half the distance from the surface of the backward curving ribs and the sternum.

353. THE FRAMEWORK OF THE CHEST IS OPEN above, where the opening is surrounded by the vertebræ, ribs, and sternum, and on the sides between and below the ribs, where the opening is almost the full size of the Chest. It has a singular outline, and is worthy of so much study as shall make it familiar.

354. THE CONSTRUCTION OF THE FRAME OF THE CHEST, and the applicability of it, is most ingenious and admirable. Its strength, from the arched form of the ribs and the condensed air when the windpipe is closed, is exceedingly great. The violence it can receive without injury is astonishing.

355. *Illus.*—A loaded wagon passed over the chest of a man without fracturing a rib.

356. Being built upon the upper part of the spinal column, THE CHEST IS PERFECTLY ADAPTED to be the support of the upper extremities, setting them out from the

---

349. How does —? 350. What does —? 351. What —? 352. How —? 353. How —? 354. What said of —? 355. Illustrate. 356. To what —?

spinal column, and adding to their own extent of mobility.

357. But THE MOST WONDERFUL THING in the chest framework is its adaptation to the peculiar motions demanded of it; for to the perfection of structure for movement in the spinal column it adds its own by means of the form of its ribs, their joints and cartilages.

358. THE CAPACITY OF THE CHEST IS INCREASED OR DIMINISHED by simply raising or depressing the ribs, and in either case the elasticity of their cartilages will at once restore them.

359. THE INTERVERTEBRAL SUBSTANCES can also be made to lend aid in increasing and diminishing the capacity of the chest, for when they are put upon the stretch by raising the spinal column, or compressed by curving it, the ribs have so much the greater extent of motion. .

360. THE EXQUISITE ADAPTATION OF THE FRAMEWORK OF THE CHEST is worthy the contemplation of every studious mind, for it is truly wonderful that so much flexibility of motion should be combined with such stability; that such complexity of results should be gained by such simplicity of construction.

361. THE CARTILAGES OF THE CHEST, like those of other parts, CAN ONLY BE KEPT IN GOOD CONDITION by alternate activity and repose; and continued pressure or disuse will despoil them of all their beautiful properties, uses, and results.

362. TIGHT CLOTHING impedes the action of some, and subjects others to constant pressure, the result of which is diminished capacity for breathing, with all its dire results to the complexion, the expression, the activity of mind, etc., and incapacity for graceful movements, that are always dependent as a *sine qua non* upon the elasticity of cartilage.

357. What in the chest —? 358. How —? 359. What — aid from? 360. What said —? 361. How —? 362. What said — ?

FIG. 167.  FIG. 168.

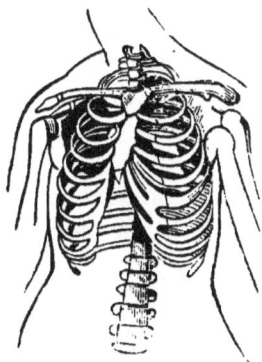
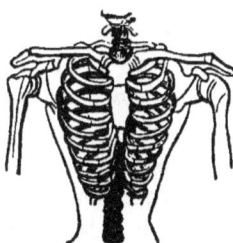

363. THE FIGURES REPRESENT a good and a constricted Chest. With the former, health and beauty, grace of movement, and a lively expression, are compatible, but are not possible in connection with the latter.

364. THE WORST MISTAKE THAT CAN BE MADE *is to try to improve personal appearance by constricting the size or action of the chest.* TIGHT CLOTHING IS A BANE TO BEAUTY.

365. *Remark.*—It does not by any means follow that a natural development of the chest will be attended with every desirable excellence. It IS NOT DENIED that some beauties may exist in connection with a compressed chest, nor that those may be found who with chest compressed are more attractive than some whose chests are not. But what IS ASSERTED, IS, that those who are at all attractive with the chest compressed will be infinitely more so with the chest properly expanded; that the true way to improve personal appearance is to develop as far as possible the elasticity of the cartilages of the chest and its mobility. It is asserted and CANNOT, either theoretically or practically, BE CONTROVERTED, that constricting the chest will deform; while the form and movements of the chest can be improved by easy clothing and appropriate exercise.

366. IT IS THEREFORE NOT ONLY SELF-TORTURE, SELF-MURDER, AND A SIN, TO CONSTRICT THE CHEST, BUT IT IS A MOST UNFORTUNATE SACRIFICE OF PERSONAL APPEARANCE.

---

363. What do Figs. 167 and 168 —? 364. What is —? 365. What —? What—? What —? 366. To constrict chest is what?

## Lower Extremities.

367. THE HIP OR INNOMINATE (no name) BONES
are two, of very irregular shape, and consolidated from
three, the *Ilium*, *Ischium*, and *Pubis*, which, together
with the Sacrum, form a bony ring or bowl without a
bottom, called the *pelvis*. (See Fig. 1, Pl. 14.)

368. THE HIP-BONES ARE STRONGLY BOUND by ligaments to each side of the Sacrum, and curving round
are united in front by an immovable joint on the central
line of the body, called the pubic symphysis.

369. THE INNER SURFACE OF THE HIP-BONES is
smooth though irregular, and adapted to support the
organs found within them; the external surface is very
uneven and irregular, adapted to give attachment to
numerous muscles.

FIG. 169.

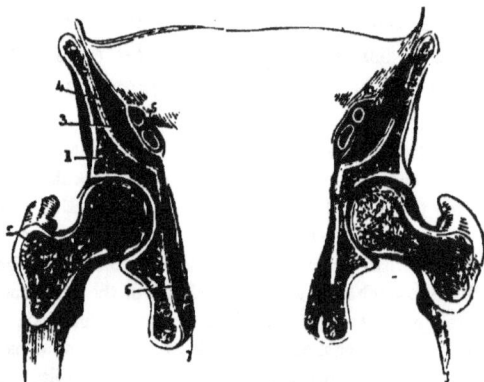

Fig. 169 represents a section of the pelvis (through its sockets) and the
heads of the thigh-bones, showing the outline of the form of the outer and
inner surfaces at those points, and 1, 2, the spongiform structure of the internal part of the hip and thigh bones.  5, Femoral artery and vein.

370. IN THE OUTER SURFACE OF THE HIP-BONES
THERE IS EXCAVATED a deep socket, the deepest in the
body, into which the thigh-bone is fastened by a central

round ligament and a surrounding capsular ligament. (See Fig. 2, Pl. 17.)

371. THE THIGH-BONE (Femur) IS CONSTRUCTED of a hemispherical head, an adjoining neck, and a long shaft that has two eminences above, called the greater and smaller trochanters, and is furnished at the knee with two enlargements, called internal and external condyles. The head and condyles are covered with cartilages. (See Figs. 3, 4, Pl. 14.)

372. THE FEMUR IS CONNECTED by a hinge-joint to the shin-bone (Tibia), enlarged very much at its upper extremity that it may assist in making a strong joint.

373. THE BONES AT THE KNEE ARE BOUND TOGETHER in the strongest manner by several ligaments; so that, though the knee is one of the most exposed parts of the body, its bones are seldom dislocated. (See Figs. 3, 4, 5, 6, Pl. 17.)

374. THE KNEE IS SUBJECT to so much forcible action, that it is supplied with a friction cartilage and an extra quantity of membrane to secrete synovial fluid. (See Fig. 6, Pl. 17.)

375. THE KNEE-PAN (*Patella, Rotulla*), is not any part of the knee-joint, but is connected with a tendon of a muscle for the purpose of allowing advantageous action.

376. BY THE SIDE OF THE TIBIA, in the lower leg, is another slender bone (*Fibula*), that is united to the tibia just below the knee, sometimes by a movable and sometimes by an immovable joint.

377. THE TIBIA AND FIBULA ARE BOUND TOGETHER throughout their extent by a broad ligament, that also presents its surfaces for the attachment of muscles.

378. THE ANKLE-JOINT IS CHIEFLY CONSTRUCTED of the tibia and ankle-bone of the foot (*Astragalus*), though the fibula is also essential, and forms the outer ankle, confining the ankle-bone in its place. (See Fig. 8, Pl. 17).

---

371. How —? 372. How —? 373. How —? 374. To what —? 375. What said of —? 376. What —? 377. How —? 378. How —?

Fig. 170.

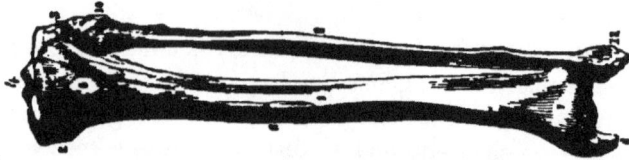

Fig. 170 represents bones of lower leg. 1, Tibia; 4, surface, joined to thigh-bone; 2, point of inner ankle (internal maleolus); 9, Fibula; 10, end attached to tibia; 11, (external maleolar process) outer ankle.

379. THE ANKLE (*tarsus*) IS CONSTRUCTED of seven bones, of which only one enters into the ankle-joint. The ankle-bone (*Astragalus*) joins and rests upon the heel-bone (*Calcaneum*) behind, and the boat-shaped (*Scaphoid*) bone in front, which is jointed in front to the three wedge-shaped (*cuneiform*) bones, the outer one of which is jointed to the cube-shaped (*cuboid*) bone that reaches back to the calcaneum.

Fig. 171.

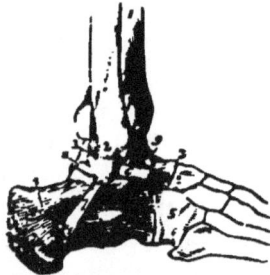

Fig. 171 represents, 1, lower end of tibia; 2, fibulæ; 3, 3, astragalus; 4, calcaneum; 5, cuboid. In front of 3, part of scaphoid; in front of it one, and part of another cuneiform. The metatarsus is seen extending from the cuboid and cuneiform; 6, 7, 8, 9, ligaments.

380. THE TARSAL BONES are half the length of the foot, and, near their middle, jointed with the leg. They are mostly composed of spongiform bone, with a thin, dense layer, having many perforations for vessels. (See Figs. 1, 2, 3, Pl. 15.)

381. THE TARSAL BONES ARE INTERLOCKED with the five bones (*metatarsal*) forming the frame of the foot, the front ends of which descend to the ground, and join

with the bones (*phalanges*) of the toe, three in number, except in the great toe, which has but two, and sometimes two in the little toe, on account of two being ossified together.

382. THE TARSUS AND METATARSUS FORM several arches, all of which are elastic on account of the cartilages with which the numerous joints are furnished.

383. Thus, THE WEIGHT OF THE BODY IS SUSTAINED in the most perfect manner, if only the perfections of the foot are developed and allowed to exhibit themselves.

384. THE PREVAILING IGNORANCE OF THE HUMAN FRAME CAUSES the sad blunder to be often committed, of cramping the foot in tight coverings, and upon high heels, which can never improve personal appearance, but must detract from it.

*Remark.*—The young lady or gentleman of low stature, who thinks to appear taller by wearing high heels, should also remember that they will thus appear more awkwardly, which will detract from, more than increased height will add to, their personal advantages.

### *Upper Extremities.*

385. THE UPPER EXTREMITIES ARE CONNECTED to the trunk by means of the collar-bones (*clavicles*).

386. THE CLAVICLES ARE short bones, one extremity of which is movably jointed and bound to the upper end of the sternum (see Fig. 2, Pl. 18), and the outer end of which is connected with the shoulder-blades (*scapulæ*), just above the shoulder-joint.

387. THE SCAPULA IS a very thin bone, often very delicately constructed, having a ridge rising from its back part and extending across it, and rising over the joint to meet the clavicle; thus forming the tip of the shoulder (*acromion process*).

388. THE FORM OF THE SCAPULA is triangular, slightly convex on its outer surface and slightly concave

382. What do —? 383. How —? 384. What does —? Effect of heels? 385. How —? 386. What —? 387. What —? 388. What —?

on its under surface, to adapt it to the form of the ribs in its natural position.

389.  THE SCAPULA IS JOINTED to the chest only as described, and merely rests on the back of its frame, some muscles intervening and being attached to it.

390.  THE INTENT OF THE CONSTRUCTION OF THE SCAPULA was, to provide a sufficient surface for the attachment of muscles adapted to move the shoulder, which is, so to speak, pivoted on the clavicle, and can be moved in a limited portion of the circumference of the sphere of which the clavicle is the radius.

391.  THE MOVEMENT OR THE POSITION OF THE SHOULDER-BLADE does not affect the size or form of the chest.

392.  *Remark.*—When the shoulders are drawn back, more of the chest is observed in front of the arms than when the shoulders are forward, hence drawing them back has been thought to enlarge the chest; but in fact the chest was so much the smaller behind the arms.

393.  CONSTRICTING THE SHOULDERS in any position will only interfere with free movement of the chest, and instead of enlarging it will only diminish the capacity of that part of the body.

394.  THE SCAPULA IS JOINTED to the upper arm-bone (Humerus) by a very shallow socket, to which the hemispherical head of the humerus is adapted, so that the arm can have a very extended motion; and when this is combined with the movements of the Scapula, and these multiplied by the motions of the chest and spinal column, the hand has an extent of motion adapted to any necessary purpose.

395.  THE HUMERUS IS WROUGHT at its lower extremity into two condyles, the internal and external, by which it is beautifully jointed to one of the bones (Ulna) of the lower arm.

396.  THE ULNA is the name of the internal bone of

the lower arm; it is furnished with a hook-like process that fits around the lower end of the humerus, and, when the bone is straightened, matches into an excavation in the back part of the humerus. It is also furnished with a process on the lower side of the joint that matches an excavation on the front of the humerus; thus, the motions of the ulna can be backward and downward nearly, but not quite, to a straight line with the humerus, and forward and upward nearly parallel to, but always at an angle with, it.

397. THE RADIUS IS THE NAME of another bone in the lower arm, situated by the side of the Ulna, and just resting against the lower end of the humerus, but forming no part of the elbow-joint; a ligament confines it to its place, but allows it to revolve partly.

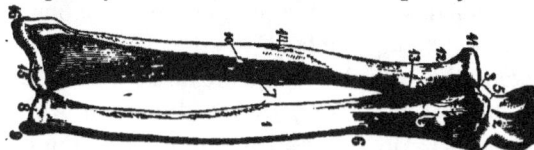

FIG. 173.

FIG. 172.

Fig. 172, ulna and radius: 4, point of elbow (olecranon); 5, process in front of joint (coronoid); 2, cavity fitting around lower end of humerus; 3, joint of 11 with ulna; 15, surface of *R* that turns over on to 8 of the *U*; 13, attachment of biceps.

Fig. 173, upper end of ulna: 1, olecranon; 2, cavity; 3, coronoid; 4, ligament binding radius in 5, the cavity in which 11 of 172 turns.

398. THE ULNA AND RADIUS are connected throughout their length by a ligament, the surfaces of which afford much space for the attachment of muscles.

399. At their lower extremities THE RADIUS IS ATTACHED TO THE ULNA in such a manner that it can turn over on to it, much as the lid of a book turns over.

400. This ROTARY JOINT OF THE LOWER ARM is the most ingenious one in the Body, and the most useful at

the same time; it is the simplest arrangement imaginable, yet how complete!

401.  THE RADIUS IS JOINTED to the wrist (*carpus*),
'that does not touch the ulna, so that when the radius
turns, the hand is carried with it.  Thus is obtained an exceedingly useful motion, as in turning a key, gimlet, etc.

402.  THE HAND IS JOINTED to the radius by a partial
ball and socket-joint, or compound hinge-joint, and thus
can be bent up and down and from side to side.

403.  THE CARPUS IS CONSTRUCTED of eight bones, in
two rows of four each.  (See the appropriate Figs. and
descriptions.)

404.  THE CARPAL BONES ARE CONSTRUCTED spongiform within and a thin dense layer at their surface.  Their
joints are supplied with cartilages, and they are bound
strongly together by ligaments.

405.  From the carpal the METACARPAL BONES extend to the bones (*phalanges*) of the fingers, of which
there are two in the thumb and three in each of the fingers, jointed by ball and socket and hinge joints, furnished with cartilages, synovial membranes, and bound
together with ligaments, limiting the bones to their appropriate motions.

406.  THE ENTIRE SKELETON, with the exception of
the U-shaped bone (*hyoid*), has been passed under notice; that bone can be better described in connection
with the organs of voice and the framework of the larynx, which will best follow the organs of respiration.

407.  THE FRAMEWORK OF SOME PARTS, the ear, tip
of the nose, the larynx and trachea, are wholly composed
of cartilage that never becomes ossified.

*Remark.*—This seems to be remarkably fortunate, for men never live
to be so old as not to be fond of thrusting their noses into other people's
affairs, and these organs would be very often fractured if they ever lost
their elasticity.

## General View of the Skeleton.

408. ALL PARTS OF THE SKELETON are connected with the commercial capital by the Blood-tubes, and with the political capital by the nerves.

409. ARTERIES AND VEINS extend into and out of all the bones by means of the numerous nutritious holes or foramina, that will be found in the surface of any bone, communicating with canals leading to capillaries in every part of it; they have the largest meshes of any in the body.

410. BY MEANS OF THE BLOOD-TUBES nutritious substances are poured through every part of the bones, and all substances that have become useless are by the same means brought away.

411. From infancy to maturity THE SKELETON ENLARGES by the gradual removal of all parts, particle by particle, and the replacing of these on an increased scale over and over again. It is impossible to conjecture how many different skeletons a person has between birth and maturity. The teeth are changed but once.

412. DIFFERENT PARTS CHANGE OR ENLARGE with varying degrees of rapidity at different periods of growth, the appropriate proportions being most wonderfully preserved, yet differing at different periods.

413. CHANGES TAKE PLACE IN THE BONES AFTER MATURITY. They become more brittle from the predominance of earthy matter, and more spongiform or cancellated, the spaces being filled with marrow.

414. IT IS NECESSARY TO HAVE THE EARTHY MATTER INCREASED to give the necessary resistance when the excavations are made, which are needed to contain marrow, useful in preventing jars from reaching the Brain, and especially in supplying a store of heat-producing material, when in age the organs cannot at times prepare it from the food directly.

---

408. What said —? 409. What said —? 410. What effected —? 411. How does —? 412. How do —? 413. What —? 414. Why —?

415. WHEN THE BRAIN DIMINISHES IN ADVANCED YEARS, the size of the entire skull correspondingly diminishes.

416. WHEN THE BONES ARE BROKEN, gristly tissue is laid down in the form of a plug within the fractured bone, and in the form of a ferule or ring (callous) on the outside of it, holding the broken ends in place, after which bony tissue is deposited between the broken ends and united with them. When this process is complete the cartilaginous plug and callous are removed so completely that at times no trace of the fracture is left.

417. THE REPAIRING OF BONE IS a long process, requiring several changes in the repaired parts, and quite an amount of nutritious material of the ossific sort. It should be attended by a hearty appetite. Fortunately, at such times, other parts may be at rest without injury to themselves or causing irregularity in the bowels, kidneys, or other organs; which is a wonderful provision, as if it was anticipated that accidents must happen.

418. THE FACTS MENTIONED IN REGARD TO CHANGES IN THE BONES SHOW that there is a unitary power superintending the action that takes place in them, increasing action and diminishing it in the various parts, according to the necessities of the part and the welfare of other parts.

419. THE NECESSITY FOR A UNITARY SUPERINTENDING POWER IS SUPPLIED by means of the nerves that extend from every part of the bones to the nervous centres.

420. THE OSSEOUS NERVES ARE DOUBTLESS of two kinds, Sympathetic and Sensatory, though only the latter can be demonstrated.

421. THE NERVES EXTENDING FROM THE BONES PRODUCE sensation only when the bones are diseased, or after they have been broken a little while.

---

415. What effect —? 416. What occurs —? 417. What said of —? 418. What do —? 419. How is —? 420. What —? 421. What do —?

422. *Illus.*—If a bone is cut across when sound, its nerves will not cause pain. Indeed, two persons stated to the author that a slightly pleasant tickling sensation was caused by the saw passing through the bone.

423. *Remark.*—A twinge of excruciating pain will be caused by cutting across a nerve of touch, if one happens, as is sometimes the case, to extend through the bone in the path of a section.

424. *Inf.*—The sooner a bone is set after it is broken, the less painful the operation.

425. THE VERY SENSITIVE CONDITION OF THE NERVES after a fracture is a wise means of compelling a perfect quiet of the parts, such as no splint can produce; for when, as sometimes happens, the nerves do not become painful, the parts are not kept sufficiently quiet to have them unite, and it is necessary to excite pain in the nerves.

426. BLOOD-TUBES DO NOT EXIST IN THE CARTILAGES OR LIGAMENTS of the skeleton, for it would not comport with the offices of those parts to have compressible Blood-tubes in them; but they exist all around those parts, which are slowly nourished from them, and tediously recover from injuries.

427. THE NERVES OF THE CARTILAGES AND LIGAMENTS, like those in tendons, are not sensitive while uninjured, nor if pricked, cut, or even burned, but if twisted ever so little are very painful.

428. *Illus.*—IF A TENDON ON THE BACK OF THE HAND, sometimes exposed by a cut, is touched with a red-hot wire, pain will not be felt, nor if it is pinched; but if the tendon is twisted in the slightest degree, pain is instantly felt.

429. *Illus.*—IF A PIN IS THRUST THROUGH THE SKIN JUST ABOVE THE KNEE, and turned downward, it may be buried in the tendon without causing pain. Boys often try this.

430. *Remark.*—There IS A CURIOUS FACT about the nerves of the hip-joint. When it is diseased, the pain caused throughout its nerves seems to be in the knee, and of that part the affected child will complain. This disease is very insidious and serious, but if taken in season

---

422. Is bone painful? 424. Inf. 425. What said of —? 426. Why —? 427. What said of —? 428. What said of —? 429. What —? 430. What —?

may usually be checked. If therefore a child complains of pain about the knee, and neither swelling nor soreness is manifested there upon examination, disease of the hip may be suspected, and skillful advice taken. A long period of repose must be allowed to the part, and particular attention paid to general health.

431. *Remark.*—FELON IS a disease of the membrane covering the bone, that becomes very painful because its dense fibres compress the blood-vessels and nerves there situated. To relieve the pain of felon, therefore, it is necessary to cut these fibres, which will be accomplished naturally by ulceration, but can better be done artificially by the surgeon's lancets. As soon as a felon is recognized, the part should be cut through to the bone.

432. *Inf.*—IT WOULD BE CORRECTLY INFERRED that *proper exercise* will circulate Blood through and around the skeleton, so as to increase its size and strength, and promote its growth, while *too much labor or exercise* will prevent its development, since the materials to form it will not be furnished, and the power needed to use them is exhausted by the exercise.

433. *Inf.*—Since WHEN THE BONES ARE BROKEN exercise cannot be taken, those parts of the Body which can be rubbed with propriety should be thus treated, to assist in more actively circulating the Blood, and producing a more speedy restoration.

434. *Inf.*—THE CHARACTERISTICS OF THE YOUNG SKELETON distinctly suggest frequent changes of position; children should neither sit, stand, or lie too long in one posture. The child should not be carried in one position long by the nurse, and should be turned frequently while it is sleeping.

435. *Inf.*—A CHILD SHOULD NOT BE PLACED UPON ITS FEET too young in order to induce or to teach it to walk, but nature should be relied upon to inspire the child to walk as soon as it is best.

436. *Remark.*—A CHILD CAN NEVER BE TAUGHT TO WALK; it will only walk when it has the requisite strength and development; setting the child upon its feet, leading it, and the use of all such means, only tends to curve the limbs or trunk. The child that is backward will walk as soon as it should, and the child that is forward will not be harmed. Of the two, pains should be taken to prevent the child from walking. Instinct in case of the child, as in the animal, is the only reliable guide in respect to walking.

437. Thus IS THE SKELETON CONSTRUCTED, a living marvel of workmanship, a thing of beauty not of dread.

---

431. What is —?   432. What —?   433. What Inf. —?   434. What said of —?
435. When —?   436. Why —?   437. How —?   How regard the skeleton?

Fig. 174.

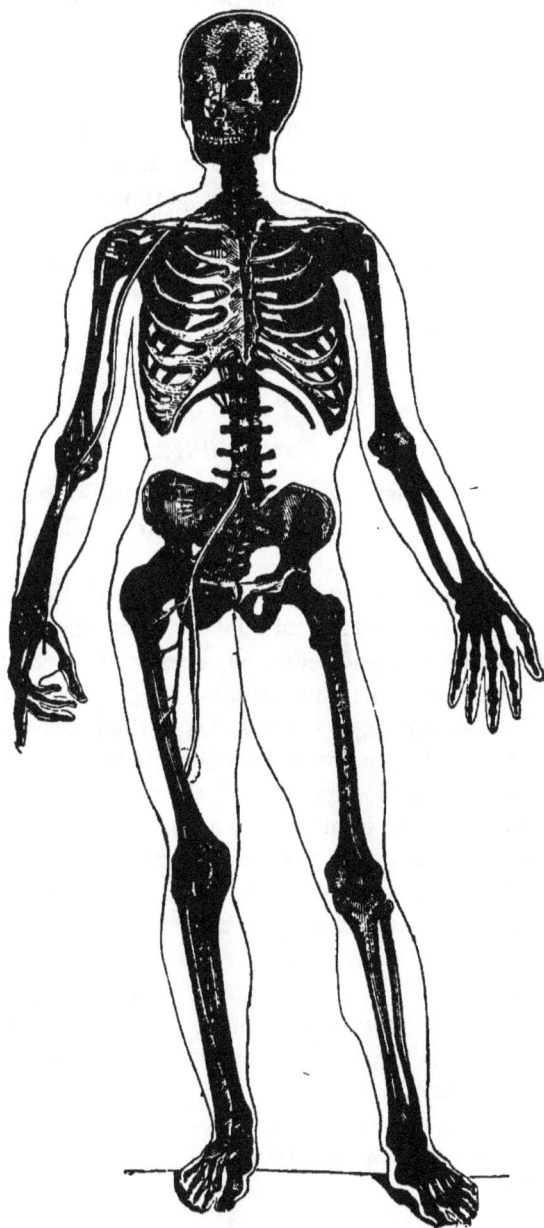

The ghastly features it exhibits to timorous ignorance, change their aspect as the intelligent mind learns to admire the superhuman wisdom that organized and so exquisitely adapted it to serve the wants of man. Whether we regard the admirable properties of the tissues of which it is constituted, or the manner in which it is constructed and preserved, mended and amended, or whether we regard the purposes to which it is devoted, its use as a support to all the organs, a protection to many, or as adapted to the production of motion, from the sole of the foot to the crown of the head, it is equally replete with edifying truths.

## SECTION II.

### *Muscles: Contractility.*

438. THE EVIDENT ADAPTATION OF THE SKELETON TO PRODUCE MOTION SUGGESTS the inquiry, What produces the motion of its parts?

439. It IS PERTINENT that a description of the muscles moving the skeleton should follow a description of it, since nothing else antecedent is necessary to a correct understanding of them; since some of them assist in the action of other organs, as the muscles of the trunk-walls in respiration, so that a description of them must precede that of the contained organs; and since other muscles produce motion of or in other organs, and will be better understood if those of the Skeleton are first described.

440. MUSCLES ARE CONSTRUCTED of two parts, the Sinewy and the Muscular or contractile.

441. THE CONTRACTILE PART OF MUSCLES CONSTITUTES what is called the lean meat of animals; and though as it is usually cut it does not seem to have any regular arrangement, it is arranged in perfect order, as may be seen by observing Pl. 2.

---

Subject of Section II. ?  438. What does —?  439. What —?  440. How —?
441. What does —?

442.  THE RED COLOR of the contractile part is owing to the Blood that is in it; hence it has different colors in different animals.

443.  *Illus.*—THE DRUMSTICK OF A FOWL EXHIBITS in an admirable manner a bundle of muscles, with their fleshy and sinewy parts.

444.  ANY PIECE OF LEAN MEAT CAN BE EASILY DIVIDED in one direction into stringy fibres, which, by the use of appropriate instruments, can be subdivided till fibrils are reached, not so large as hairs, composed of an exceedingly delicate sheath, like basement membrane, enclosing still more delicate fibrillæ composed of cells, or minute particles of a beaded form placed end to end.

445.  Commencing with the delicate cellular fibrillæ, A MUSCLE IS CONSTRUCTED by enveloping a small bundle of them in a membranous homogeneous sheath, and forming a fibril.  A bundle of fibrils placed side by side, and enclosed in a sheath of sinewy membrane, forms a fasciculus.

446.  ALL THE FIBRILLÆ in a fibril and all the fibrils in a fasciculus are parallel, but fasciculi may not be, indeed seldom are, perfectly so.

447.  *Inf.*—FASCICULI MAY BE CONSIDERED as the elements of muscles, and a single one may constitute an entire muscle.

FIG. 175.

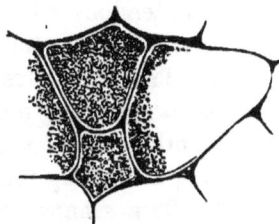

Fig. 175 represents several fasciculi cut across, their contained fibrils, and the containing sheath, each fibril being composed of several fibrillæ too small to be represented.

448.  WHEN MUSCLES ABOVE THE SMALLEST SIZE ARE TO BE FORMED, fasciculi are gathered into larger or

442. What said of —?  443. What does —?  444. How can —?  445. How —?
446. What said —?  447. How —?  Describe Fig. 175.  448. What said —?

smaller bundles, and are enclosed by a sinewy membrane somewhat thicker than that of the fasciculus, and this may be a muscle. These bundles may be again made up and enclosed in a still thicker sinewy membrane, while several muscles themselves may also be surrounded by an enclosing membrane, called a fascia. Sinewy fibres are also woven among the bundles in the form of areolar texture.

449. *Illus.*—THE MUSCLES OF THE THIGH are enclosed in a common external sheath or fascia beneath the skin, and bundle within bundle of the constituents of the thigh can be analyzed till the ultimate cellular fibrillæ, composed of albuminous membrane and contractile musculine, is reached.

450. AROUND THE FIBRILLÆ, but not entering them, is woven a beautiful network of very fine capillaries that communicate with arteries on one hand and veins on the other, so that a plentiful supply of Blood is constantly poured around the contractile elements of the muscle.

451. UPON OR IN THE SHEATHS OF THE FIBRILLÆ nerves terminate and commence, precisely how is not known; some suppose by loops and others by points.

452. Through SOME OF THE NERVES, CALLED MOTOR, influences are brought to the contents of the fibrillar sheath that causes them to contract, and through OTHER NERVES, CALLED SENSORY, influences are sent to the nervous centres, causing sensations and other effects.

453. THE SHEATHS OF SOME OF THE FASCICULI extend but a little distance beyond their contents and produce very short tendons, while in other cases they extend and are condensed in the form of long tendons.

454. THE CONTENTS OF THE FASCICULI are of nearly equal lengths in some muscles, making their ends abrupt, while in others the contents are very unequal in length, producing a tapering muscle.

455. THE FIBRES OF THE FASCICULAR SHEATHS are woven diagonally to their contents, but where they ex-

tend they are side by side, producing a very compact, hard, strong cord or expanded membrane, called, in the former case, tendon, in the latter, aponeurosis.

456. WHEN THE MUSCLE IS TO ACT UPON SOFT PARTS, the sheaths of its fasciculi blend directly with the part which is to be acted upon, and no tendon is necessary.

457. THE FASCICULI MAY ALSO BECOME ATTACHED by one extremity to the enclosing sheath of the muscle they form.

458. DIVISIONS OF THE TENDON frequently extend into the muscle, making divisions of it, to each side of which the fasciculi can be attached.

459. THE TENDONS ARE ATTACHED TO THE SKELETON by having their fibres become continuous with those of the fibrous or sinewy membrane (periosteum) covering the bones, or with the fibres of the ligaments.

460. THE TENDONS SOMETIMES EXTEND along in grooves in the bones, or turn around ligaments, or move over places where friction is likely to occur.

461. BURSÆ ARE small bags or cells composed of sinewy membrane, and lined with serous, secretory tissue placed at different points, to prevent the friction that would otherwise be caused by tendons. Their form is in accordance with the place they are to occupy.

462. *Illus.*—Behind the third finger, at the edge of the palm, (1, Fig. 176) under the skin and a ligament, A BURSA OF AN HOURGLASS FORM IS LOCATED; by pressing on the front part of it, its contained fluid is pressed back, and produces a little swelling easily noticed and felt.

463. *Remark.*—The bursa, from some cause, sometimes becomes distended by fluid, feels very hard, and is sometimes mistaken for an enlargement of a bone. Weakness of the part is the usual attendant. It is called WEEPING SINEW, or sometimes a ganglion. It is easily cured. The best thing to do at first is to bind something hard upon it for several days, when, if it do not disappear, recourse should be had to a surgeon, who will, by a slight operation, remove it.

---

456. What said —? 457. How —? 458. What said of —? 459. How —? 460. How do —? 461. What —? 462. Where —? 463. What is —?

FIG. 176.

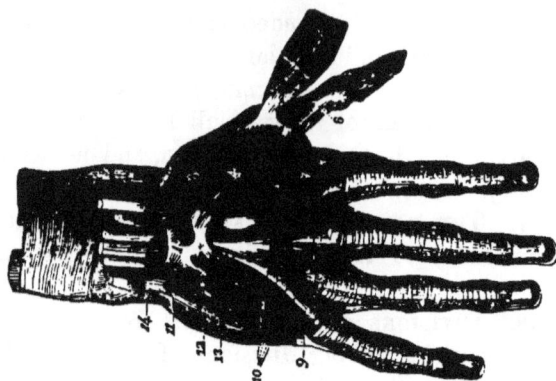

464. THE TENDONS OF EITHER EXTREMITY OF A MUS-CLE may be attached to one or to several different points, or to a continuous surface. In one case they are said to have one or several heads or origins, and in the other to have one or several insertions. The head is that extremity that is usually firm, while the insertion is the point or extremity to be moved.

465. THE USE OF A MUSCLE is, by contraction, to produce motion.

466. THE CONTRACTION OF A MUSCLE cannot be continued for any considerable length of time; it must in a short time relax, or lengthen.

467. THE REASON WHY CONTRACTION CANNOT CONTINUE is, that decomposition is going on as long as contraction continues, and as it shuts out the Blood recomposition cannot take place; relaxation must therefore occur in order that the Blood may have an opportunity of pouring into the muscle, carrying away the used and useless substance, and supplying new material.

468. *Illus.*—By placing one hand on the front part of the upper arm, and raising and lowering its fore arm, the action of the muscles can be appreciated. When the forearm is raised the muscle shortens,

becomes broader and harder; when the fore arm is lowered, the muscle returns to its former condition.

469. Contraction of a muscle could not be continued as long as it now is IF THE ENTIRE MUSCLE CONTRACT-ED at once.

470. Only A PORTION OF THE CONTENTS OF A FIBRIL contract at a time, when another portion takes it up, then another, then another. Thus there is a vibratory contraction of a muscle during the time that it appears to contract.

471. ANOTHER PECULIARITY OF MUSCLE IS TO BE NOTICED: if it is cut it gapes. The muscle must therefore be constantly contracted to a degree. This is called its tone or its tonicity, and exists without reference to the will.

472. THE TONE OF THE MUSCLES WILL DIFFER in different persons, and in the same person at different times. Whether it is the same as its contractility, only in a lesser degree, is not known. It can be suspended or diminished by the action of various medicines, for instance, those that sicken the stomach.

473. It is the tone of the muscles that balances THE ACTION OF ONE MUSCLE AGAINST THAT OF ANOTHER, and preserves an equilibrium of the parts of the Body.

474. WHEN MOTION IS PRODUCED BY THE CONTRAC-TION OF ONE OR MORE MUSCLES, the relaxation of one or more must take place at the same time. Contraction and relaxation must always be harmonious.

475. A DOUBLE INFLUENCE MUST ALWAYS BE EXERT-ED in producing motion, one controlling the action of the muscle or muscles directly producing the motion, and another controlling the action of the relaxing or opposing muscles; and the requirement for the one is as imperious and requires as nice adjustment as the other.

476. SOME MOTIONS ARE PRODUCED by the direct action of fasciculi, others by the combined action of sev-

<hr>

469. What —? 470. What said of—? 471. What is —? 472. How —? 473. What said —? 474. What said—? 475. Why —? 476. How —?

eral fasciculi, neither of which contracts in the line of motion; others by the successive action of fasciculi, and others again in a direction opposite to the direction of fascicular action, produced by a changed direction of the tendon. (See 9, in succeeding figure.)

FIG. 177.

Fig. 177 represents a section of the socket of the eye and that organ with its muscles in situ. 4, the muscle that elevates the lid, a part of which, with the lashes, has been cut off and left with the muscle; 10, points to the loop of ligament round which 11, the tendon of the muscle, 9, turns, a bursa being interplaced to prevent friction. A farther description of the muscles of the eye will be given when that organ is described.

477. Sometimes ALL THE FASCICULI OF SEVERAL MUSCLES will be required to produce a desired motion, and again all those of one and only a few of another.

478. Thus by THE VARIED DIRECTION OF THE FASCICULI, and by combining the action of muscles or parts of muscles, all kinds of desired motion can be produced.

479. THE EXTENT OF MOTION DEPENDS UPON the length of the fasciculi, and the strength of motion upon the number of them.

480. MOST OF THE MUSCLES ARE ATTACHED to the skeleton so as to produce rapid motions at the expenditure of strength, for they are usually attached near the joint. As a hand near the hinge of a gate must use great exertion to swing it, so must the muscle to move the part to which it is attached.

Describe Fig. 177. 477. What said of —? 478. What said of —? 479. — what? 480. How — ?

481. THE HAND THAT SWINGS A GATE BY ACTING NEAR THE HINGE, moves but a very short distance; so the muscle contracts but a little to produce much motion.

482. A MUSCLE CAN CONTRACT but a small part of its length with profit, while the exertion necessary to produce the effect can be made in an instant, and much time saved in producing motion.

483. *Inf.*—THE WHOLE BODY IS CONSTRUCTED *with reference to saving time* in doing that which is merely mechanical.

484. IT ALSO IMPROVES PERSONAL APPEARANCE to have muscles in many instances attached near the joint, since thus a compact and graceful form is exhibited.

485. *Remark.*—We may now hastily take a view of the muscles of the different members of the Body, for the purpose of noticing anything particularly interesting, and for the purpose of illustrating some of the facts already stated. But it will not be worth while to enter into a detailed description of all the muscles, as nearly all of them are illustrated by the plates, and the description of them in connection with the plates is sufficient for all ordinary purposes.

SPECIAL MUSCLES.

## Muscles of the Head.

486. THE MUSCLES OF THE HEAD ARE DIVIDED into those of the Cranium, Facium, and Organs of Sense.

*Remark.*—Those that move the head upon the spinal column are classed with those of the neck.

487. THE OCCIPITO FRONTALIS is the name of a broad thin muscle that extends from the back of the cranium to the front of it, having two fleshy parts, one at the back and the other in the forehead, the connecting part being a broad aponeurosis, some of the fibres of which closely connect with those of the scalp, on account of which this muscle can move the scalp. The contraction of the front part of it wrinkles the forehead. It has a line of tendon extending from front to back,

through its centre, dividing into right and left; hence, though named as if single, it is double. It assists in giving expression.

FIG. 178.

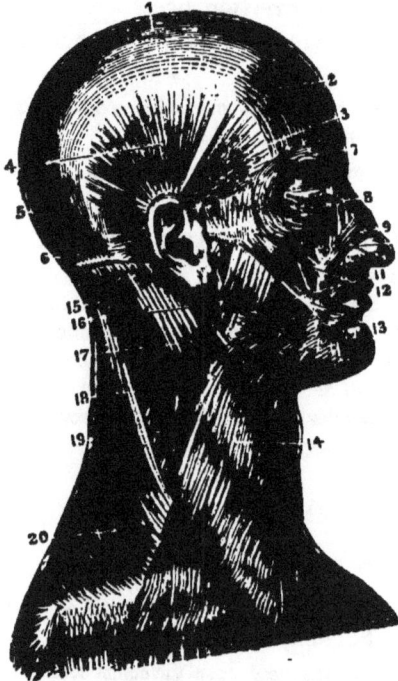

Fig. 178, side view of muscles of Cranium, Facium, and Neck. 1, tendon; 2, front, 5, back of Occipito frontalis; 3, 4, 6, muscles of external ear. Beneath 3, temporal faintly shown; 7, Orbicularis; 8, levator; 9, compressor; 10, levator; 11, buccinator (trumpeter); 12, Zygomatic; 13, ditto and orbicularis; 16, masseter; 14, platysma myoides; 17, Sterno Mastoid; 20, trapezium.

488. THE SUPERCILIARY are situated beneath the eyebrows, extending from the ridge above the nose outward. When they contract they corrugate, and draw down the skin covering them.

489. *Remark.*—The two preceding muscles are the only ones of the Cranium, and as they are only useful in giving expression, they might better be associated with those of the facium.

490. THREE MUSCLES, ABOVE, IN FRONT, AND BACK OF THE EAR, extend from it to the cranium, moving it when contracted.

Describe Fig. 178. Which muscles belong to cranium? face? neck? 488. Describe —. 489. What said in Remark? 490. What said —?

491. *Remark.*—PERSONS USUALLY CANNOT CONTRACT THESE MUSCLES, not because they are wanting in any case, but from not being used, as any one can prove by frequently making an effort to contract them with a determination, and giving the mind direction toward them. Not a few other muscles through disease become equally indolent and deaf to the calls of the mind. Thus by tight dressing and supports the muscles of the chest not only become infirm, but absolute nullities.

492. THE TEMPORAL can readily be felt while chewing, by putting the fingers on the temple (so called from *tempus*, time, because first showing gray hairs), to the bone of which, and a fascia stretching from its ridge to the zygoma, muscular fasciculi are attached, and from which converging, their lower extremities terminate in a tendon that passes under the yoke or zygoma (an arm or process of bone extending in the line between the ear and eye), and seizes upon the upper part of the lower jaw, just in front of the joint, and assists in drawing it up when chewing; hence the perceptible action of the muscle at such a time.

493. THE MASSETER can be felt working at the same time on the side of the face; it extends from the yoke and the cheek (malar) bone down to the lower jaw.

494. THE PTERYGOID, internal and external, are two muscles upon each side, within the jaw, against the masseter, that assist in raising the jaw, drawing it forward and from side to side, as in grinding.

FIG. 179.

Fig. 179 represents a portion of the lower jaw removed to expose 1, the internal pterygoid. A portion of the zygoma and the temporal bone in front of 2 is removed to expose 2, 3, the external pterygoid.

491. Why —? 492. What said of —? 493. What said of —? 494. What said of —? Describe Fig. 179.

495. *Inf.*—Four strong muscles upon each side assist in mastication, several of minor importance assisting in drawing the jaw downward.

496. THE ORBICULARIS PALPEBRARUM is composed of fasciculi, that extend from the inner angle of the eye, sweeping round the lids of that organ and by contraction closing it, most of the contraction taking effect upon the upper lid, which is raised by a special muscle extending from the bottom of the socket.

497. THE OTHER MUSCLES OF THE FACIUM, as their illustrations show, extend from various points to the nose or to the mouth, producing motion of those parts.

498. THE ORBICULARIS ORIS (circular of the mouth) is usually described as if its fasciculi entirely surrounded the mouth. It would be better if it were described as extending from the angles of the mouth, where it is blended with the fibres of other muscles, to the central line of the lips, where there is always a thin partition tendon dividing the muscle distinctly into right and left.

499. *Remark.*—THERE ARE OTHER MUSCLES within the mouth, eyesocket, and ear, to be described in connection with those organs.

### *Muscles of the Neck.*

500. THE MUSCLES OF THE NECK MAY BE CLASSED as adapted to move the head and neck on the spinal column, to raise the shoulders, to control the action of the mouth and throat, and to act upon the larynx.

501. THE STERNO-MASTOID (neck-cord) is the most conspicuous muscle of the neck, extending from the sternum to the mastoid process just back of the ear. Its contraction draws the ear toward the sternum and moves the head accordingly. Both muscles, acting at the same time, draw the head directly forward. (See Fig. 178.)

502. *Remark.*—The other muscles bending the head are small, and, for the most part, lie close to the column, as shown by the illustrations.

---

495. How many muscles raise jaw? 496. Describe —. 497. What said of —? 498. Describe —. 499. Where —? 500. How —? 501. Describe —.

503. THE TRAPEZIUS (or monk's cowl) hangs down from the back of the head, a thin muscle, the fasciculi of which reach out and take hold of the shoulder-blade (Scapula), which is raised toward the head by the upper part, drawn back by the middle part, and downward by the lower part, and moved in a rotary manner by the successive contraction of the fasciculi. (1, Fig. 1, Pl. 22.)

*504. THE TRAPEZIUS WILL ILLUSTRATE the effects of culture in producing graceful motions; for if its fasciculi are called into action successively, a curvilinear motion will be exhibited, while if whole clusters of them act at once, angular and abrupt motions must be the consequence. The successive action of parts of a muscle, and of several muscles, can only be obtained by much culture, and is one of the greatest accomplishments.

505. *Illus.*—THE OCCUPATION OF ALMOST EVERY MECHANIC CAN BE RECOGNIZED in his every-day life when away from his shop, the muscles being so apt to act in accordance with the habit of exercising them.

506. THE LEVATOR ANGULI SCAPULÆ, (5, Fig. 1, Pl. 21,) is used to shrug the shoulder, and hence sometimes called the Frenchman's muscle.

507. *Remark.*—WHEN THE SHOULDERS ARE MADE FIRM, the preceding muscles will move the head and neck, as will some of those adapted to move the head assist in raising the ribs if the head is held firm; observe that action of the neck-cord.

508. THE SUBCUTANEOUS CERVICAL is a broad thin lamina that originates in the tissue of the upper part of the chest, and extends up just beneath the skin, to which some of its fibres are attached, to the lower jaw, melting into the parts covering it. Its contractions assist in giving expression (also called Platysma Myoides, Fig. 178.)

509. THE DIGASTRIC MUSCLE, as its name signifies, has two fleshy parts, with an intervening tendon that slips through a loop on the side of the hyoid bone, from which the front part of the muscle extends up to the

çentre of the inner surface of the lower jaw, and the back part extends up to the temporal bone. This muscle can therefore draw down the jaw if the hyoid bone is firmly held, or if the jaw is firmly held the hyoid will be raised. It is a curious muscle on account of its central tendon and its arrangement.

510. *Remark.*—Other neck-muscles may best be described, so far as they need to be mentioned, in connection with the throat and larynx.

## Muscles of the Trunk-walls.

511. THE MUSCLES OF THE TRUNK-WALLS MAY BE CLASSED as those adapted to act on the upper extremities, to elevate the ribs, to depress them, to compress the organs of the abdomen, and to bend or sustain the spinal column.

512. THE MUSCLES ADAPTED TO ACT ON THE UPPER EXTREMITIES MAY BE CLASSED as those acting on the blades and on the arms, and both may be classed as front and back, or anterior and posterior.

513. THE PECTORALIS MAJOR, as its name signifies, is the largest of the breast, and is worthy of notice from the curious arrangement of its fasciculi, where they are attached to the arm (14, Pl. 20, Fig. 1). The lower ones extend up under the upper ones, to be attached as near the joint as possible, since their use is to draw the arm downward, while the upper fasciculi extend over the upper ones and are attached as far as possible down from the joint, a very extraordinary and ingenious arrangement for gaining important results by simple means. This muscle forms the inside of the armpit.

514. THE SERRATUS MAGNUS (Fig. 1, Pl. 21) is a broad, not very thick muscle, that extends from several of the ribs under the shoulder-blade, to the under surface of the back edge or base of which it is attached;

510. What said of other neck-muscles? 511. How —? 512. How —? 513. Describe —. Why is this muscle curious and interesting? 514. Describe —.

by its contractions it draws the blade downward and forward.

515. THE LATISSIMUS DORSI (broadest of the back) extends from the back of the hips and the region of the loins upward and outward to be attached to the arm, and by contraction draws it down and backward. Its tendon forms the back side of the armpit. (2, Fig. 1, Pl. 22.)

516. THE RHOMBOID has the shape its name signifies, and extends from the upper part of the spinal column down and out to the back edge or base of the blade, which it raises up and draws back.

517. It therefore appears that THE SHOULDER-BLADE is imbedded in muscles, by which it is suspended in its place and moved in every direction, the use of it being to furnish surface for the attachment of muscles to act through it upon the shoulder and arm; for the shoulder-blade is never to be moved except to move the shoulder-joint and extend the use of the. arm. The blade lies upon the chest, its position and projection being determined by the chest, but not affecting its form or size.

518. *Remark.*—MOST OF THE MUSCLES OF THE UPPER PART OF THE TRUNK-WALLS AND OF THE NECK may assist in raising the ribs, as may be noticed by drawing a deep breath.

519. THE INTERCOSTALS, internal and external, are two layers of fasciculi extending obliquely between the ribs. By their oblique direction they are longer than if direct, and if they contract one third their length, will almost or quite bring the ribs together, while a contraction of one third their length, if direct, would only move the ribs to an equal extent. (20, 21, Fig. 1, Pl. 21.)

520. AGAIN, IT IS FOUND, WONDERFUL TO TELL, that from their arrangement one set of these fasciculi raise the ribs and the other depresses them, and neither can do the work of the other.

---

515. Describe —. 516. Describe —. 517. What said of —? Which effects the other, the blade or chest? 518. What said of —? 519. Describe —? 520. What —?

Fig. 180.

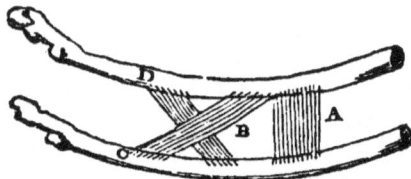

Fig. 180, a plan of ribs, to show direction of intercostals and how they cross.

521. THE RECTUS ABDOMINALIS (straight of the abdomen) extends from the front lower part of the chest down in front of the abdomen to the pelvis. It is divided by tendons into the upper, middle, and lower parts; for if the fleshy part had been continuous, its contractions would have produced inconvenience on account of the prominence its central part would have exhibited. (26, Fig. 1, Pl. 20.)

522. THE USE OF THE RECTUS is to depress the chest, and when moderately distended it can act with greatest advantage; hence, a person can speak or use the expiratory organs with greatest ease after eating a reasonable quantity of food, if not to soon.

523. THE SIDES OF THE ABDOMEN ARE WALLED by three layers of Muscles, the external and internal oblique, and the transverse. They are fleshy on the sides, and form a dense sinewy aponeurosis over the centre of the abdomen. The rectus passes down through this about two thirds its length, and then passing between the fibres of the transverse, extends behind them and those of the oblique to the pelvis. (15, Fig. 2, Pl. 21.)

524. THE USE OF THE THREE LAYERS is to draw down the ribs and compress the organs of the abdomen, and thus doubly assist in expelling the air in common with the rectus.

525. THE LUMBAR QUADRATUS is a stout muscle that extends from the upper edge or crest of the hip to the lowest rib, and assists in depressing the chest. (Fig. 181.)

526. *Remark.*—MOST OF THE MUSCLES ACTING ON THE CHEST, shoulders, neck, and head, can assist in bending or sustaining the spinal column. The muscles of the front walls of the abdomen, especially the rectus, are used to bend the spinal column forward.

527. NUMEROUS SMALL, LONG, DELICATE, BEAUTIFUL MUSCLES ARE PACKED in the hollows upon each side of the spinous processes, and assist in bending the spinal column from side to side, raising it up and sustaining it. (See Figs. 3, 4, Pl. 21.)

528. TWO MUSCLES WORTHY OF NOTICE ARE FOUND within the abdomen: the *Illiac* is attached to the inner surface of the upper part of the hip-bone; the *Psoas* arises from the bodies of the lumbar vertebræ and their substances, and extends down to unite its tendons with that of the former muscle, when they both extend out

FIG. 181.

Fig. 181. 1, lower sternum; 2, costal cartilages; 3, lumbar column covered with ligament; 4, crest of hip; 5, 6, 7, under or concave surface of *Diaphragm* (see Fig. 66); 8, 9, lower back border of D.; 10, 11, central attachments called crus, curiously braided around openings, 12, for (aorta) large artery; 13, for Œsophagus, and 14, for large vein (vena cava); 15, front part of lumbar quadratus; 16, small, 17, large Psoas; and 18, Illiac muscle, with lower part removed, as is lower part of 4; 19, 20, part of the transversalis and tendon.

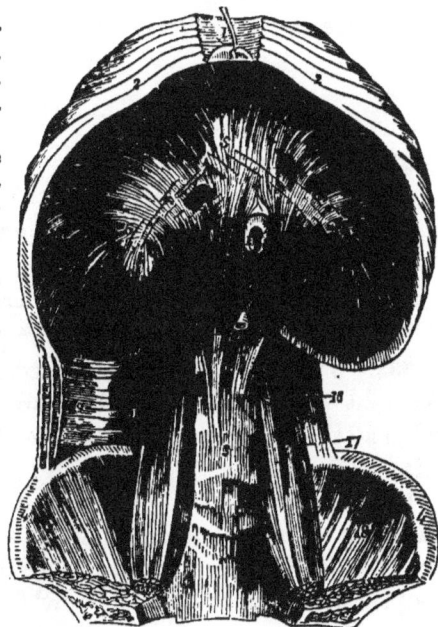

upon the lower part of the front margin of the pelvis, and connect with the upper part (lesser trochanter) of the thigh-bone (Femur).

*Remark.*—The Psoas are the only muscles that extend across the Pelvis. With these exceptions, the Pelvis is the origin of motions both upward and downward. It is the circle of support of all the upper parts of the body and the portion of the body that walks upon the parts below. It is the " dead point " in the body.

529. THE USE OF THE ILLIAC AND PSOAS MUSCLES is to bend the column and trunk forward upon the thigh-bones, or to draw the leg up and turn it slightly. They also essentially assist in maintaining the trunk erect on the lower extremities.

530. THE GLUTEI ARE very large strong muscles on the back of the hip-bones, inserted by their lower tendons in the thigh. They are the chief means for supporting, in a backward direction, the trunk upon the lower extremities. Several small muscles assist them.

### Muscles of the Lower Extremities.

531. THE MUSCLES OF THE LOWER EXTREMITIES may be classed as those that move the thigh on the hip; the lower leg on the thigh; the foot on the leg; and the toes on the foot.

532. *Remark.*—THE MUSCLES THAT MOVE THE THIGH UPON THE HIP are numerous : some already noticed are very large, but most are small.

533. THE SARTORIUS (tailor's) extends from the upper front point of the hip diagonally down the thigh, and is attached just below the inside of the knee to the tibia. (Fig. 2, Pl. 23.)

534. THE USE OF THE SARTORIUS is to raise the lower leg over the other, as when a tailor sits. This muscle is very interesting from being the longest in the Body, slender, graceful, with parallel fasciculi, prominent

near the surface, having peculiar uses, and particularly because its inner edge at the middle of the thigh is a guide to the position of the great femoral artery that lies between it and the thigh-bone. (See Fig. 182.)

535. SOME MUSCLES OF THE THIGH, besides assisting in moving it, bend the lower leg upon the thigh and straighten it, or are flexors and extensors of the lower leg.

536. FOUR MUSCLES OF THE FRONT PART OF THE THIGH unite their tendons below and connect with the kneepan, from which a ligament extends fastening it to the tibia (shin) at the upper part. (Fig. 2, Pl. 23.)

537. THE USE OF THE KNEEPAN (Patella) is to throw its tendon out from the knee and allow the muscles to act with increased advantage. A Bursa is placed below the patella.

538. THE LARGE MUSCLES ON THE BACK OF THE LOWER LEG unite their lower tendons to form the heel-cord (tendo Achilles), in which the fibres are, so to speak, braided, to produce the strongest possible tendon. (Fig. 2, Pl. 22.)

539. THE MUSCLES UPON THE FRONT PART OF THE LOWER LEG, as the figures show, send down their tendons to the top of the foot and outward to the toes, passing under a ligament that extends from the outer to the inner ankle, and binding down the tendons, between which and it bursæ are placed to prevent friction.

540. THE LONG PERONEAL is situated in the outside of the leg: its tendon extends down through a groove at the back of the ankle, turns through another groove beneath the foot, which it crosses, and becomes inserted into the bone back of the great toe (1st Metatarsal). The grooves are lined with bursæ. It has the most curiously arranged tendon in the Body. (Figs. 3, 4, Pl. 23.)

---

535. What said of —? 536. What said of —? 537. What is —? 538. What said of —? 539. What said of —? 540. What said of —?

### Muscles of the Upper Extremities.

541. *Remark.*—NEARLY ALL THE MUSCLES THUS FAR DESCRIBED may assist in giving latitude of motion to the upper extremities: those that act on the scapulæ are especially to be classed with those of the arms.

542. THE MUSCLES OF THE UPPER EXTREMITIES are to be classed as those acting upon the shoulder-blades, and moving it upon the chest and its fulcrum, the clavicle; upon the upper arm, and moving it upon the scapula; upon the lower arm, and moving it upon the humerus; acting upon the wrist, and moving it upon the lower arm; and upon the fingers, moving them upon their joints.

543. SEVERAL MUSCLES EXTEND FROM THE SCAPULA, and are attached to, the humerus, producing its movements.

544. THE BICEPS arises, as its name signifies, by two heads from the scapulæ, extends down the front of the humerus, and is inserted in the radius, which, it will be remembered, does not at all form the elbow-joint, being fastened to the ulna near the joint; but to its lower end the wrist is jointed, and when the hand is to be lifted it is better to have the biceps take hold of the radius than of the ulna. (Fig. 6, Pl. 24.)

545. THE TRICEPS arises from the scapula, extends down the back of the humerus, and is inserted into the point of the elbow. (Fig. 5, Pl. 24.)

546. THE USE OF THE BICEPS AND THE TRICEPS is to raise and lower, or flex and extend, the forearm.

547. THE MUSCLES OF THE FOREARM EXTEND, some from one bone to the other, to roll the radius, and make the hand prone or supine; some to the wrist, to bend it upon the arm; and again others to the very finger-ends, passing under a ligament like a bracelet, at the wrist, provided with numerous bursæ to prevent friction.

548. MUSCLES EXTEND FROM THE WRIST and from

the ligaments surrounding it to the thumb and fingers, and to some of the tendons extending to the fingers' ends small muscles are attached, extending to the sides of the fingers; and again from between the bones of the frame of the hand, muscles spring up and are inserted into the finger-bones. (Fig. 4, Pl. 25.)

549. *Remark.*—THE FINGERS ARE THUS INSURED powerful, various, and rapid motions, to the utmost extreme that can be demanded of them, and they can express almost as many shades of emotion as can the muscles of the face, or the flexible organs of speech. It is only necessary to watch the nimble fingers of a skilful performer fly over the piano-keys, with all the varied touch that can be given them, to be convinced that the hand is a masterpiece of workmanship, and could only be made for, as it can only be perfectly used by, a well-developed mind, which it also serves to develope.

550. THE MUSCLES OF THE LARYNX may best be described when the organs of speech are considered.

## *General View of Muscular Action.*

551. *Remark.*—Having observed the special uses of the various muscles, it WILL NOW BE PROPER TO NOTICE the general effects of the action of all—what is for their good, and how they are likely to suffer.

552. THE MUSCLES, VIEWED AS A WHOLE, excite our admiration on account of the compact manner in which they are built up, and the graceful outline that they bestow upon every part of the Body.

553. *Remark.*—IT WOULD AT FIRST THOUGHT APPEAR that the muscles must somewhat interfere with each other and produce friction.

554. FRICTION OF THE MUSCLES, against each other, is PREVENTED by the beautiful arrangement of the areolar texture, which is wrought around and between the muscles, becoming continuous with their sheaths, which in fact are only dense forms of the same tissue. This texture also loosely connects the muscles with the all-sur-

---

549. What are —? 550. What said of —? 551. What —? 552. What said of —? 553. What —? 554. How is —?

Fig. 182.

rounding skin, and affords a nidus at various points for the deposit of fat, to serve as a packing, or to round the limbs and give perfection to the beautiful forms already moulded by the skeleton and muscles.

555. *Remark.*—It IS UTTERLY IMPOSSIBLE TO EXPRESS the delightful emotions the mind experiences as it contemplates the developing beauty that begins to clothe the uncouthness of the skeleton, in which was really the basis and the purpose of all that is lovely in the human form or graceful in its motions, though unseen by the unprophetic eye of ignorance.

556. In the midst of, upon, and sometimes through, the muscles, THE BLOOD-TUBES AND NERVES EXTEND, from their centres to their terminal points; the larger arteries being buried deeply beneath muscles, and thus protected from injury and loss of heat.

557. Thus EVERY MUSCLE BECOMES a heart to other Blood than that which it receives, since it presses upon all the vessels in its neighborhood, and, owing to valves in the veins, has the effect to keep the Blood moving through its circuit.

558. *Illus.*—When a person is bled from the arm-vein, he is directed to grasp something and alternately loosen his grasp, when each contraction of his muscles spirts out the Blood.

559. *Inf.*—A PERSON MAY CORRECTLY BE SAID TO HAVE as many voluntary hearts as he has voluntary muscles.

560. THE POWER WITH WHICH MUSCLES CONTRACT is almost beyond belief, the influence which controls them acting so very easily.

561. *Illus.*—A person walking against anything, or swinging his hand against it unexpectedly, strikes with a force that surprises him. A person raising or holding a hundred pounds' weight in his hand, eighteen inches from his elbow, exerts a force with the muscle (Biceps) upon the front part of his upper arm of no less than eighteen hundred pounds, or almost a ton, since the lower end of the muscle is attached not more than an inch from the centre of the joint or fulcrum. When a person stoops and raises himself, the power exerted by the muscles

straightening his trunk and head upon his lower extremities is equal to many thousand pounds.

562. EVERY EXERTION OF MUSCULAR POWER IS ATTENDED WITH a corresponding amount of waste in the material of the muscle, as is proved by the fact that the dark red color of the Blood increases with continued muscular exertion, as seen in the case of animals hunted to death, and by the fact that exertion too long continued decomposes the muscle so much that it will never recover.

563. *Illus.*—THE AUTHOR IS ACQUAINTED WITH A CASE where, by over-exertion at a fire, a person's muscles became incapable of recovery, and for years he was unable to move.

564. THE DECOMPOSITION OF THE MUSCULAR SUBSTANCE renders it necessary to have large respiration and great muscular exertion go hand in hand, since oxygen must be furnished to assist in removing the decomposed substance from the muscle, while carbonic acid must also be removed from the blood.

565. *Inf.*—THIS ACCOUNTS for a full development of the chest and muscles always existing together; the former is a *sine qua non* of the latter.

566. *Inf.*—THIS ALSO ACCOUNTS FOR the fact that a higher activity of the muscles is attainable in cool than in hot weather.

567. *Inf.*—It FOLLOWS ALSO that the circulation should be more active through active muscles than through those that are inactive.

568. EVERY MUSCLE IS a heart, especially in respect to itself, since it assists in the circulation of so much Blood, at least, as flows through itself.

569. EVERY MUSCLE, WHEN IT CONTRACTS, presses out the Blood it contains, chiefly onward into the veins, and when the muscle relaxes (since from the valves in the veins the Blood cannot flow back, and since there is a pressure in the arteries crowding the Blood forward), it gushes through every part of the muscle, supplying it

with nutritious material and the ever-active scavenger, oxygen.

570. *Inf.*—Since time is required for the Blood to course through the relaxed MUSCLE, IT SHOULD NOT BE CONTRACTED AGAIN INSTANTLY; if it is, it cannot renew itself as rapidly as it decomposes, and is soon exhausted.

571. *Illus.*—A horse driven rapidly with a light load, day after day, falls off; while a horse driven slowly, drawing heavy loads, improves. Stage-horses improve during bad going, and fall off when the roads are good.

572. *Inf.*—Light work rapidly performed is more wearisome than laborious work.

573. *Remark.*—Such work as sewing is very wearing if long continued, from the frequent contraction of muscles which, though small and few, are not allowed sufficient time for relaxation; from the continued contraction of some; from the entire inactivity of others; from being done in close, uninviting rooms, and under few of the stimulating emotions, hope, good pay, etc. Relief can and should be given in most of these respects. (See Ap. L.)

574. SINCE THE MUSCLES REQUIRE A LARGE FLOW OF BLOOD, it follows that tight clothing of the chest, hands, or feet, will not only prevent free muscular contraction in those parts, producing stiff and awkward movements, but also injure the muscles by cutting off their supply of Blood.

575. Particularly WHEN PERSONS (or animals) ARE GROWING, the muscles ought not to be constrained, and should not be over-tasked, but should be allowed considerable repose, and a free circulation promoted.

576. WHEN A FREE CIRCULATION IS PERMITTED, and alternate contraction and relaxation of the muscles, with proper periods of repose, are allowed or caused, they will increase in size and vigor wonderfully.

577. *Illus.*—A BLACKSMITH'S ARMS become large, hard, and strong, from the effects of exercising them, but the right is no healthier than the left one.

---

578. WITHIN CERTAIN BOUNDS THE EXERCISE OF
THE MUSCLES PROMOTES their own health and that of
other parts of the Body ; beyond this, exercise is exhaustion.

579. THE PERSON WHOSE BRAIN MUST BE VERY
ACTIVE, cannot digest food enough for that and for constantly exercised muscles. To be a Cicero is incompati-
ble with being a gladiator.

580. THE MUSCLES OF THE YOUNG CRAVE light,
sportive exercise, and they should be indulged; as, if
they are, it will prevent many a mischievous outburst.

581. THE YOUNG, IN PARTICULAR, SHOULD NOT BE
CONFINED long to any one position, either by means of
improper clothing or on any other account.

582. EXERCISES UNDER THE INFLUENCE OF MUSIC
are most healthful as well as pleasant. Military exercises are admirable for the health and for mental effect.

583. EXERCISES OUT OF DOORS are better than indoors. Horseback-riding is one of the best exercises,
giving sweet converse with Nature, and building up
both body and mind at the same time.

584. IT MUST ALSO BE REMEMBERED that muscular exercise—every contraction of a muscle—is attended with
waste of material of a very expensive character'; therefore no more muscular exercise should be caused than is
necessary for maintaining health and accomplishing the
object in view, for all excess of exercise is costly.

585. *Inf.*—EVERY MECHANIC should have his tools handy, for he not
only loses time, but loses money, by the unnecessary muscular action.

586. *Inf.*—EVERY FARMER should have everything convenient; each
unnecessary step destroys other labor as well as that. Let the head
save the heel.

587. *Inf.*—In particular, EVERY HOUSE and its furniture, especially the *kitchen furniture*, should be made convenient. How many
thousands of unnecessary steps do over-worked women take in the year
for want of convenient arrangements! how many thousands of other

inconvenient and unnecessary exertions! This matter may with great profit be improved. (See Ap. L.)

588. PROPERLY EXERCISED, THE MUSCLES ARE great sources of enjoyment through their own sensatory nerves, and by the sensations they induce through other organs by increasing the activity of the circulation, and the volume of fresh air which is inspired.

589. *Remark.*—There would be no difficulty in inducing the inactive to take requisite exercise, if they could only appreciate the truth of the previous paragraph.

590. There SEEMS TO THE AUTHOR NO DOUBT that the digestory canal or some of its glands is an eliminatory organ to the Muscles, which by action furnish the material that in the most healthy manner stimulates elimination. This should be a very powerful motive to take proper exercise.

591. THE EXERCISE OF THE MUSCLES MAKES A DE-MAND for food, producing appetite—one of the greatest of good things for which to be thankful.

592. EXERCISE IMPROVES THE COMPLEXION by the results mentioned in the two preceding paragraphs (for there is nothing more evidently intimate than the condition of the skin and the digestory canal), by circulating the Blood freely to the skin, increasing the heat of the Body and starting the perspiration, and particularly by causing the inspiration of large quantities of óxygen. A beautiful paint, coming from within, is thus delicately spread under the influence of buoyant emotions, those matchless painters, that will challenge the admiration of even the envious.

593. THE MUSCLES ARE THUS SHOWN TO BE our friends as well as our servants, our entertainers as well as our dependents, demanding a support, yet, if properly cared for, merrily repaying their cost with interest; they are the poor man's necessity, the rich man's comfort, the

physician to good health; they give beauty to the graceful, and may be a mine of wealth to all, that princes cannot buy nor untold riches equal.\*

SECTION III.

*Nerves.*

594.  It is very easy to understand the structure,
general arrangement, and purpose of the nerves, but
precisely how they perform their duties is one of the
mysteries of life from which the curtain has not yet
been lifted.

595.  Nerves are either white or reddish gray
pulpy cords, about the consistence of new-made cheese,
of various sizes, extending either between nervous centres, when they are called commissures, or between a
nervous centre and some other part of the Body.

596.  Nerves large enough to be seen with the
naked eye are bundles, covered by a thin sinewy
sheath.  Hence a nerve readily splits up into delicate
fibres.

597.  Nervous fibres are constructed of fibrillæ, having three parts, an outer envelope or sheath nucleated like basement membrane, inclosing a layer of oily
substance, called the Medullary sheath, in the centre of

\* The author has not thought it necessary in this work to speak of any
particular calisthenic or gymnastic exercises, since there are so many good
works upon those subjects.  The classes of motion are few, and may be
traced by the divisions or classes of muscles laid down.  Any kind of exercise that brings into action all the muscles *gracefully* (and that is best
done under the influence of music), without great exertion, and excites
active respiration, is sufficient, while exercises that cause the "holding of
the breath," or those that put a person "out of breath," are not advisable.
Indeed, the chief benefit of any exercise is derived from its causing increased respiration.  Therefore, muscular exercise, to be advantageous,
must be taken in pure air, and with the chest perfectly free to move.

Foot-note. What not necessary?  What sufficient?  594. What —?  595. What
—?  597. How —?

which is the part called the axis of the nerve. The two latter substances are called nervine.

598. THE NERVOUS FILAMENTS OR FIBRILLÆ differ in size in different cases, but each one is uniform in diameter, and continuous from end to end.

599. THE EXTREMITY OF A NERVE, CONNECTED WITH ITS CENTRE or ganglia, is called its origin, inner, central, or centripetal extremity, while THE OTHER extremity is called its termination, outer, or centrifugal extremity, though in case of sensation the influence acting through the nerve begins at what is called the termination of the nerve.

600. IN THE COURSE OF THE NERVES numerous instances occur of filaments or bundles of them passing across from one nerve to another, particularly in the region of the neck. This constitutes what is called a plexus.

FIG. 183.

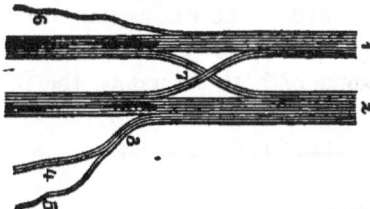

Fig. 183 represents a plexus, 7, formed by side nerves, 1 and 2, interchanging two fibres; 3, a branch of three fibres; 4, another, of two; 5 and 6, two branches of one each.

601. THE USE OF A PLEXUS is to allow nerves that are extending from different centres to the same part to be enclosed in a common sheath.

602. There are THREE DIVISIONS WITH WHICH THE NERVES CONNECT: the Brain, the spinal cord, the Sympathetic Ganglia; hence,

603. THE NERVES ARE CALLED the Cranio-spinal nerves, and the Sympathetic.

604. THE SYMPATHETIC NERVES are of two kinds, a white kind, finer than the Cranio-spinal, and a much greater number of a reddish gray character.

---

598. What said of —? 599. What said of —? 600. What occurs —? Describe Fig. 183. 601. What is —? 602. What —? 603. How —? 604. What said of —?

605. **The spinal nerves may be divided** into two classes: 1st, those that have their origin or inner extremities in the spinal cord, where they connect with the ganglionic cells of the cord; and 2d, those that extend up through the spinal cord to the Brain, and which are truly Cranial nerves.

606. *Inf.*—The spinal cord is partly a nerve, or bundle of nerves, and partly constructed of nervous centres or ganglia.

607. **The cranial nerves,** including the latter division of the spinal, are divisible into two classes, the Motor and the Sensatory.

608. **The Motory and Sensatory nerves** are alike except that on the latter, quite close to their origin, there is a ganglion, the use of which is not understood.

609. **It is also observable** that the Motor nerves are connected with the front part of the spinal cord and the Sensatory with the back part. (See Figs. 76 and 69.)

610. **All the spinal nerves** have two roots or origins, and some fibres of both commence in the cord, and some of both extend to the Brain. Some of the nerve-fibres commence from the cells of the ganglia on the roots of the sensatory nerves.

611. **Some of the Cranial nerves** proper have two roots, but most of them but one, and those are either all motory or all sensatory.

612. At their outer extremities **some of the nerves** lose their external sheath and **terminate** by their axis-part on the sheaths of the muscular fibrillæ.

613. **Nerves also terminate** in three peculiar bodies or corpuscles.

614. **The Pacinian bodies** are composed of a number of layers of sinewy tissue, the spaces between which are filled with a colorless liquid. The axis only of the nerve extends along the centre. The use or mode of action is unknown.

Fig. 184 repre-
sents, 8, a cluster
of Pacinian bod-
ies : 1, much mag-
nified body; 2,
pedicle ; 3, por-
tion of nerve ; 4,
several nerve-fi-
bres ; 5, sinewy
sheath ; 6, nerve
filament ; 7, its
axis.

FIG. 184.

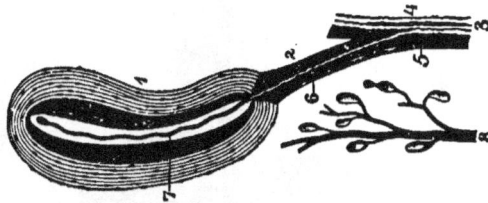

615. TACTILE CORPUSCLES are very small, composed of membrane filled with granular matter, connected with, usually, two, but sometimes only one, or even three or four, nerve filaments. They exist in the papillæ of touch of the hands and feet.

616. THE CORPUSCLES OF KRAUSE are still smaller spherical bodies filled with a transparent soft substance, with one or two nerve filaments entering them.

617. ALL THESE BODIES are supposed to be concern-ed in the production of sensation, but how, is a question.

618. *Remark.*—It is not certain that all the methods of termination of the nerves are known.

619. THE NERVES EXTEND BETWEEN THEIR ORIGIN AND TERMINATION by the most direct course, which is shown by the figures.

620. THE NERVES WILL TOLERATE considerable ex-tension without injury, are readily repaired when in-jured; and when even a section is made, or a portion removed of considerable extent, they will be repaired in a short time.

621. THEY ARE PASSIVE AGENTS, like telegraph wires, for the transmission of influences ; and whether the influence in passing through them necessitates a change in their constitution, or leaves them unaffected, is not known, nor can it be ascertained at present how

rapidly they change nor how great is their requirement
for nutrition. From the small number of Blood-tubes

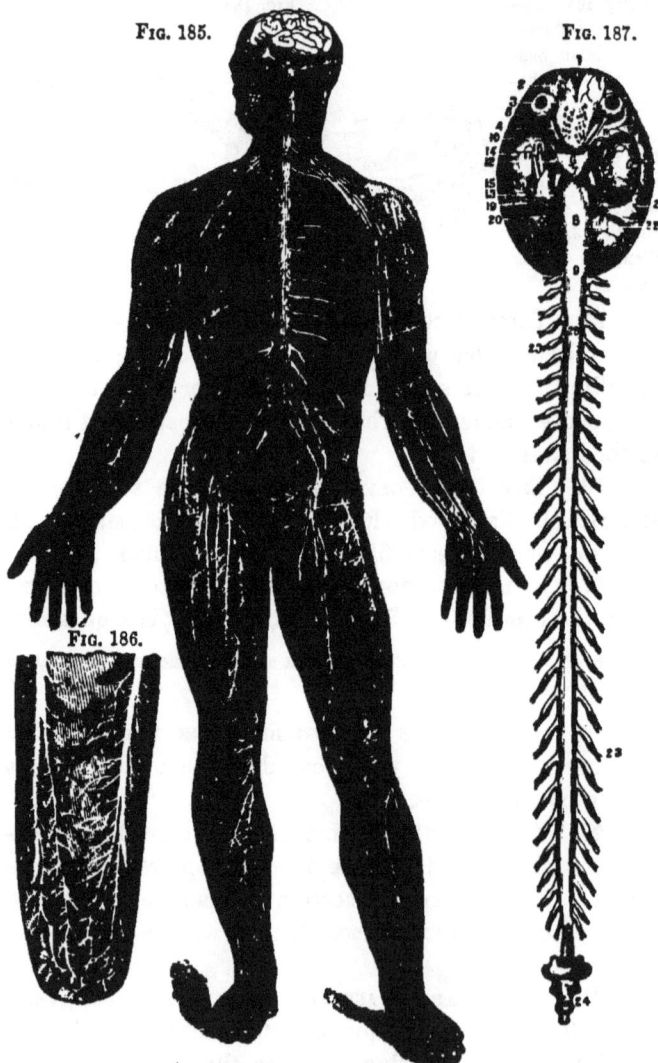

FIG. 185.    FIG. 187.

FIG. 186.

Fig. 185 represents front, side, and under view of brain, front view of
cord and nerves leading from it, the large branches being correctly, the
small ones ideally, drawn. Fig. 186 shows the end of a finger, natural
size, the nerves enlarged. Fig. 187, a front view of the brain-cord and
roots of nerves enclosed in the membranes of the cranio-spinal cavities,

found in large masses of white fibres, it is not supposed to be great.

622. *Remark.*—The special nerves may best be described as parts of apparatus.

## SECTION IV.

### *Ganglia.*

623. GANGLIA are organs designed to excite activity (of mind, of muscle, of the secretory tissue), the essential part of which is cells, in which the power of exciting activity is developed.

624. THE CELLS OF GANGLIA ARE PRODUCED in the midst of a granular substance, and are connected with the nervous fibres, through which the cells act and are acted upon.

625. How THE CELLS ACT, or how they are acted upon, is not understood; it is said to be by nervous influence, but what that is, or how produced, or how it acts, is not known; it is so called because it acts through the nerves, and they are concerned in and essential to its action.

626. Whether THE NERVOUS INFLUENCE IS ALWAYS ONE THING, or whether it varies according to the circumstances under which it acts, is not known; it is only certain that in some way it produces different effects in different cases.

627. It IS ALSO CERTAIN that the exertion of this influence, either upon or from the cells, is attended by a change in them, requiring their nutrition in harmony with their action.

628. THE GANGLIA MUST ALSO BE CONSTRUCTED of fine fibres of sinewy tissue woven through and around them, and if they have any free surfaces liable to friction they must be covered with secretory tissue; and to

623. What are —? 624. How are —? 625. How —? 626. Is —? 627. What is —? 628. How must —? What tissues in —?

supply nutrition to the very active gray part, an almost
infinite number of capillaries must be woven through
it. Thus Ganglia require three tissues for their construction:

$$\text{GANGLIA} = \begin{cases} \text{Nervous } T. & \begin{cases}\text{Gray,}\\ \text{White.}\end{cases}\\ \text{Sinewy } T.\\ \text{Secretory } T. \end{cases}$$

629. THE RELATIVE POSITIONS OF THE GRAY AND
THE WHITE TISSUES is not a uniform or essential matter; sometimes the gray is external, sometimes internal,
sometimes interlaid; thus the color of the Ganglia will
differ.

630. THE SIZE OF GANGLIA varies from that of a pin's
head to the weight of a pound, and their form is equally
variable, evidently depending on the position in which
it is convenient to have them located.

631. SOME OF THE GANGLIA HAVE free surfaces,
and some of them are merely surrounded by a connective, sinewy tissue, supporting and retaining them
in place.

632. Nothing is uniform, nothing therefore IS ES-
SENTIAL IN GANGLIA except a proper number of perfectly
constituted nerve-cells connected with nerve-fibres, and
supplied with a sufficient quantity of good Blood to sustain their nutrition, upon which their continued action
depends.

*Remark.*—We will now describe the different ganglia, when the
great facts of the preceding paragraph will become still more apparent.

633. THE GANGLIA ARE CLASSED as Cranial or Brain-
al, Spinal, and Sympathetic.

*Remark.*—THE GANGLIA ARE OFTEN CLASSED as Cerebro-spinal and
Sympathetic, and the former subdivided: this is well except as to name,
which should be Cranio-spinal, for all the ganglia do not belong either
to the Cerebrum or Spinal cord, some forming the Cerebellum, &c.

## *Cranial Ganglia.*

634. IF THE UPPER PART OF THE SKULL AND ITS LIN-
ING BE CUT THROUGH AND LIFTED FROM THEIR PLACES, the
brains will be presented to view, standing up, a jelly-like,
tremulous mass, with a surface perfectly smooth, glairy,
and well fitted to the portion of skull removed. (See
Figs. 45 and 46.)

635. AN EXAMINATION WILL DETERMINE that the
surface under view is an exquisitely beautiful specimen
of secretory tissue, composed of a single layer of thin,
scale-shaped, but not dry cells, upon a basement membrane supported by a mere shadow, so delicate is it, of
sinewy fibres.

636. This SPIDER'S WEB MEMBRANE (ARACHNOID)
extends over the general surface of the brains, without
following its smaller indentations, passing across from
point to point, being parallel to the inner surface of the
lining of the skull, with which in life it is in contact. At
the indentations of the brain the end of a small tube can
be adroitly inserted, beneath, and air blown under it,
raising it up and bringing it into view very beautifully.

637. *Remark.*—THE OUTER SURFACE OF THE BRAIN AND THE INNER
SURFACE OF THE LINING OF THE SKULL are precisely alike, and at various
points will be found to be continuous; and the idea is usually expressed
by saying the covering of the brain is reflected on the lining of the
skull, or the inner layer of the lining of the skull is reflected over the
brain.

638. WHEN THE ARACHNOID IS REMOVED, the surface
below is found to be very uneven, corrugated, or convo-
luted, like the surface of a peach-stone, the eminences
being called convolutions, and the indentations being
called anfractuosities. Near the surface THE ANFRACTU-
OSITIES ARE GROOVES, but a little deeper their sides
touch, as represented in the Figs.

639. THE SURFACE OF THE CONVOLUTIONS AND AN-

FRACTUOSITIES is a network of minute capillaries supported by a small number of delicate sinewy fibres called the pia mater.

640. FROM THE *pia mater* STILL SMALLER CAPILLARIES EXTEND INTO THE STRUCTURE BELOW, supplying it still more intimately and abundantly with nutritious blood. On the other hand, veins lead from the *pia mater* directly into the sinuses or large veins of the lining of the skull, in order that the Blood may flow rapidly away, and do not follow back a more tortuous path alongside the arteries, as is the case elsewhere; this shows the importance of a free passage from the brain of its waste material.

641. A VIEW OF THE BRAIN FROM ABOVE SHOWS it divided on the central line by a deep fissure, in which is the membrane called the falx (36, Fig. 160), into what are called the right and left hemispheres, or right and left Brains.

642. AN EXAMINATION OF THE SIDES OF THE FISSURE lying against the falx, DETERMINES that the convolutions and anfractuosities exist there as well as externally.

643. THE CONVOLUTIONS ARE NOT EXACTLY ALIKE in the two sides, much less in two persons; and thus is sustained what was said, that form, etc., are not essentials.

644. By removing the side of the skull, and VIEWING THE BRAIN AT THE SIDE, IT WILL APPEAR to be constructed of two parts, one, about one eighth the size of the whole, being situated under the back part of the other, and separated from it by a partition or shelf. The larger is called the cerebrum, and the smaller the cerebellum.

645. THE CEREBELLUM has a more even external surface than the cerebrum, and a shallow groove at its back part, on the middle line, occupied by a membrane corresponding to the falx.

---

640. Why do —? 641. What does —? Describe falx, Fig. 160. 642. What does —? 643. Where are —? 644. What —? 645. What said of —?

646. THE CEREBRUM, VIEWED AT THE SIDE, EXHIBITS by its form three divisions, called the anterior, middle, and posterior lobes, only superficial, not corresponding to any uses, and only serving as convenient designations.

647. Below, about the centre of the cerebrum, and at the front of the cerebellum, the spinal cord (Fig. 45) is seen, called, while in the cranium, THE MEDULLA OBLONGATA (oblongated marrow).

648. IF THE CRANIUM HAD BEEN INVERTED, AND THE LOWER PART REMOVED, many nerves, their sheath, and arteries must have been severed, and the spinal cord cut across, and instead of the general spherical form of the upper surface, a very uneven general surface of the cerebrum would have been presented, as in the subjoined figure, representing the base of the brains turned upward and the cushion (subarachnoidean texture) removed.

FIG. 188.

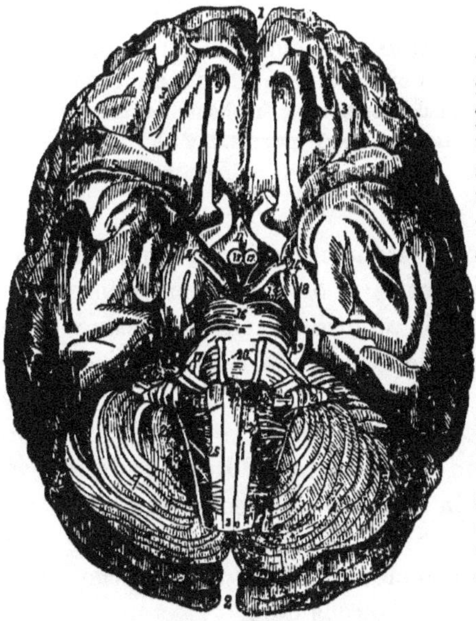

Fig. 188. In this view, the cerebellum, 7, is seen above the back part (posterior lobe) of the cerebrum and the shelf between them; the oblongata, 30, is in view; also the pons, 16, and the commencements of 12 pairs of nerves; 1, 2, is the anterior and posterior part of the deep fissure nearly dividing the cerebri; 3, anterior, 4, middle lobes; the convolutions of the under surface of which are evidently numerous, as they also are above the cerebellum.

646. What does —? 647. What is —? 648. What would have been presented —? Describe Fig. 188.

649. The contents of the Cranium, viewed on all sides, present the right and left cerebrum, cerebellum oblongata, Pons, and commencement of twelve pairs of nerves.

650. The Cranial ganglia, commonly called "the Brain, the seat of the intellect, the will, the sensations, and the emotions" (Leidy), viewed externally, form a large oval, spheroidal mass, shaped much like a duck's egg, with a third of the under part of the smaller end removed, and, with the enveloping membranes, completely fill the Cranium.

651. The Brain varies in size and form in different persons and at different ages; "the average weight in males is fifty ounces; in females, forty-five; the maximum sixty-four, and the minimum twenty."—*Draper.*

652. *Remark.*—It does not necessarily have the same form when having the same size, for the parts of it are differently proportioned in different cases. The cerebellum averages one eighth the weight of the whole; it may be one third or one twelfth. Nor, if the proportional weight of the parts and that of the whole are the same in two brains, is it necessary that the particular forms of each should be the same. It must be remembered THAT THE ONLY ESSENTIAL THING FOR EFFICIENCY IN NERVOUS CENTRES are cells, communicating fibres, and a copious supply of good Blood.

653. In early life the elasticity of a frame renders other protection against jars of the Brain unnecessary; but as life advances, in addition to the increasing quantity of marrow in the bones, the Brain requires something more.

654. The arachnoid membrane beneath the Brain increases in strength by addition to its sinewy fibres, which, also in the form of areolar texture, grow in between the arachnoid and pia mater, forming what is called the subarachnoidean areolar texture, the areolæ of which are filled with serous fluid.

---

649. What said of —? 650. Viewed externally, what are —? 651. How —? 652. What are —? 653. What does brain require —? 654. What said of —?

655. Upon the subarachnoidean tissue, as upon a hydrostatic bed, THE BRAIN IS PROTECTED from the least effect of the jars sent up through the body. In old age, in the thickest part, this has become an inch thick. (See *a*, adjoining Fig.)

656. THE BRAIN IS FULLY PROTECTED by the curved, arched, and irregular form of the skeleton, and by the flexures of the joints by which jars are dispersed; by the cartilages of the joints; by the spongiform structure of the bones and the marrow they contain; by the subarachnoidean cushion; by the arched

FIG. 189.

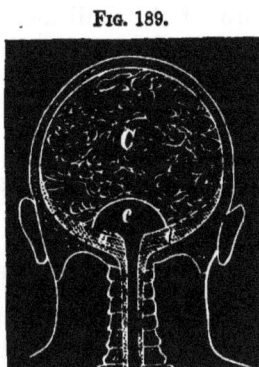

FIG. 190.

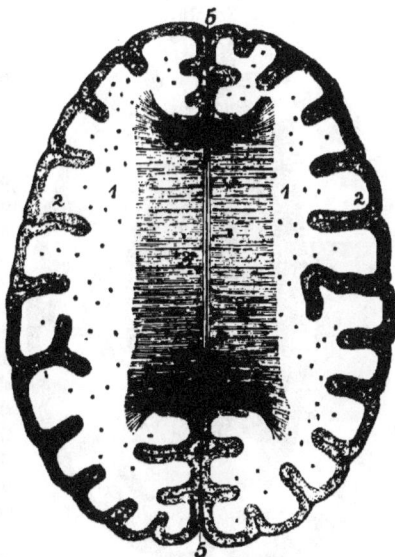

Fig. 190 represents the cerebri sliced down to the upper surface of 4, the bridge (corpus callosum) of fibres that extends across from one to the other. 5, 5, are the deep fissures behind and before, that extend up between the parts removed; the bottom of the middle part of the fissure is represented lengthwise the centre of 4. This fissure is occupied by the falx, the lower edge of the central part of which touches 4, the ends sinking down at 5, 5. 1 is the white tissue, its fibres being interlaced by fine sinewy fibres; the dots show sections of a few capillaries extending among them; 2, the gray tissue, the dark line indicating the anfractuosity or the division of capillaries between touching portions of the surface of the gray Tissue.

Describe Fig. 189. 655. From what—? 656. Enumerate the means by which—. Describe Fig. 190.

12*

form of the skull and the different constitutions of its three tables; by the membranes within and on the outside of the skull; by muscles; by the skin, and by the hair.

657. IF A PORTION OF THE UPPER PART (Cerebrum) OF THE BRAIN IS SLICED OFF, the gray external tissue is found to be only about an eighth of an inch thick, within which the white tissue alone is found.

658. IF THE GRAY SUBSTANCE SHOULD BE REMOVED FROM THE WHITE and spread out, it would present a figure several times larger than it now does; that is, the pia mater is several times the extent of the arachnoid.

659. IT WILL BE OBSERVED that in the fissures the convolutions are proportionately as numerous and the anfractuosities as deep as described in ¶ 642.

660. THE PROBLEM SOLVED BY THE ARRANGEMENT

FIG. 191.

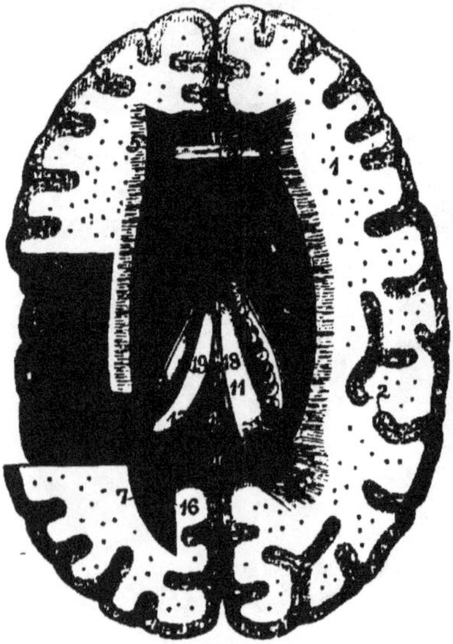


Fig. 191 represents the brain sliced a little lower than in the preceding figure. 1, the white tissue; 2, the gray; 3, 4, the front and back portions of the bridge or corpus callosum, the middle portion being removed and exposing the ventricles, in which, and forming the sides and floor of which, ganglia are seen. Portions of the brain are cut out to show 7 and 6, extensions of the ventricles; 8 is one of the largest ganglia, and from the white and gray substance being in alternate layers, it appears striped, hence its name, striated body (corpus striatum); 19, portion of thalamus; 14, hippocampus; 18, a plexus of capillary vessels (the choroid plexus).

---

657. What —? 658. What extent of surface could be shown —? 659. What — in fissures? 660. What is —? Describe Fig. 191.

OF THE GRAY TISSUE evidently is to pack in the smallest space a large amount of gray tissue, the cells of which shall be in connection with white fibres.

661. IF A HORIZONTAL SECTION THROUGH THE LOWER PORTIONS OF THE CEREBRI BE MADE, or if a perpendicular section of the brains be made on the line of the fissures, or if the corpus callosum be cut out, numerous small ganglia will be revealed as located in the lower central portions of the Brain and in front of the cerebelli.

662. THE CENTRAL GANGLIA CAN ALSO BE EXPOSED by raising up and turning forward the back part of the cerebri, and cutting a few fibres that confine them.

663. THE ARRANGEMENT OF THE GANGLIA CAN BE BEST UNDERSTOOD by constructing the Brain from below; beginning with the spinal cord, where it enters the cranium, ganglia of small size are added to it, enlarging its size into the oblongata; to the back part are added the cerebelli, from the right to the left of which, around the upper part of the oblongata, extends the pons (16, Fig. 188), above which other ganglia are placed, some of the fibres from the cord and nerves extending into them all, and into the cerebelli, while the remainder extend upward and forward into the great ganglia, the cerebri.

664. ACCORDING TO THE SUPPOSITION, THE CEREBRI WOULD BE LIKE two great pear-shaped masses, and the whole chain of ganglia would present a very irregular outline, not susceptible of protection, and altogether uncouth.

665. Hence THE MASSIVE CEREBRI ARE, so to speak, FOLDED BACK, over, upon, and by the side of other parts, slightly overhanging the cerebelli, and quite covering in the central ganglia.

666. WHERE THE SURFACE OF THE CEREBRI COME IN CONTACT WITH THE CENTRAL GANGLIA, they do not adhere to them, but both surfaces are free, and moistened

---

661. What —?  662. How —?  663. How —?  Construct the Brain from below by successively adding ganglia.  664. What —?  665. How—?  666. What said of —?

with the same fluid as if they had not been thus placed. These places are called ventricles.

667. *Remark.*—Dropsy of the Brain is apt to be a collection of the fluid in the ventricles; it may be outside the Brain, or it may be in both places. If outside the Brain, it can be reached by tapping through the skull, which is a simple operation.

668. WHITE FIBRES (commissures) EXTEND between the cerebri, cerebelli, and other ganglia, and between different parts of all the ganglia, by which communication is established and influences exerted, the whole being so woven together that *intellections, emotions, sensations,* and *volitions,* can harmoniously and reciprocally act on each other, as occasion may require.

669. THE CEREBRI are the organs of the Intellect, but by what method they act is not known.

670. The central ganglia of the brain are organs through which sensations are caused, and hence they are called THE SENSORIUM. A part of these ganglia are probably concerned in producing emotions, since they can be excited despite the will by certain substances swallowed or inhaled; and substances that excite emotions produce unusual circulation of Blood about some of these ganglia.

671. THE CEREBELLI are smoother at their surface than the cerebri, and in this respect are like the central ganglia. Cut across, they present in their centre, upon section of either side, the white tissue branched like a tree, hence called tree of life (arbor vitæ).

672. THE USE OF THE CEREBELLI is to harmonize the action of all the muscles of the body, as is especially necessary in walking; and by some they are also thought to be the sensatory ganglia of muscular sensations; but this is not certain, as they may perform the office of influencing contractions without necessarily exciting sensations.

## Spinal Ganglia.

673. THE SPINAL CORD IS CONSTRUCTED of an external sheath, the surface of which is a continuation of the arachnoid of the under surface of the Brain. Adjoining this is found a continuation of the subarachnoidean areolar texture, filled with the same fluid as that of the Brain, next to which is a continuation of the pia mater. In other words, the spinal cord is but a prolongation of the lower portion of the Brain, which is an enlargement of the cord.

674. THE SPINAL CORD IS CONSTRUCTED INTERNALLY of white tissue enclosing gray, the latter, when cut across, showing the form of a quarter moon, the horns being near the surface of the cord, where the nerves enter or leave, thus dividing the cord on each side into three parts, called the front, middle, and back columns, while two grooves at the centre of the front and back face of the cord indicate the line of its division into its right and left halves, or right and left cords.

675. THE GRAY CENTRAL PART OF THE CORD CONTAINS millions of cells, with some of which only the white fibres that enter and leave are connected: other cells are connected with white fibres that enter and leave, and also with those extending up to the cranial ganglia, while others again are connected with fibres that either enter or leave, and with those that extend to other cells in the cord or in the Brains.

676. THE WHITE PART OF THE CORD contains some fibres that extend to or from the Brain, and some that extend to cells in the cord, and some that extend between cells of the cord.

677. ON ALL THE FIBRES CONNECTED WITH THE BACK PART OF THE CORD there is found, a short distance from the cord, but within the canal, a small quantity of gray

---

673. How —? 674. How —? 675. What does —? 676. What said of —?
677. What found —?

tissue, called the ganglion of the posterior or sensatory root of the nerve. On all the sensatory nerves or sensatory parts of nerves extending from the cranium, a similar ganglion is found. Hence, these ganglia are thought essential to sensation, but in what manner is not known: some of the nerves terminate in them.

678. THE USES OF THE SPINAL CORD are twofold: 1st, it is a large bundle of nerves, connecting the Brain and various other parts of the body; and 2d, it is a chain of nervous centres, causing action in the various parts with which it has connection.

679. *Illus.*—IF THE SPINAL CORD OF A FROG IS CUT ABOVE HIS HIPS, and his hind-foot is pricked, it will be immediately moved; but no pain is caused, since he makes no effort to remove his body, as he will do if his fore-foot is pricked. Accidents have injured the spinal cord in cases where, if a feather was drawn across the sole of the foot, motion would be caused; yet it was not by the will of the patient, for he felt neither the feather nor the motion. The feather through the nerves affected the centres in the cord, and they sent down an influence that caused the muscles to contract.

## *Sympathetic Ganglia.*

680. THE SYMPATHETIC GANGLIA DIFFER FROM THE CRANIO-SPINAL in these respects: the latter are protected within a bony case, while the former are not; they are very numerous, but all very small, while some of those of the Cranium are very massive.

681. THE SYMPATHETIC GANGLIA, WITH THEIR NERVES, FORM a double chain on each side of the spinal column, extending, in fact, from the head to the extremity of the coxcyx. Several are found in the head, three in each side of the neck, and one near each exit of the nerves from the remaining length of the spinal column.

682. THE SYMPATHETIC GANGLIA CONNECT with each other, and with the spinal and cranial nerves; they

Fig. 192.  Fig. 193.

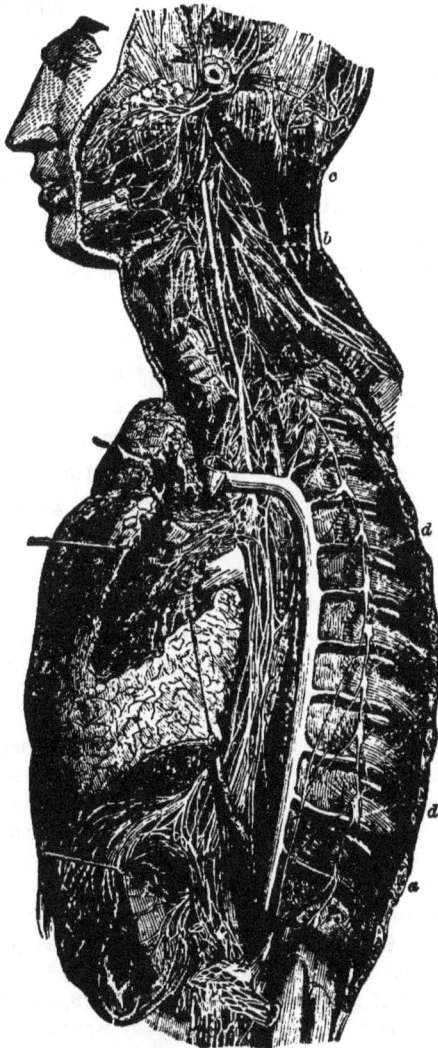
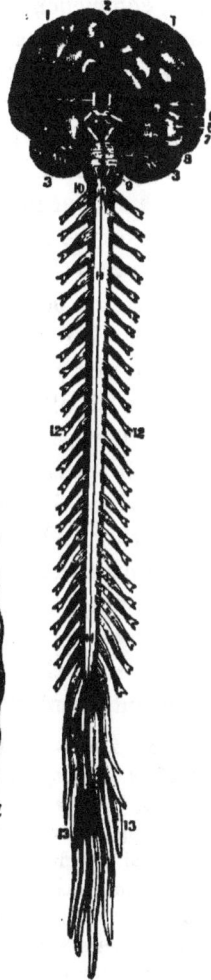

Fig. 192 represents a side view of the face, neck, and upper part of the abdomen. *a*, section of the ribs, being made at the point of their greatest curvature backward; the left lung, *L*, and stomach being cut open and drawn forward and to the right to exhibit *d, d,* sympathetic ganglia and nerve, the divisions of *b*, pneumo-gastric, or tenth nerve, and the plexuses they form; *c*, facial nerve.

Fig. 193, the brain and nerves, with ganglia on their posterior roots.

also, commencing with those of the neck, give out nerves that lead from each side downward and toward the centre till, meeting, or nearly meeting, they form inextricable plexuses, associated with ganglia, called the central ganglia, and having specific names, cardiac, solar, &c.

683. FROM THE CENTRAL GANGLIA and plexuses, nerves extend in great numbers, especially following the course of the large arteries to the organs of the trunk, and, as it is thought, extending with the arteries into all parts of the body.

684. It CAN NOW READILY BE COMPREHENDED that with so many nervous centres, associated by millions of fibres with all parts of the body, all the combinations and sequences of action that can be desirable, are produceable; that action without and with sensation can be provided for; and that the body is, to a degree, a self-regulating and also superintended machine, so arranged that a cause acting on one part may, through a common centre, affect a very distant part, which again may exert influences upon half a dozen parts, all of which may concentrate their influences on the first part, or still other intervening steps may be necessary.

685. *Illus.*—A PARTICLE OF DUST IS INHALED, and proves annoying by influence through a sensatory nerve; but that is not sufficient for its removal. Another centre is at the same time acted upon, and influences from it gush down upon the muscles of inspiration, contracting them, succeeded by influences closing the mouth and windpipe, which acts exert their influence, and the expiratory muscles, contract, followed by the sudden relaxation of those closing the windpipe, and continued contraction of those closing the mouth, and the air is forced rapidly through the nose, sweeping away the obnoxious particles. At the same time other influences or branches of the same influence are exerted upon the gland that supplies the tear-fluid, which it pours copiously over the eye. The fluid, coursing rapidly to the inner angle, finds its way down the ducts into the nose and helps wash away the obnoxious matter. Thus horseradish, mustard, and other volatile, irritating substances, act on the lining of the back part of the nose, and produce a corresponding series of effects.

## SECTION V.

## *Organs of Sense.*

**686.** *Remark.*—After a description of the Ganglia, it is natural to take up a description of those organs by which the ganglia are acted upon or excited to action.

**687.** The organs of sense are, figuratively in case of all of them, and literally in case of touch, the hands by which the ganglia reach out and grasp the various properties of the objects constituting the external world, and obtain the elementary knowledge needed by the mind to work out the facts of the physical constitution of the universe.

**688.** The organ of sense is not first formed, and afterward its nerve, and then its ganglion, but inversely; the ganglion is formed first, and reaches out an appropriate nerve, and the organ of sense is found at its extremity.

**689.** *Remark.*—In case of deformities, parts will be defective in the order stated: first, the organ of sense; secondly, the nerve; and lastly, the ganglion. If it is wanting, the other two will surely be.

**690.** *Remark.*—The simplest of the organs of sense should be first described, particularly as it will be found an element in some of the compound organs of sense.

## *Muscles as Organs of Sense.*

**691.** The Muscles are primarily organs for the production of mechanical motion, without reference to the direct acquisition of knowledge. But their action in that respect could not be regulated without the degree of every contraction was instantly known by the mind; hence it is necessary that every voluntary contraction be attended by a corresponding sensation.

**692.** *Illus.*—What clumsy work people would make in their speech unless every contraction of the muscles of speech, after being ordered, and while taking place, was reported to the mind! It would be necessary to wait till the voice was heard before it would be known whether the required contraction had exactly taken place.

693.   THE ACTION OF THE MUSCLES IS SO IMPORTANT to their health, and that of all parts of the body, it is ALSO NECESSARY that they should produce sensations of discomfort when improperly used, either too much or too little, and sensations of comfort when they are properly used or allowed their required rest.

694.   The Muscles are doubtless, or at least can be, greater SOURCES OF COMFORT OR DISCOMFORT than all other parts of the body taken together.

695.   But THE MUSCLES, AS ESPECIAL ORGANS OF SENSE for the acquisition of knowledge, are also of more advantage than all the other organs collectively.

696.   *Remark.*—In fact, it will be seen that two of the most important portions of knowledge obtained through the eye and ear are dependent on sensations excited by muscles used in their construction.

697.   THE MUSCULAR SENSE IS USED in ascertaining the densities of objects, their force, configuration, size, and the adhesiveness of surfaces, to measure distances, angles, etc.

698.   *Illus.*—The muscles of the arm will point out the position where an amputated part of the body would be, if it had remained, as accurately as the most exact measurement would do.

699.   THE MUSCULAR SENSE COMES INTO USE the earliest of the senses, and fades the earliest: the knowledge it gives us being our first requirement, and most rudimentary, its utility is soonest completed.

700.   HOW THE NERVES OF SENSE COMMENCE IN THE MUSCLES is not yet a settled question, nor how they are acted upon.   Whether the simple pressure of the contracting muscular fibre upon the nerve-fibre is all that is necessary, or whether the commencement of the nerve-fibre is furnished with a corpuscle that is acted upon by

693 — what — ?  694. What are — ?  695. What are — ?  697. How is —?
698. What can muscles of arm do?  699. When does —?  700. How do —?

the contracting muscular fibre, and then exerts an influence through the nerve, is not known.

701. *Remark.*—It is not positively certain that there are two kinds of nerves connecting between a muscle and the centres, but it is supposed there are, because it does not appear probable that the motor influence can be acting toward the muscle, and the sensatory from it through the same nerve-fibre at the same time; and if the back roots of the spinal nerves be cut, reflex motion cannot be excited by acting on the muscle. This seems conclusive, although all the nerve-fibres in the muscle appear alike; motory and sensatory cannot be distinguished there.

702. All that is known with certainty is that every contraction and relaxation of any part of even the smallest fibrilla of the voluntary muscles can instantly produce a sensation, and also other exhilarating or depressing effects upon, or through the effect of, many nervous centres, waking the whole body to life and joy, or overwhelming it with gloom and misery.

703. It is our duty to study and observe the effects of over and under exercising the muscles, the fatigue, exhaustion, and dragging sensations produced by over toil, and the discomfort, unrest, uneasiness, and impulsiveness, that is caused by a want of muscular exercise. A regularly, healthily exercised boy is very little inclined to mischief.

### The Skin as an Organ of Sense.

704. Forming the surface of the body, it is necessary that the skin be endowed with the power of exciting sensations. It must be an advance picket-guard, to report upon the presence of objects in contact with the body, and upon the temperatures to which it is exposed, and must excite a sudden alarm if injured in any way.

705. The Skin is constructed of three layers,

cuticle, basement membrane, and dermis, sometimes called the true skin.

706. THE DERMIS IS CONSTRUCTED of sinewy fibres, woven most densely near its surface and quite loosely below, where, in fact, the fibres pass into the form of areolar tissue, in the areolæ of which, and in some of the larger meshes of the skin, fat-cells are found more or less abundantly.

707. THE DERMIS is that part of the skin that is termed leather, being tanned dermis, the hair, cells, nerves, etc., being removed in the process.

708. THE SURFACE OF THE DERMIS IS RAISED in ridges of small points called papillæ, in the most sensitive parts, as may be seen at the ends of the fingers and elsewhere.

709. THE MESHES OF THE DERMIS ARE OCCUPIED by glands, lymphatics, blood-tubes, a loop of which extends into every papilla, and nerves which terminate in, or more properly commence in, nervous corpuscles in the papillæ.

FIG. 194.

Fig. 194 represents the capillaries near the surface of the dermis. The loops that are in the papillæ are easily recognized.

710. THE BASEMENT MEMBRANE IS SPREAD over the dermis and all its eminences, and from it soft cells constantly grow up, producing a deliciously soft, elastic cushion to protect the exquisitely sensitive nerves from too great pressure; the slightest touch on them exposed is exceedingly painful.

711. THE THICKNESS OF THE DERMIS VARIES in dif-

706. How is —? 707. What is —? 708. How is —? 709. How are —? Describe Fig. 194. 710. How is —? 711. How does —?

ferent parts from a twenty-fourth to a sixth of an inch; the basement is one thing everywhere; the thickness of the cellular layer (cuticle) varies very much, as does the dryness of its cells; in some places, as the palm, heel, &c., they become quite horny by the time they are at the surface, while in other places the cells retain considerable of their moisture and pliability at the very surface.

*Remark.*—CORNS ARE PRODUCED by pressure upon cells at some point, by which the fluids they contain are caused to exude, leaving them one under another mere horny scales matted together, pressing on the nerves and even down into the dermis. A sharp-pointed instrument can be used to take them out without pain, as cutting the cuticle devoid of nerves cannot cause pain; and, if the instrument is carried no deeper than the corn, it cuts no nerves. If the pressure is not renewed the corns will not return; if it is, they will.

712. TOUCH IS the name given to the sense, when used in a general or passive manner; when used intentionally, and for special purposes, it IS CALLED TACT, and also implies some training and acquired skill.

713. TACT AND THE MUSCULAR SENSE often work together to get knowledge for the mind, that neither alone could well obtain.

714. THE ORDINARY SENSATIONS PRODUCED THROUGH TOUCH are neither particularly pleasant or unpleasant. They may be called neutral, since they are neither essential nor injurious to our welfare; when they are either, the skin rouses itself and is the cause of intense sensations; entrancing the mind with pleasure, as when it receives a refreshing bath, or the gentle breeze of summer is wafting away its accumulating heat; or irritating and vexing the soul most unpleasantly, as when the cold, drizzling, raw air chills it through; in all cases proving equally a precious friend and worthy of being maintained in the highest condition of health and activity.

715. OUR WELFARE REQUIRES that attention be given to provide seats that are easy, beds that are comfort-

able, apartments of a proper temperature, neither too cool nor too warm, baths for the skin, brisk rubbing to circulate the blood freely through it, and clothing appropriate for all seasons and all kinds of weather; that we should walk and ride and exercise in various ways in the open air, where the skin can receive agreeable influences and produce numerous and lively sensations.

*Remark.*—The skin and the care it should receive will again be a proper subject for instruction when the eliminating organs are considered.

## The Mouth as an Organ of Sense.

716. THE MOUTH IS PRIMARILY an organ for preparing the food for the digestory action of the stomach, but that duty is changed into a pleasure by the addition of the nerves of taste.

717. NERVES OF TOUCH EXTEND from every part of the mouth to the nervous centres, in the same manner as they do from the skin, generally; they are called nerves of common sensation.

718. THE NERVES OF TASTE COMMENCE in the papillæ, on the sides and tip of the tongue, in those about its roots, from some of those in the sides of the back part of the mouth, and from some of those of the soft palate.

719. THE NERVES OF TASTE OR TOUCH MAY BE PARALYZED without the other being affected, or either may be in either half only.

720. SOME SUBSTANCES do not have the power of affecting the nerves of taste, and are called tasteless; some produce taste in some persons and not in others; some produce slight effects on some and strong taste in others.

721. *Remark.*—Whether substances produce the same identical kind of sensation in all persons who taste them is uncertain; it is not probable they do.

722. THAT SUBSTANCES MAY BE TASTED they must be

---

716. What is —? 717. From what part of the mouth do —? 718. Where do—? 719. How may —? 720. What effect on taste of —? 722. What necessary —?

dissolved and act through the cellular and basement layers that cover the papillæ, upon their tactile corpuscles, in which the nerves of taste commence.

*Remark.*—How they act, and why one substance produces a different effect from another, is not known.

723. THE TEETH SERVE AN ADMIRABLE PURPOSE for grinding the substances to be tasted, and the SALIVA is equally applicable for dissolving them.

724. THE SENSE OF TASTE IS in some way ASSOCIATED with that of smell, since while the nose is closed many substances become tasteless. The *rationale* of this is not understood.

*Remark.*—The value of this sense will be further illustrated when the apparatus is made up.

## The Nose as an Organ of Sense.

725. THE NOSE IS PRIMARILY for the passage of air, and nerves of common sensation extend from all its parts, the nerves of smell being superadded to certain portions.

FIG. 195.

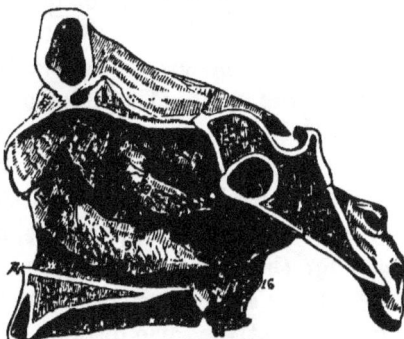

Fig. 195 represents a perpendicular section of the bones of the nose a little to the right of the central partition, exhibiting, 14, the front point of the floor, and 4, the roof of the nose, between them being the passage through, in the outside of which is hanging 9, the inferior turbinated or spongy bone, slightly coiled, as its name signifies; 8, the middle; 7, the superior turbinated or spongy process. At 3, a sinus or cavity in the sphenoid is seen opening at 10; 4 is the frontal sinus, opening into the upper part of the nasal cavity; 12 is the opening into a sinus in the upper jaw-bone directly above the eye-tooth. (Other cavities not shown are in the roof of the nose, opening into it.)

---

723. For what do —? What is the use of the —? 724. How is —? 725. What is —? Describe Fig. 195.

726. THE CAVITY OF THE NOSE is formed of parts of several bones complemented by cartilages. A partition extends through the centre, dividing it into two cavities or nostrils, the central sides of which are smooth, but upon the outsides of each a bone and two processes (Turbinated) are constructed, that partially fill each nostril, and very much increase the extent of surface presented to the action of air; especially in some animals that hunt their prey, in which the bone is even coiled.

727. CAVITIES CALLED SINUSES, in four different bones, open into each nostril, as indicated in the description of the preceding figure. Their use is unknown. They are lined by a delicate membrane that sometimes pours out a watery fluid, which, when the head is bent forward, will be freely discharged into the nose. Sometimes the fluid is, or will become, viscid. Sometimes the membrane lining these cavities becomes inflamed, as when a person has a cold in the head, and in other ways proves troublesome.

728. THE SKIN LINING THE NOSE is a continuation of that covering it, only thinner, and the cuticular cells are changed to ciliated epithelial in the lower part, and to thin, pavement-shaped cells above, where the nerves of smell are found. (The lining of the sinuses is continuous with that lining the nose.)

729. THE NERVES OF SMELL DESCEND INTO EACH NOSTRIL in two bundles, through the cribriform or sieve-like holes in the roof of the nose, one bundle being destined for the lining covering the turbinated processes (see 1 in the succeeding figure), the other bundle extending for about the same distance down the lining of the partition. (See Fig. 196.)

730. ODOROUS PARTICLES dissolved in the air and floated or snuffed through the nostrils, striking upon the delicate surfaces of their upper walls, permeate the thin

---

726. How is — constructed?    727. What said of —?    728. What said of —?
729. How do —?    730. What effect produced by —?

Fig. 196.

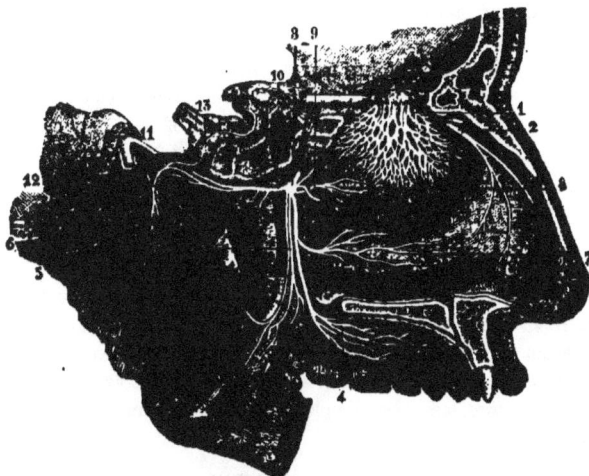

Fig. 196, a section of nose a little to the left of central partition; 2, nerve of smell; 2, 9, 8, 7, 4, nerves of common sensation; 5, 6, motor nerves of soft palate; 8, speno-palatine ganglion; 12, deep petrous connection between the ganglion and the carotid plexus, and thus with the sympathetic.

layer of cells that protects the corpuscles in which the nerve-fibres of smell commence, and some change that is wrought in them produces, as an ultimate result, a sensation of smell.

731. From the DIFFERENT EFFECTS PRODUCED BY DIFFERENT SUBSTANCES, as in case of taste, their characteristics may be learned, and they may be distinguished.

732. MUCH ENJOYMENT IS ALSO ATTAINABLE THROUGH THE SENSE OF SMELL by judicious treatment. It is the province of the sense of smell to enjoy the sweet odors of flowers, the freshness of the cultivated fields, and the fragrance of the wildwoods. Nor are the arts of the perfumer to be disesteemed. The savor of preparing food quickens the appetite, heightens the pleasure of eating, and facilitates digestion.

Describe Fig. 196. Where do the nerves of common sensation, 2, 9, 8, 7, 4, terminate? 731. What said of —? 732. How is —?

13

733. *Inf.*—It is of great consequence to have food prepared in a savory manner: many a plain dish may thus be made exceedingly attractive.

734. *Inf.*—Rooms should be furnished not only with flowers, bouquets, and nosegays, but, as this name implies, they should be fragrant. A party should not be given to gratify the sense of taste only or chiefly; but the other senses should also be addressed, and flowers, sweet-scented flowers, should abound, as a source of refined entertainment.

## The Eye.

735. *Remark.*—The eye is usually thought to be a complicated organ in structure and action, because many acts have been attributed to it that it does not perform, and because it has not been examined in a natural manner. The process of seeing, so far as the eye is concerned, and its structure, are very simple and easily understood, if correctly examined.

736. The eye is constructed of several distinct classes of parts, each adapted to its peculiar purposes, that should be distinctly described.

737. The parts of the eye may be classed, 1st, as the nervous; 2d, as those adapted to cause the light to act upon the nervous part; 3d, as those promoting the movements of the eye; 4th, as those moistening the eye; 5th, as those adapted to protect the eye. Each of these classes of parts may be subdivided.

### Nervous Structures of the Eye.

738. The nervous parts of the eye are constructed of the dura mater, the pia mater, and the ganglionic structure, called the retina, which will be best understood if described in the following manner:

739. The Optic nerve, or nerve of sight, extends from its ganglion in the Brain, through an appropriate opening in the skull at the bottom of the eye-socket. It is formed externally of the dura mater lining the skull,

which forms a sinewy sheath to the millions of nervous fibres it encloses.

740. WITHIN THE SHEATH OF THE OPTIC NERVE EXISTS a very thin layer of pia mater, derived from that of the Brain, and WITHIN THE PIA MATER ARE the true nervous fibres, in bundles, having a small quantity of sinewy tissue wrought around and between them.

FIG. 197.

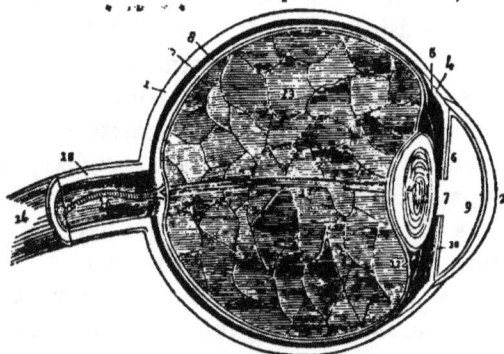

741. THE NERVE (14, Fig. 197) EXTENDS FORWARD an inch or more into the socket, where it expands into the form of the larger part of a sphere, nearly an inch in diameter.

742. THE SHEATH OF THE NERVE (15, Fig. 197) is here thickened and made very dense by the addition of sinewy fibres, and becomes (1, Fig.) the white or sclerotic (hard) coat of the eye, in the front part of which an opening is left, through which the colored part of the eye (6) shows.

743. THE PIA MATER OF THE NERVE is also thickened into (3, Fig. 197) an innumerable multitude of capillary Blood-tubes, supported by a small number of sinewy fibres, having in their meshes a large number of colored cells, intensely black at the inner surface, where the

740. What —? What does Fig. 197 represent? 741. How does —? 742. What does — become? 743. What does — become?

light acts freely upon them, and being a deep brown near the sclerotic. This part of the eye is called the choroid coat.

744. THE CHOROID DOES NOT REACH FORWARD AS FAR AS THE SCLEROTIC, is loosely connected with it throughout by Blood-tubes and sinewy fibres, that become more dense at the front part of the choroid, and more intimately connected with the sclerotic; thus forming the ciliary body (4), in which the capillary vessels are found in greatly diminished numbers. (See ¶ 746.)

745. THE PROPER NERVOUS STRUCTURE OR RETINA OF THE EYE exists in five layers: 1st, a layer of columnar cells (membrane of Jacobi), resting against the choroid; 2d, a granular layer (like that of the ganglia); within that is, 3d, a layer of caudate cells, connected with, 4th, the nervous fibres, covered with, 5th, a layer of structureless membrane, like basement membrane.

746. THE RETINA does not EXTEND QUITE AS FAR AS THE CHOROID, the capillary vessels of which diminish rapidly in number as soon as no longer necessary in large numbers to furnish nutriment to the retina.

747. THE NERVOUS FIBRES OF THE RETINA are millions in number, some terminating almost the instant they enter the eye, in caudate cells near the entrance, others extending to the last row of cells at the front border of the retina.

748. *Remark.*—THE NERVE-FIBRES TURN BACK to connect with the cells forming the layer next behind the fibre-layer. They do not present their points toward the concavity of the eye, as for certain reasons has been represented in several of the figures illustrating the eye, nor do they terminate in points, but in or at the caudate cells, with which they are continuous.

749. ALL THE LAYERS OF THE RETINA are transparent, as can be judged from the fact that the pupil of the eye appears black, owing to our looking through the

pupil quite back to the black choroid. Hence, light can readily pass through the retina to the choroid, and all that reaches it is neutralized, absorbed, killed, and therefore none is reflected; and that which affects the nervous structure, must do it while passing through the structure or by the reaction of the choroid upon it.

750. *Inf.*—There must be A SPOT IN THE EYE WHERE THE FIBRES ENTER, where cells and other parts of the retina, except fibres, do not exist; this spot is devoid of sight, showing that the fibres alone are not sufficient for sight, and that cells are necessary.

751. *Illus.*—Close the left eye, and with the right one look at the left-hand spot, at ordinary reading distance; the right-hand spot will not be seen if the distance is correct; if seen, move the book nearer or farther, and the spot will vanish, and reappear at a greater or less distance. The reason is, that the light from the spot falls on the blind point in the eye. Close the right eye, and with the left look at the right-hand spot, and the same effect will be produced.

752. *Remark.*—THE NERVOUS STRUCTURE OF THE EYE, the protecting sclerotic, the supplying choroid, and the visual retina of five layers, are now complete for the reception of light.

*Remark.*—THE RETINA APPEARS TO BE merely a ganglion of peculiar form, adapted to receive the influence of light.

## Nature of Light.

753. There are THREE PRIMARY KINDS OF LIGHT, red, yellow, and blue, by the proper proportional mixture of which all shades of all colors can be produced. (Pl. 8.)

754. *Illus.*—*Three* parts red and five *yellow*, produce Orange, *eight* parts; *three* red and *eight* blue, produce Purple, *eleven* parts; *five* yellow and *eight* blue, produce Green, *thirteen* parts; while *three* red, *five* yellow, and *eight* blue, produce White, *sixteen* parts. These parts, colors, and proportions, are the natural ones, those of the rainbow, pure and

most beautiful. Black is a sensation produced when no light is acting on the organ of sight. It is not, however, a negative, but a positive sensation.

755. SOME OBJECTS give off, others transmit, and others reflect, one or two, or all kinds of light, while some do neither.

756. *Illus.*—THE SUN is the only source of all kinds of light in perfect proportions. Many objects, candles, etc., give off all kinds, but not in perfect proportions. Some give off but one or two varieties. Glass transmits all kinds well, but the front parts of the eye most perfectly. Some objects transmit one or two, and not more. Looking-glass, a sheet of water, polished surfaces, etc., reflect all kinds. All objects, except " good reflectors," and black, reflect some kinds, but not others. Grass reflects blue and yellow; blood reflects red chiefly.

757. Those objects that give off, transmit, or reflect any kind or kinds of light, are correspondingly CALLED WHITE, RED, GREEN, ETC., OBJECTS. That an object may appear black, it is only necessary that it should neither give off, transmit, nor reflect any light. (See Figs. 10 to 17, Pl. 8.)

758. OBJECTS THAT PRODUCE OR REFLECT LIGHT, EX-CEPT GOOD REFLECTORS, throw it off in all directions in straight lines.

759. GOOD REFLECTORS AND THOSE OBJECTS THAT TRANSMIT LIGHT, throw the light only in certain directions, depending upon their form, their density, and the direction in which light has fallen upon them.

760. *Illus.*—A SUN-GLASS (convex lens), properly exposed to the sunlight, will bend (refract) the light as it enters, and again, as it leaves, in such a manner that all which enters will be brought, just beyond the lens, to a single, very bright point, called a focus.

761. As LIGHT COMING FROM TWO DIFFERENT POINTS to the lens must come from different directions, the lens must bend them to different foci; and every point which can throw light upon the lens will be represented by a focus on the other side of the lens.

762. *Illus.*—If TWO LIGHTED CANDLES are placed at a little distance apart, and a lens before them, the light from them, striking on the lens, will be reflected to two foci, and if a hundred lighted candles were similarly placed there would be a hundred foci.

763. *Remark.*—A CANDLE, though the best, is NOT A FAIR ILLUSTRATION, since its flame is composed of millions of points, from each of which light passes to the lens, through which millions of foci are formed so near together as to cause the points of the flame to appear contiguous; yet each focus is independent in its cause and all its conditions.

764. IT IS THEREFORE POSSIBLE FOR THE EYE TO BE FURNISHED WITH transparent media, by passing through which the light thrown into the eye from any minute point will be gathered to a corresponding focus.

765. *Remark.*—THESE POINTS, FROM WHICH LIGHT COMES TO THE EYE, and which are the true objects of vision, are the minutest distinguishable, and will be called visual points, or visual objects. They are to be carefully distinguished from general objects that, to be sure, may be, single visual points, but are usually composed of myriads of them. A tree, a limb, a twig, a leaf, or a spot upon it, is called, in general terms, an object. Again, visual points, examined under the microscope, are found to be composed of still smaller points, that can thus be distinguished. Still again, some eyes appreciate points smaller than those seen through other eyes, for some can see blue and yellow where others see green: some painters will be found to declare those colors well mixed, when others will see them streaked, and require that they be still more intimately mixed before they will allow that a good green has been produced. We are now prepared to describe

## The Media of the Eye.

766. THE MEDIA OF THE EYE are four: the *Cornea*, the *Aqueous, Humor*, the *Lens*, and the *Vitreous Humor*.

767. THE FRONT OF THE EYE IS CLOSED by a transparent window, 2, called the cornea (horny part, though not composed of horny matter). It occupies about one fifth of the sphere, is more convex than the sclerotic, or is like a section of a smaller sphere inserted in a larger

---

762. What said of —? 763. Why —? 764. What is —? 765. What said of —? 766. What said of —? 767. How is —?

one. The edge is bevelled, to fit the sclerotic, though they are continuous, and the cornea cannot be removed without section.

768. THE CORNEA IS CONSTRUCTED of sinewy fibres, arranged in many layers, so as to be perfectly transparent, devoid of blood-vessels, but containing many cavities, wrought like the lacunæ in the bones, that the white nutriment of the blood may reach all parts. There is a canal around its edge, lined, and serving as a vein.

769. THE CORNEA IS COVERED with the delicate serous membrane lining the lids, which upon the eye becomes very thin and transparent, its surface-cells being scale-like or squamous.

770. THE CORNEA IS LINED with an exquisitely delicate serous membrane, that also covers the front surface of the iris, or colored part of the eye.

771. BACK OF THE CORNEA IS a space lined with serous membrane, just described, divided in two chambers, the anterior, 9, and posterior, 10, by a partial muscular partition, 6, surrounding an opening, 7, the pupil. Both chambers in common are filled with a fluid, the aqueous (watery) humor, that is the second medium of the eye.

772. THE IRIS is the colored part of the eye, with a round opening in its centre through which light passes. It corresponds to the choroid, as the cornea does to the sclerotic, and the covering of the back part of it is filled with dark pigment-cells, which give color to the iris, according to their own number and the transparency of that portion of the iris in front of them.

773. THE IRIS IS CONSTRUCTED of muscular fibres, some forming rings around the pupil, and, by contracting, lessening it, and some radiating from the rings, and, by contracting, enlarging the pupil. The muscles and their action can be seen by watching the eye as a light is brought near it and removed.

---

768. How —? 769. How is —? 770. How is —? 771. What is —? 772. What is —? 773. How is —?

774. THE USE OF THE IRIS is to admit more or less light; hence it will be large in a feeble and small in a bright light.

775. THE CRYSTALLINE LENS, 11, is the third medium of the eye, shaped as shown, and enclosed in its capsule.

776. THE CRYSTALLINE LENS IS CONSTRUCTED of sinewy fibres arranged in layers like coats of an onion, and also in several sections. It is very beautiful; no diamond of the choicest water equals its transparency; and could it be taken from the eye and consolidated with all its primitive beauty, no gem from Golconda's mines would equal its price.

777. THE FOURTH MEDIUM OF THE EYE, 13, the vitreous (glassy) humor, fills all the back part of the eye between the lens and the retina.

778. THE VITREOUS HUMOR IS CONSTRUCTED of the hyaloid (glass-like, from its transparency,) membrane, that entirely encloses the vitreous, lying against the retina; it intersects the vitreous, dividing it into small spaces filled with a watery fluid. Through the centre the central artery, 17, extends to the capsule of the lens. The artery, as shown, is enclosed by the hyaloid.

779. THE FRONT SURFACE OF THE VITREOUS surrounding the crystalline is indented; the indentations, 60 or 70 in number, receiving folds of a membrane, called the ciliary processes, triangular in form, that with the spaces between them occupy the side, 5, of the posterior chamber of the eye. They are also attached to the front part of the choroid, and to the iris near its attachment. Their surfaces and the intervening surface of the hyaloid are pervaded by black pigment-cells.

780. IN THE FOLDS OF THE CILIARY PROCESSES muscular fibres are found, extending backward and forward, which, by their contraction, will draw forward the lens, and also affect the densities of both humors of the eye.

---

774. What is —? 775. What is —? 776. How is —? 777. What is —? 778. How is —? 779. What said of —? 780. What found —?

13*

781   *Inf.*—THE MUSCLES IN THE EYE are three in number: the circular and radiate of the iris, and the ciliary.

## *Effect of the Eye and Light upon each other.*

782. THE USE OF THE MEDIA OF THE EYE is to cause the light coming from different visual points through the pupil to be brought to corresponding points or foci in the retina. The cornea has the greatest effect on the light, the lens considerable, the aqueous and vitreous humors but little.

783. THE EFFECT OF THE LIGHT BROUGHT TO A FOCUS IN THE RETINA is to produce changes in the cells acted upon, different in case of different kinds of light; resulting in corresponding sensations, and also in the perception that the light has come from a certain direction.

784.   *Illus.*—Red light causes a sensation of redness; blue and yellow acting together cause a sensation of green, etc. If a point in the lower back part of the eye is acted upon, it causes a perception that the light comes from above; a point in the right side of the retina causes a perception of light coming from the left, etc.

785.   *Illus.*—IF ONE EYE IS PRESSED WITH THE FINGER, the point on which the light will be focused will be changed, and the direction of the object from which the light comes will seem to change; the object will also appear to move, and, if the other eye is open, will appear double. If both eyes are pressed at the same time, the object will appear double, and seem to move in two directions.

786. THE USES OF THE EYE, then, are to gather the light that from minute or visual points comes into its pupil, to corresponding points in the retina; to excite corresponding sensations of color; and to cause perceptions of the directions of the minute or visual points. These are the complete work of the eye, the only data furnished by it, and all that are necessary; at the same time very pleasurable sensations are excited in connection with the fulfilment of these offices.

787. *Remark.*—A larger office HAS FREQUENTLY BEEN ATTRIBUTED TO THE EYE than it is possible or desirable for it to perform, and when an attempt has been made to explain what of course was unexplainable, it has been found that the explanation itself must be explained. The elements of knowledge gained through the eye are few, very simple, yet complete, and quite sufficient. The mind should not be embarrassed, when studying the eye, with previous theories and dogmas, such, for example, as that there are images in the eye; that they are inverted, etc. (Pl. 8.)

788. THE PERFECTION OF SIGHT DEPENDS UPON having a healthy nervous structure, and upon having the light from each visual object brought with the utmost exactness to its corresponding visual point in the eye.

789. *Experiment.*—IF TWO OR MORE LIGHTS ARE PLACED IN A STRAIGHT LINE before the lens, their foci will not be found in a straight, but in a curved, line, and the degree of curvature will depend upon the distance of the lights.

790. *Inf.*—THE RETINA MUST HAVE an exact curvature corresponding to the foci naturally produced, and the curvature must be changeable to suit objects at different distances.

791. *Experiment.*—IF A LIGHT IS HELD BEFORE A CONVEX LENS, AND ITS FOCUS OBSERVED, it will be found to vary in its distance from the lens, either by the motion of the light toward or from the lens, or by the movement of the lens toward or from the light. If the lens be exchanged for one more or less convex, the distance of the focus will also be correspondingly changed.

792. *Experiment.*—But IF THE LENS IS MOVED IN THE SAME RATIO AS THE LIGHT, the focus will be at the same point; or if the screen or object on which the focus is received is moved correspondingly with the light, the focus will of course fall upon it; or if both the screen and lens be moved according to the needed adjustment, the combined effect will preserve the focus at the same point on the screen.

793. LIGHT FROM DISTANT AND FROM NEAR OBJECTS cannot have its focus at the same point of the retina in the same eye without being changed in some respects.

794. THE MUSCLES OF THE CILIARY PROCESSES are thought to be the means of regulating the position of the lens and the conditions of the humors in such a

manner, that the light from objects within reasonable
distance shall reach a focus at the retina.

795. *Remark.*—THE CURVATURE OF THE RETINA certainly, and perhaps also its distance from the lens, is affected by the external muscles
of the eye, yet to be described.

796. THE EYE CHANGES, WHEN PERFECT, from that
of short to that of long sight without effort, but in long-
sighted persons it cannot be changed to short-sight, and
in the case of the short-sighted it cannot be changed to
long sight.

797. *Remark.*—THE WORDS SHORT-SIGHTED AND LONG-SIGHTED do
not convey a correct idea, for neither is anything more than a perfect
eye, but fails in being something less. The short-sighted is no more so
than any perfect eye, but is not long-sighted.

798. *Remark.*—It is, however, seldom that LONG OR SHORT-SIGHTED
PERSONS ARE RESTRICTED to a single distance, their eyes nearly always
having some power of adaptation, but not sufficient for perfect sight.

799. *Remark.*—In eyes called good there is A GREAT DIFFERENCE
IN THE DEGREE OF ADAPTATION: one person can see at a few inches' or at
miles' distance, while another has a more restricted range. All eyes have
their limit, and those only in case of which it is inconvenient are noticed.

FIG. 198.

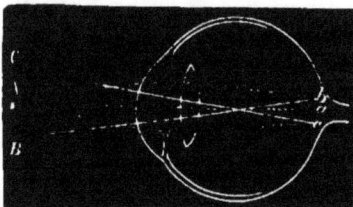

800. IN THE CASE OF
THE LONG-SIGHTED at ordi-
nary distances for reading,
the light from *A*, Fig. 198,
is not brought to a focus at
*a*, as it would be in a perfect eye, but spreads over
the surface from *b* to *c*, on account of which *A* will seem
to occupy all the space from *B* to *C*, since, from the necessary constitution and action of the eye, the light acting on *b* will appear to come from *B*, whether it does or
not, and the light acting at *c* will appear to come from
*C*, and the light acting on all the visual points between
*b* and *c* will appear to come from an equally numerous
series of objects between *B* and *C*.

801. IN THE CASE OF THE SHORT-SIGHTED, at ordinary distances the light is refracted too much, and reaches a focus before it reaches the retina, as in Fig. 4, Pl. 8. But as there is nothing to check the light it passes on, spreading beyond the focus, till it reaches the retina and acts over the visual points from *b* to *c*, and the same sensations and appearances will result as in the case of the equally long-sighted.

802. *Remark.*—Figs. 3 and 4, Pl. 8, represent in colors the same ideas, only more fully, as the light from three points is represented, and it is seen that they overlap each other on the retina, producing confused sensations of color as well as of direction. To a long or short sighted person a point appears to occupy much space, and objects appear blended, indistinct, and of course not in their true colors.

803. *Experiment.*—It is found that the light passing through the outer part of a lens is not brought exactly to the same focus as that passing through the centre.

804. ANOTHER USE OF THE IRIS is to cover the circumference of the lens and also shut out light coming through the circumference of the cornea.

805. It is also found that some colors are not bent by the same medium quite as much as others, and that by putting together SEVERAL MEDIA, each operating differently on different kinds of light, the evil is nearly corrected, and in the eye most effectually.

806. *Remark.*—As the eye, in one position, can take in but a very limited range of vision, it is important that it be moveable to a very large extent.

## *Parts moving the Eye.*

807. AROUND THE EYE a deep, protective, and otherwise useful socket has been formed: narrow at its inner extremity, it widens outward by being funnel-shaped.

808. Part of THE SPACE BETWEEN THE EYE AND ITS SOCKET is occupied by muscles, blood-tubes, etc., but it

is chiefly packed with a firm fat, on and against which the eye rests and turns very securely.

FIG. 199.

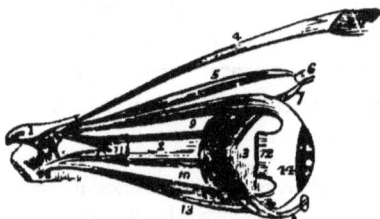

809. SIX DELICATE MUSCLES commence near the inner point of the socket. Four, called the straight (recti), extend, one to each side of the eye, to which they are attached, a little back of the cornea. The fifth (the superior oblique) extends to a loop in the upper, inner, front part of the socket. The round tendon of the muscle, enveloped in a bursa, passes through the loop, turns backward to one side, and downward over the eye, to be attached to the outside of the back part of it under the external straight muscle. The sixth, or elevator muscle, extends above the eye and is inserted in the upper eyelid.

810. THE SEVENTH MUSCLE OF THE SOCKET, the sixth that moves the eye (the inferior oblique), is attached, by one extremity, to the lower inner part of the socket near the nose, and extending outward, backward, and upward under the eye, is attached to it near the same point with the superior oblique.

811. THE USE OF THE OBLIQUE MUSCLES is to draw the eye forward and partially turn it on its axis.

812. THE USE OF THE STRAIGHT MUSCLES is to turn the eye, as their direction signifies, and by excessive contraction to roll the eye.

## The Apparatus for Moistening the Eye.

813. THE MEANS FOR MOISTENING THE EYE may well be called an apparatus, since several different organs are

constructed to work together harmoniously for a given common result.

814. THE TEAR (Lachrymal) APPARATUS IS CONSTRUCTED of a gland to form the tear-fluid, ducts to conduct it to the eye, and ducts to conduct away what does not evaporate. The lids and the Meibomian glands might also be counted as necessary to distribute the fluid over the eye, and also to prevent it from running over the lids.

815. THE LACHRYMAL GLAND IS CONSTRUCTED of two parts (Plate 4), the rounded and the flat, situated just within the socket above the eye, and to the outside of it. The rounded part is about the size of a sparrow's or robin's egg, the flat or thin part being smaller. Six to a dozen minute tubes lead from all parts of the gland, within which an immense number of minute cryptæ or pouches surround the tubes, and pour into them the tear-fluid secreted from the blood. The tubes or ducts open at their lower extremity upon the under surface of the upper part of the lid, and pour the fluid into the eye, over which it is spread by the motion of the lids.

816. THE TISSUES OF THE GLAND are the sinewy in very small quantity, the secretory chiefly, and the nervous. The Blood-tubes of the gland are very large in proportion to its size, though not exceeding a large knitting-needle in diameter: the amount of fluid the gland can secrete in a short time, under strong mental influences, is astonishing.

*Remark.*—The only creature that weeps is Man. Other creatures have tear-glands for supplying fluid to moisten the eye, and in case of injury it sometimes overflows and trickles down; but the flow of tears proper, under the influence of emotions, is only human.

817. TWO BLACK POINTS, TO BE SEEN AT THE INNER ANGLE OF THE EYE by turning out the lids, are the openings of two minute tubes that curve around and connect

---

814. How is —? 815. How is —? Place the finger over the opening of tubes from L duct. 816. What are —? 817. What are —?

with a large one, called a sac, the lower end of which
leads into the lower part of the nose.   (See Pl. 4.)

818.   It is not unusual for A COLD transiently or permanently to stop the ducts leading from the eye.  Pressure
upon the upper part of the sac will sometimes clear the
passages again.  If they are permanently closed, a small
silver tube can be inserted, that will draw the fluid off
from the eye and prevent its overrunning, and obviate
the attendant wiping, that is likely to injure the eye.

FIG. 200.

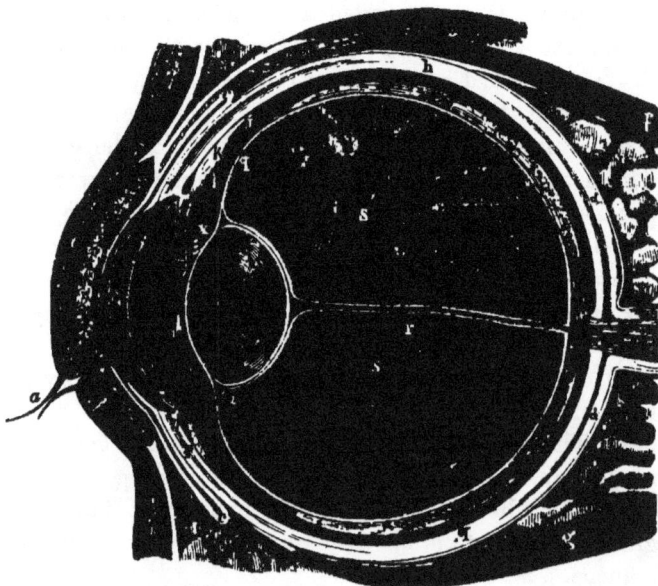

Fig. 200 exhibits a magnified section of the eye.  *a*, the lashes of *b*, the
upper, *d*, the under lid, in the middle of which the orbicularis is to be
seen, and in the back part the Meibomian glands opening at the edges of
the lids; *c*, the upper and lower points of the lining of the inside of the
lids; *i*, the cornea; *d*, the sclerotic; *h*, the choroid ; *j*, its division; *k*,
ciliary ligaments, *b*, body, *l*, muscle; *m*, iris; *v*, anterior chamber; *u*,
posterior chamber; *t*, lens; *o*, retina; *r*, central artery; *p*, hyaloid; *s*,
vitreous humor; *e*, muscle raising lid; *f*, superior rectus; *g*, inferior rectus; *y*, fat.

---

818. What frequently the effect of —?  What does Fig. 200 represent?  Class
the parts of the eye represented under their proper heads.

819. THE FLUID FURNISHED TO THE EYE contains an appreciable quantity of salt to render it transparent. When this is not supplied the eye soon inflames; in such cases, to wash it frequently with a solution of salt in water will be highly serviceable. Too much salt in the blood, and in the tear-fluid, on the other hand, tends to inflame the eye.

## Protections of the Eyes.

820. THE PROTECTIONS OF THE EYE are the lids, the Meibomian glands, the orbicularis muscle, the lashes that fringe the lids, and the eyebrows that shed off the perspiration and catch particles of dust that might incommode the eyes.

821. *Remark.*—WHEN DUST DOES ENTER THE EYE, one of the best ways of removing it is to draw the upper lid down over the lower lid, and let it slide up. This experiment is almost always successful, and particularly valuable on the cars.

## Review of the Eye.

822. A REVIEW OF THE EYE SHOWS that perfect vision depends upon a very large supply of Blood to its nervous part, to sustain the great activity of the retina; upon a proper form and density of the Media of the eye; and upon the reception of a proper supply of light; while incidentally the front surface must be kept moist and protected from dust.

823. THE EYE MUST HAVE repose, temporarily as well as at long intervals.

824. *Illus.*—If the eye be acted upon continuously by any color, in a short time the object of regard will begin to appear dark, on account of the exhaustion of the eye, no longer able to respond to the light. If then another color be presented to the eye, it will be discerned with great pleasure; if it be white, the complementary part of it only will be effective. For example, if the eye regard red till fatigued, and look upon white, it will see green, that is, the blue and yellow of white.

---

819. What said of —? 820. What are —? 821. What to be done —? 822. What does —? 823. What must —? 824. Illustrate the idea.

825. The eye is reposed temporarily by receiving different kinds of light in succession, or if all three kinds act together, by not receiving any for an instant.

826. *Inf.*—Black and white are the best colors for a reading-book (and black letters upon a white ground are better than white letters on a black ground would be), since the retina that corresponds to black is perfectly rested, and that acted upon by white has the influence of all colors at once; while any other colors would give constant activity to the retina. The proportion of white and black should be better adjusted in school-books than is usual, through an incorrect idea of cheapness. What injures the eyes is never cheap.

827. *Inf.*—The art of pleasing through the eye consists in presenting before it a succession of colors, the aggregate of which will be white. The complexion, clothing, adornments, furniture, and everything colored must be arranged upon this principle.

828. *Illus.*—A moss rosebud pleases, because its red and green are in proper proportions. A Violet (Viola tricolor), five sixteenths yellow and eleven sixteenths purple, will please, their proportions being right. Blue is becoming to blondes, because in their hair and complexion orange predominates, which with blue produces white.

829. The retina is sometimes blind entirely, and sometimes to only one or two kinds of light, when it is called color-blind. Sometimes one eye is blind to one color and the other to another.

830. The author has seen a person who recognized a green object as blue with one eye and yellow with the other, one eye being blind to blue, the other to yellow.

831. Blindness is sometimes produced by an imperfection in the character of some of the Media of the eye. This blindness is of three classes: 1st, want of transparency; 2d, too little, and 3d, too great, refractory power.

832. The medium that is wanting in transparency can sometimes be removed, and its place supplied with a glass that will, in a measure, restore sight.

833. Long sight is produced either because the Media of the eye are not sufficiently convex, or because

the means for adjusting them are imperfect. When this occurs in youth, as it sometimes does, an effort should be made to bring into action the adjusting means, by looking at near objects. When long sight is the result of age, sight can only be assisted by convex glasses.

834. SHORT SIGHT IS PRODUCED either by too convex media or imperfect means of adjustment.

835. *Remark.*—THE FREQUENCY OF SHORT SIGHT IN CHILDREN can be accounted for, if it is remembered that as the eye grows one part may change its form and size more rapidly than another, and the media of the small eye are of course more convex than those of the large one ought to be.

836. *Inf.*—CHILDREN AND YOUTH should accustom themselves and be directed to read, and look at objects, at different distances, and, if growing near-sighted, much pains should be taken to often exert the eye in looking at the most distant objects that can be discerned.

837. SHORT SIGHT, DEPENDENT ON TOO GREAT CONVEXITY OF THE MEDIA, can be assisted only by wearing concave glasses, which should be carefully adjusted, and a little weaker rather than stronger than is essential for clear sight, and used only when necessary.

838. *Remark.*—Do not let any near-sighted child or youth grow up without enjoying the assistance of glasses in viewing nature. Often a person is to a degree near-sighted, and is not aware of it, losing all the delight conferable by the magnificence of the starry heavens and the widespread beauty of Nature.

839. ARTIFICIAL LIGHT should be of the same kinds, and as nearly as possible in the same proportions, as in sunlight; should be abundant, diffused, and come from above the eye.

840. *Remark.*—LARD gives the best light, "gas" next best, of ordinary materials; wax gives a beautiful light, but is too costly. STUDENTS SHOULD RATHER CLUB THEIR MEANS, and secure a brilliant light, than separately use an inferior one. LIGHTS SHOULD NOT BE SHADED, so as to make a bright light before the eyes and darkness elsewhere, as the eye must change too suddenly when directed from one object to another.

---

834. How — ? 835. What said — ? 836. How treat the eyes of — ? 837. What said of — ? 838. Should glasses be used? 839. What said of — ? 840. What said of — ?

841. AT THE INNER ANGLE OF THE EYE is a small red eminence enclosing a cluster of sebaceous glands, and furnished with minute hairs. Sometimes they grow out and are troublesome; they are then merely to be extracted. The fleshy part sometimes extends over the cornea; if it obscures vision or is troublesome, it should, as it can very easily, be removed.

842. COLDS, WINDS, AND DUST annoy the eye, and may cause inflammation. From the latter it can be protected; the former should be avoided.

843. THE EYE IS SOMETIMES PERMANENTLY TURNED inward (cross-eyed), or outward (wall-eyed); usually because it is weak, or in some respect imperfect.

844. *Remark.*—If the imperfection is outgrown, and yet the direction of the eye not corrected, as is sometimes the case, the operation of cutting one of the muscles may be tried, with the expectation of a favorable result; but if the weakness of the eye continues, the operation will be worse than useless.

845. THE EYE IS A COMPOUND ORGAN: the eye proper is constructed of sinewy, secretory, muscular, and nervous tissues, and in its accessories gristly and bony are also found.

846. *Remark.*—Thus in STRUCTURE, in MODE OF ACTION, in the CAUSES THAT EXCITE IT, and THEIR REGULATION, and in all that is necessary to the preservation of the eye, there is a remarkable simplicity that cannot fail to excite admiration and enlist the attentive study of so important an organ. Some other exceedingly interesting facts will be presented, when it is, with other organs, made up into the apparatus of sight.

### The Ear.

847. THE EAR is one of the most delightsome and precious as well as useful of the organs of sense, for it is not only attuned to all the sweet and varied music of nature, but, properly cultivated, it enraptures the mind with the marvellous productions of art. No other organ

approaches it in power to sway the mind. The persuasive tones of eloquence through the ear exert more power to stir or stay the passions of men than all the arguments the ablest reasoner can address to the judgment. "Lend me your ears," said the ancient orator; and when his hearers had done that, and they were filled with the bewitching cadences of his voice, they were the unwitting slaves to his will. Yet as a physical organ of hearing the ear is exceedingly simple: its wondrous power consists in its nervous relations to the mind.

FIG. 201.

Fig. 201 is a beautiful view of the Ear; 1, external, 2, middle, 3, inner ear; 13, a section of the air-tube, which section extends through the front of 2, and the middle of 19, a tube called Eustachian, leading to back part of nostril. If a person close the nose and mouth, and blow air from the lungs, it will press through 19 into 2, and produce a sound. 14, bottom of air-tube, a vibratory membrane through which *force* is determined; 22, semicircularcanals, through which *quality* is determined; 24, cochlea, through which *pitch* is determined; 18, bones of ear.

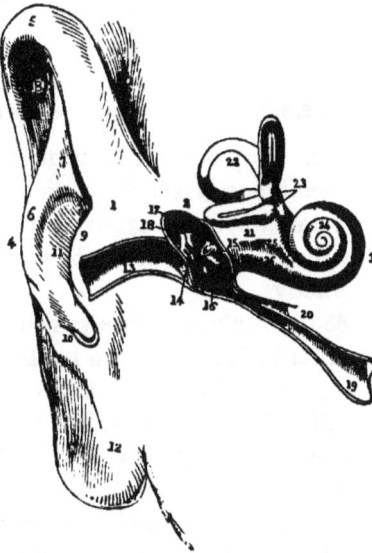

848. THE EAR IS CONSTRUCTED in three parts, called the external, middle, and internal ears.

849. THE EXTERNAL EAR IS CONSTRUCTED of the auricle, or ear, and the meatus, or tube.

850. THE AURICLE IS CONSTRUCTED of a framework of gristle and sinewy fibres covered with skin, in which there is some fat and numerous oil-glands.

FIG. 202.

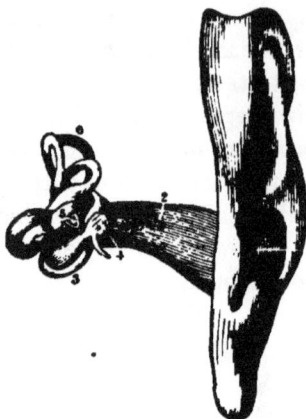

851. THE MEATUS IS CON-STRUCTED of a frame partly gristly and partly bony, lined with skin, in which there are numerous glands secreting the ear-wax. The meatus is about an inch in length, inclining a little downward and forward, and closed by the skin lining the meatus stretching across.

Fig. 202, 1, mouth of; 2, meatus extending to 3, bony ring; 4, hammer attached to membrana stretched upon 3; 5, 6, 7, labyrinth; the anvil closes the ovale near 5.

852. THE BOTTOM OR END OF THE MEATUS IS CALLED the membrane of the drum (membrana tympani). It is not parallel with the external openings, nor is it stretched directly across, but is slightly concave or funnel-shaped externally.

853. THE USE OF THE EXTERNAL EAR is to conduct the vibrations of the air down to the membrane of the drum, which is thereby thrown into corresponding vibrations.

854. THE MIDDLE EAR IS CONSTRUCTED of the drum (tympanum), Mastoid cells, and Eustachian tube.

855. THE DRUM is a cavity about as large as a kidney-bean, placed flatwise toward the external ear. It is excavated in the temporal bone. In the outside of the drum is the membrane previously mentioned. In its opposite or inside bony walls are two openings, called the round and oval windows (fenestra rotundum and ovale), the former covered by the lining of the drum, and the latter closed by the stirrup-bone and its ligament.

856. A CHAIN OF BONES, the smallest in the body, stretches across the drum. The *hammer* is attached by

one arm to the external membrane, and by its head is jointed to the *anvil*, which by one arm is jointed in early life, but is soon consolidated with the *round* bone that is jointed to the *stirrup* attached to the oval window.

857. Three ligaments SUSPEND THESE BONES in their places, and three delicate muscles CONTROL THEIR MOVEMENTS, tightening or relaxing the chain of bones, and of course the opposite membranes to which they are attached.

858. THE DRUM IS LINED by a very delicate membrane, that is continued over its chain of bones as well: the inner surface of the membrane is constructed of cells.

859. THE MASTOID CELLS are numerous small cavities in the Mastoid process, just back of the ear, opening into the back part of the drum, and lined with a similar membrane. Their use is not known.

860. THE EUSTACHIAN TUBE is one in which the lower part of the drum terminates: it narrows till not larger than a crow-quill; then enlarging, opens into the side of the upper and back part of the nostril, and of the pharynx.

861. THE USE OF THE EUSTACHIAN TUBE is to allow air to pass into the drum; and when it is closed by a cold or disease, hearing is lost. It also serves as a drain to the drum and mastoid cells, and is sometimes closed by collections from them.

862. THE INTERNAL EAR (labyrinth) IS CONSTRUCTED of a vestibule, semicircular canals, and cochlea.

863. THE VESTIBULE is an irregular, small, oval cavity, wrought into the bone; into it the oval window would open if the stirrup should be removed. It is directly within the drum. It is lined with a delicate membrane, adhering to the bone on one side, and finished with cells on the other, and is filled with serous fluid called peri-lymph.

---

857. What —? 858. How is —? 859. What are —? 860. What is —? 861. What is —? 862. How is —? 863. What is —?

864. THE MEMBRANOUS VESTIBULE is a sac nearly filling the bony vestibule, floating in the peri-lymph, and itself also filled with similar fluid, called endo-lymph.

865. THE SEMICIRCULAR CANALS are three small tubes at right angles to each other, opening out of the vestibule (two, however, conjoining) by five orifices. Each canal has an enlargement near one end. The bony canals are lined with the extended lining of the vestibule and filled with peri-lymph.

866. THE MEMBRANOUS VESTIBULE is ALSO EXTENDED through the canals, and correspondingly filled. Both surfaces of this membrane are formed of cells.

867. WITHIN THE MEMBRANE are small collections of minute crystalline particles of carbonate of lime (otolites) that adhere together and are connected with the filaments of nerves that terminate in the fluid contained in the sacs of the membranous labyrinth.

868. THE COCHLEA is, as its name signifies, like a cockle-shell in shape, and is constructed with a spiral cavity like that in the cockle, turned nearly three times round its axis, and rising with each turn.

869. THE SPIRAL CAVITY IS DIVIDED nearly in its centre by a partition stretching across from its axis outwardly, about one third being bone, one third membrane, and the other third muscular. The partition does not extend quite to the top. Here both cavities unite, forming what is called the cupola. The spiral cavities are called scalæ.

870. THE MOUTH OF ONE OF THE SCALAE COMMENCES at the round window; the other opens from the vestibule, but does not communicate with the inside of the membranous part. Both spirals are lined with the extended lining of the bony vestibule, and filled with the same peri-lymph.

871. NERVES EXTEND UP THE AXIS OF THE COCHLEA,

and out in the spiral partition between the scalæ, and terminate in numerous nerve-cells that are there situated in several rows.

FIG. 203.

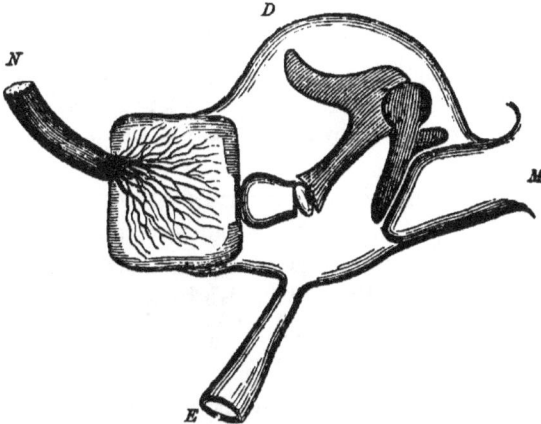

Fig. 203, ideal ear. *M*, the Meatus or external tube; *D*, drum, with a chain of bones stretching across it from the external membrane to the internal membrane, covering an opening into the labyrinth, filled with fluid, in which the nerve, *N*, is suspended. *E*, Eustachian tube.

872. THE GENERAL MODE OF ACTION OF THE EAR MAY BE UNDERSTOOD by observing the preceding figure. The various vibrations of air pass down the tube *M* and throw its membrane into vibrations; through the chain of bones the internal membrane is caused to vibrate; the fluid in which the nerve is suspended is thus acted upon, and the nerve consequently affected.

873. THREE MODIFICATIONS OF THE GENERAL MODE OF ACTION FOR HEARING will be necessary, in accordance with the three classes or characteristics of vibrations, by which they manifest *intensity*, *pitch*, and *quality*.

874. THE INTENSITY OF VIBRATION IS DETERMINED by the space it moves over in a given time; and that can very well be measured by the action of the vibration on the membrane of the drum, and the muscular

Describe Fig. 203. What is the difference between an ideal and real Figure? 872. How may —? 873. What —? 874. How is —?

14

action needed to keep it tense under the influence of the vibration.

*Remark.*—Thus muscular action and sensation is again used to determine force, as it always is in the body. According to the action of the muscles of the middle ear, so are the sensations they will cause.

875. THE PITCH OF VIBRATIONS IS DETERMINED by the number of vibrations in a given time.

876. *Illus.*—If a string of a certain tensity is vibrating a certain number of times in a second, one of half its length at the same tensity would vibrate twice as many times in a second, and its sound would be an octave higher.

877. IN THE SPIRAL PARTITION BETWEEN THE SCALÆ OF THE COCHLEA exist all the requisites for appreciating the rapidity of vibrations; for from bottom to top the scalæ and their partition diminish by a perfect gradation corresponding to every possible period or pitch of vibration of the air.

878. *Illus.*—If a musical instrument is sounded near a harp or a piano, a string of these instruments, corresponding in proportionate length, and being of course in harmony with the instrument sounded, will also be thrown into vibration, and cause a sound.

879. THE QUALITY OF VIBRATIONS is determined by some as yet undistinguishable characteristic of the vibrations, different in case of almost everything causing sound.

880. *Remark.*—Herein is a wonder of wonders—that there is something so perfectly evident, and yet eluding our grasp most effectually! It is also still more strange that this can be so when there are so many millions of different qualities in sounds and of course vibrations, in every human voice not only, but that of every bird and even of every insect, as well as in all the noises of dumb nature.

881. THE NERVES CONNECTED WITH THE VESTIBULE AND THE SEMICIRCULAR CANAL are constructed for the purpose of appreciating the qualities of vibrations, the effects of which reach them; for there is no other office

---

875. How is—? 676. Illustrate. 877. What exist —? 878. Illustrate. 879. How determine —? 880. What wonderful —? 881. What said of —?

for them to perform. That office is performed by some part, and many arguments drawn from the structure of the ear in lower animals go to establish the statement.

882. THE EAR MIGHT BE CONSTRUCTED by tracing out the nerves of hearing from their ganglia till an appropriate place is found in the outer walls of the head for suspending the nerves and their cells in contact with an enclosed fluid; in front of which appropriate media, the middle and outer ears should be constructed for receiving and transmitting the extremely delicate vibrations of the most sensitive of all material things, the air, capable of vibrating in three modes, expressive of intensity, pitch, and quality, through a great range of each.

883. IN THE STRUCTURE OF THE EAR all six (*B.*, *G.*, *Si.*, *M.*, *N.*, *Se.*) tissues are required.

884. *Remark.*—There ARE FEW CAUSES OF DERANGEMENT OF THE EAR. Ear-wax is sometimes allowed to collect and harden in the tube; a little sweet oil dropped into it several times and washed out with castile soap and warm water will suffice to remove that difficulty. Colds are very apt to affect hearing, through the Eustachian tube; they must be very carefully avoided if they affect hearing. If it is thus permanently stopped the opening can sometimes be regained by passing a tube into it. If that cannot be done, sometimes hearing can be regained by making an opening in the external membrane. If a watch placed against the head can be heard, there is hope. Throat diseases, scarlet fever, etc., are very liable to leave children with a hardness of hearing. To prevent that result, and to prevent still worse effects, great care should be taken during recovery from such diseases to avoid taking cold; and if it occurs, and care is taken, the trouble will usually be outgrown, or, if it is not, it will ingrow deeper.


SECTION VI.

*Circulatory Organs.*

885. The Circulatory Organs are of five kinds: Capillaries, Veins, Arteries, Hearts, and Lymphatics.

---

882. How —? 883. What tissues required —? 884. What are to —? What the effect of taking cold upon hearing? 885. Class circulatory organs.

## *Capillaries.*

886. CAPILLARIES are exceedingly minute tubes differing slightly in size in different parts, but uniform in size in a given part, united with each other so as to form a network, the meshes of which vary in size and arrangement in different parts, according to their structure and requirements in respect to Blood. The capillaries are the most essential parts of the circulation, always existing first and acting last in life, the other parts being added in order to pass the blood from one set of capillaries to another.

FIG. 204.

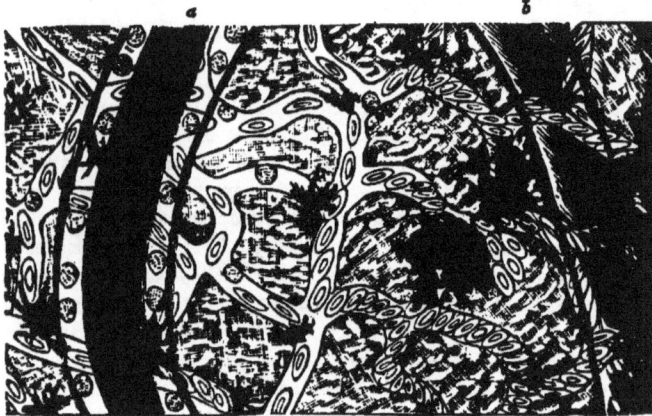

Fig. 204 represents a portion of frog's foot, during life, very much magnified: *a*, an artery; *b*, a vein deeper than *c*, a network of capillaries near the surface. In the network the two kinds of blood-cells are noticeable; the red ones are largest, and, from compression in the minute capillaries, somewhat elongated. The white cells are also noticeable at 1, in the artery, near its walls, the red cells pouring along in the centre so rapidly as not to be separately discerned, and only as a dark current, as represented. Above *c*, in the centre, a capillary has been ruptured and cells are crowding into the adjoining mesh. The irritation thus caused has retarded the circulation in the neighboring capillaries, and the cells are crowding each other. The irregular black spots are pigment-cells that give color to the part.

886. What are —? What different figures represent capillaries? Describe Fig. 204.

887. *Remark.*—THE ARRANGEMENT OF THE CAPILLA-

FIG. 205.

RIES is so exactly in accordance with the structure to which they belong that an examination of a small portion of a capillary network would determine whence it came. In Fig. 205, 5, 5, represents a network of capillaries around the root of a hair, 3.

888. THE CAPILLARIES ARE CONSTRUCT-ED of a homogeneous membrane, like basement membrane, with nuclear spots in its sides, shown in *c*, Fig. 206, capillaries, into which *a*, an artery, opens.

889. THE CAPILLARIES OPEN from the arteries on the one hand and into veins on the other, as a rule, but there are ex-

FIG. 206.

ceptions, 1st, in which capillaries only are the means necessary for a very limited circulation, and, 2d, where capillaries open from and into veins,—for example in the liver.

890. THE USE OF THE CAPILLA-RIES is to bring the Blood into very near relations with the parts upon which it is to act, or with the substances that are to act upon it.

891. There DO NOT APPEAR to be any apertures or pores IN THE SIDES OF A CAPILLARY more than there are in the sides of a soap-bubble, but the substances it contains pass through it at any point, and substances pass into it without any rupture. Where one occurs, as in Fig. 204, the entire contents burst out.

892. THE SUBSTANCES ABOUT TO PASS OUT FROM THE CAPILLARY approach its inner surface from the current that is more quickly flowing through its centre, then hesitate, are quiet for an instant, and, in the twinkling of an eye, appear on the outside.

887. What said of —? 888. How are —? 889. How do —? 890. What is —? 891. What —? 892. What said of —?

# 318 DETAILED SYNTHESIS.

893. *Remark.*—It is evident, from Fig. 204, that there is consider-able space between the capillaries, filled with the substructures of the part; and the substances having made their exit from the capillaries, must travel on to their destined place of use. So also must those that have ful-filled their purpose travel back a greater or less distance to the free moving current in the capillaries.

894. *Inf.*—Thus there is a circulation outside the circulation in the Blood-vessels, a great commotion of substances moving to and fro, the whole living body being a mass of activity; and rubbing the body and muscular contraction promotes this essential activity, this internal circulation, in a still more important degree than they do the general or external circulation in the Blood-vessels.

895. *Remark.*—It is not settled whether the capillaries are mere-ly passive tubes or also adapted to secrete substances. The appearance of the wall of the capillary is like that of secretory tissue; and while it acts, to a certain degree, passively, in allowing certain substances to pass through it, without change, it appears even then to have a selecting in-fluence, or else it is exerted through the wall by the structures outside. In other cases, as in the ganglia, there does not seem to be any reason to believe the capillaries do not secrete the granular substance which is found around them, and is doubtless the material from which the cells are formed.

## Veins.

896. Veins are continuations of the capillaries, the wall of the latter being the basement membrane of the former. It is lined with squamous (scale-like) or pave-ment-cells, and is surrounded with sinewy fibres, partly white or inelastic and partly yellow or elastic. By these means the vessel is strengthened and thickened. In the larger branches a few unstriated muscular fibrillæ are found.

897. The larger veins of the extremities, the walls of the Trunk and Head, and the superficial veins of the neck, are furnished with valves.

898. *Illus.*—The positions of some of these can be seen by rub-bing the finger down upon the skin over the veins in the back of the hand, since the blood is pressed against the valves, closing them, and dis-

---

893. What is —? 894. Is there —? 895. What —? Do capillaries secrete in gan-glia? 896. What —? 897. What —? 898. How positions of vein-valves seen?

tending the vein. If the finger is carried below the valve, the blood is pressed away from it, and the vein between the valve and finger will be empty.

FIG. 207.

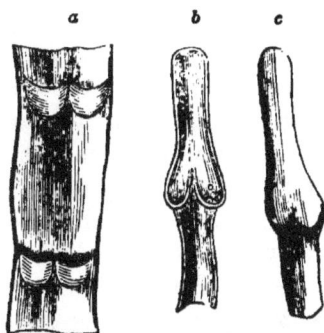

Fig. 207 represents, *a*, vein, laid open by a section of one side, showing two sets of valves, in pairs. *b*, same vein, in perpendicular section, through one set of valves. *c*, same vein, entire, showing the distension of the vein behind the valves.

FIG. 208.          FIG. 209.

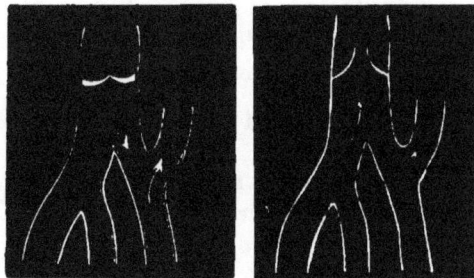

Figs. 208, 209, are plans of veins, showing how the blood flows into side branches below the valves when they are closed.

899. THE USE OF THE VALVES is to prevent the Blood, when pressure is made upon the veins, from flowing back into the capillaries.

900. *Remark.*—The valves of the lower extremities are very numerous, yet the severe and continued pressure upon the veins there not unfrequently distends them till the valves fail to close them, and the Blood flows back, distending the veins below to a greater degree, causing a knotted appearance of the veins, called varicose. In this case, as far as possible, the pressure should be removed, and an elastic bandage worn over the entire extremity below the upper part of the varicose condition. Standing positions are apt to produce this condition.

901. *Inf.*—The arrangement of the valves teaches that when the

Describe Fig. 207. Describe Fig. 208. Describe Fig. 209. 899. What is —? 900. What are varicose veins? 901. What do the valves teach?

surface is rubbed, the outward movement should be made with gentle pressure, and that toward the centre of the body with vigor.

902. THE VEINS EXTEND FROM THE CAPILLARIES of the body toward the chest in two courses, one near the surface, and another deep, near the arteries, most of which have a vein on each side. These two courses frequently communicate, both directly and indirectly. In both courses, but particularly in the superficial one, the veins unite nearly at right angles, as seen on the back of the hand. They at last unite at the auricles of the heart, into which they pour the Blood drained from every part of the body.

903. THE USE OF THE VEINS is merely that of passive strong tubes, allowing the flow of Blood from one part of the body to another. The force of its flow through the veins is derived from or through the capillaries, the action of the muscles about them, and temporary pressure upon the surface of the body.

904. *Inf.*—Rubbing the surface must promote circulation of Blood in veins.

905. VEINS ARE CONSTRUCTED OF THREE TISSUES, a framework of sinewy, a lining of secretory, and a little muscular.

### *Hearts.*

906. HEARTS MAY BE CONSIDERED as merely veins, with walls thickened with muscular tissue; for, if in the walls of *a*, Fig. 207, considerable muscular tissue should be introduced, a heart would be constructed.

907. *Remark.*—The form of the heart is not an essential point, a muscular pouch of any form, with valves at its two openings, allowing the Blood to move in only one direction, is all that is necessary. Hearts like that suggested in the preceding paragraph exist in some animals, and the human heart is the same in general principles, its modifications adapting it to its peculiar offices.

---

908. THE HUMAN
HEART IS merely a hol-
low, heart-shaped mus-
cle, with several com-
partments and valves,
lined with a continua-
tion of the lining of
the veins (the serous
coat), and covered with
a similar coat; three
coats in all. It weighs
about ten ounces, is
about five inches in
length by three and a
half in diameter, and is

FIG. 210.

Back view of Heart.

enclosed in a heart-case (pericardium), the inner surface
of which is a continuation of the membrane forming the
surface of the heart.

909. THE HEART IS SUSPENDED to the spinal column
in the upper part of the chest by the blood-vessels and
ligaments connected with its base (that is above), and
extends downward, forward, and slightly to the left, behind the breast-bone; the place where it is felt beating
being against its extreme left and lower point.

910. WHEN THE HEART CONTRACTS, its point is
thrown forward and strikes against the chest-walls; this
point being supposed to be over the centre of the heart,
it is thought to be lower and more to the left than it is.

911. *Illus.*—Speakers will, unawares, not unfrequently indicate
their *true* feelings by referring to the heart, and placing the hand more
nearly over the stomach than over the region the heart occupies.

912. *Remark.*—Instances have occurred in which the heart inclined
to the right rather than the left side.

913. THE HEART IS SEPARATED FROM THE LUNGS,
between which it hangs, by its heart-case, and by the
partitions 17, Fig. 211, that pass across from the breast-

908. What is —? 909. How is —? 910. Effect when —? 911. What is a frequent blunder of speakers? 913. How is —?

14*

Fig. 211.

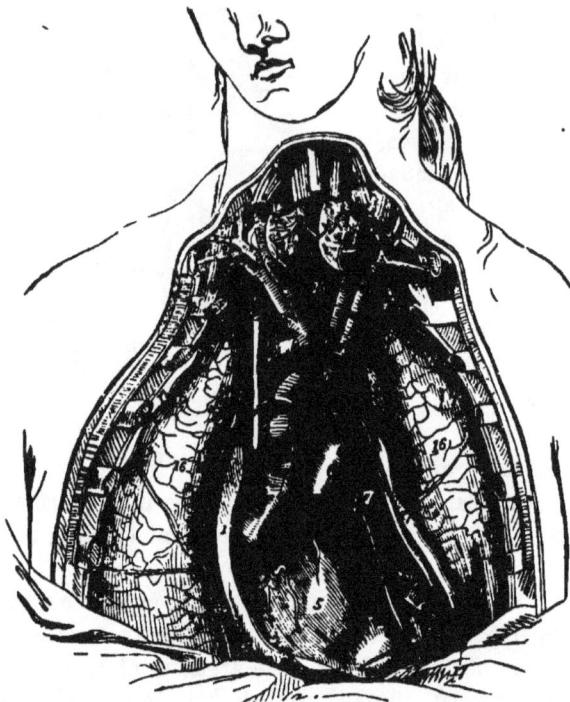

Fig. 211 represents the chest opened. 16, the lungs, have receded a little, a section of 17, the membrane, lining the chest having been made. A portion of the pericardium, 1, has also been removed, exposing the heart, 5, 8, one auricle, 4, and the commencement of its arteries, 6, 7, 9, and 3, the inner extremity of one vein, seen above at 2; 10, the continuation of 9, which is attached to 8, heart, behind 6; 11, 12, 13, branches; 14, Thyroid gland; 15, windpipe.

bone to the spinal column, curving round the heart, and forming a wall to enclose the cavity of each chest. These partitions and the heart-case are formed of sinewy, and their free surfaces of serous secretory, tissue.

914. THE COMPARTMENTS OF THE HEART are four, in pairs, named ventricles and auricles, right and left, pulmonary and systemic.

915. *Remark.*—Rightly speaking, the auricles are the enlarged inner extremities of veins, and should be considered thus ; the hearts being constructed properly of ventricles.

916. THE PAIRS OF VENTRICLES AND AURICLES have not any communication, and therefore they may be considered as forming two distinct hearts, that are adjoining, because they can thus assist each other, be thus formed of less material, and because the place adapted for one is perfectly adapted for the other.

917. EXTERNALLY THE HEART APPEARS TO BE a unit, but internally, physiologically, it is double.

918. *Remark.*—In some animals, as in the Dugong, Fig. 212, the hearts are nearly distinct externally.

FIG. 212.

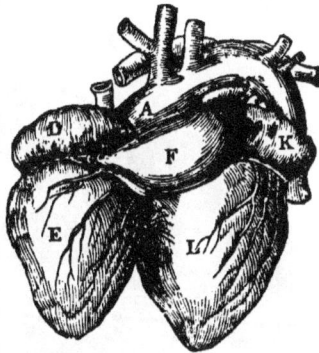

919. THE AURICLES OF THE HEART are very similar. Their name is derived from an appendage or extension, that, particularly in the case of the right auricle, appears a little like a dog's ear. They are irregular-shaped cavities of small size, and compared with the ventricles have thin, but compared with the veins thick, walls.

920. THE AURICLES ARE CONSTRUCTED of a continuation of the veins, that terminate in them ; or in fact, they are the inner extremities of the veins, with some muscular tissue inserted in their walls, that, from the uniting of several veins, have changed from a tubular to an irregular, pouched form. The muscular tissue reaches out into the walls of the adjoining veins a varying distance.

921. THE AURICLES HAVE several openings on their sides toward the veins, at which there are no valves, and one opening toward the ventricle with valves through

915. How should auricles be considered? 916. What said of —? 917. What —? Describe Fig. 212. 919. What said of —? 920. How —? 921.

which, in case of the right auricle, blood can flow back
if much force is exerted; but in case of the left auricle,
none can flow backward, or the lungs might be endangered.

922.   THE VENTRICLES ARE ALSO SIMILAR, the right
being a little more capacious, and its walls thinner, than
the left.

923.   THE INNER SURFACE OF THE VENTRICLES is
smooth; at the same time it is uneven, many bundles of
fibres (*columnæ carnæ*) extending across the cavity,
some of them connected with the points of valves, as
shown at 9, Fig. 213.

FIG. 213.

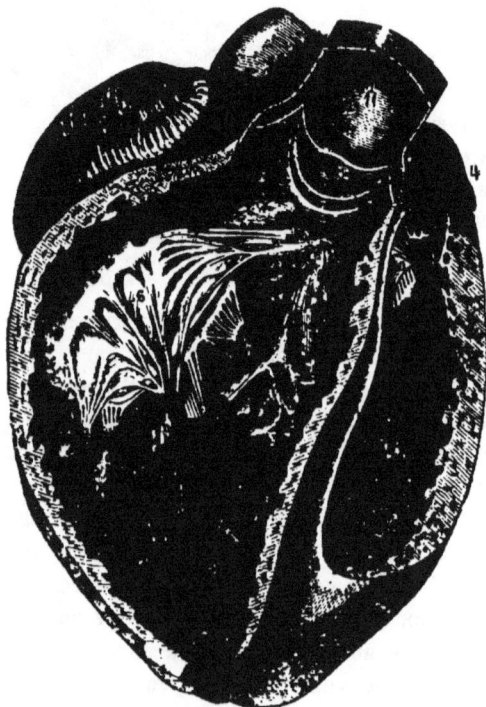

Fig. 213, heart,
the front view;
1, right, 2, left,
or systemic ven-
tricle, with por-
tion removed; 3,
R. auricle; 4 L.
auricle; 5, artery
opening from 2;
6, 7, 8, tricuspid
valves, connect-
ed by muscles, 9,
to inner surface
of heart; 10, bi-
cuspid valves;
11,    semilunar
valves opening
into 11, artery,
leading to lungs
(pulmonary).

924. EACH VENTRICLE HAS TWO OPENINGS, one into and the other out of it, both situated at the upper part of the ventricle, and both furnished with valves.

925. THE VALVES AROUND THE OPENING INTO THE VENTRICLE have an irregular, fringed edge, are constructed of sinewy tissue, and so arranged that when the heart contracts, and the fibres attached to them relax, the contents of the ventricle, crowding against the inner surface of the valves, presses them together, and their position being regulated by their muscles, the orifice is closed. In the right, or pulmonary ventricle, the valves having three main divisions are called tricuspid; in the left ventricle, the valves having two general divisions are called bicuspid.

926. THE VALVES AROUND THE OPENINGS FROM THE HEARTS have a half-moon shape, and are called semilunar, each one partially overlapping its fellow; and at the centre of each there is a little point, situated in such a manner that, when the valves are down, the opening is perfectly closed.

927. THE USE OF THE VENTRICLES is to receive the Blood from the auricles, and, by contraction, pour it out into the arteries.

928. *Inf.*—The auricles and ventricles should not contract at the same time; the auricles should first contract, and then the ventricles.

929. THE AURICLES OF BOTH HEARTS CONTRACT at the same time, occupying one fourth of the time of a full contraction of the heart, thus resting three fourths of the time.

930. THE VENTRICLES OF BOTH HEARTS CONTRACT at the same time, occupying one half the time, and resting the remainder of a full beat.

931. THE CONTRACTIONS OF THE HEART IN HEALTH, AT MATURITY, are from 60 to 80 per minute, more frequent in woman than in man, and in early than in late

---

924. What has —? 925. Describe —. 926. Describe —. 927. What is —? 928. Order of heart's contraction. 929. How do —? 930. How do —? 931. What are —?

life, varying also according to health, exercise taken, state of mind, etc.

932.  THE CAPACITY OF EACH VENTRICLE is at least between one and two ounces, and if 80 beats per minute are made, and only one ounce is received and thrown out, it will equal five pounds per minute, three hundred pounds, or more than a barrel, per hour!

933.  *Inf.*—The contemplation of the statistics of the labor performed by the heart, and of the great change that must go on in the body to render such action necessary, must convince every one of the importance of properly exercising and rubbing the body, and indeed of doing everything that will exert a favorable influence on the circulation.

934.  *Remark.*—IF EXERCISE QUICKENS THE HEART'S ACTION ten beats per minute, that circulates ten ounces more blood per minute, or forty pounds more per hour; an immense advantage in the way of promoting changes in the body.

935.  THE HEARTS ARE CONSTRUCTED of three tissues: Sinewy, Muscular, and Secretory.

936.  THE NECESSITY FOR TWO HEARTS arises from the fact that all the Blood of the body, as soon as it has passed through the capillaries, needs to be acted upon by the air, and must be therefore forced through the lungs as soon as it can be drawn together from the capillaries. The delicate structure of the capillaries of the lungs would not tolerate the force necessary to drive the Blood through them and then out into all parts of the body, or, if they would, such a force would drive the Blood through them too rapidly; therefore, after the Blood has been passed through the lungs, it must be received by another heart.

937.  *Inf.*—Since the Blood loses more than it gains in the lungs, THE SECOND HEART MENTIONED NEED NOT BE SO CAPACIOUS AS THE OTHER; but as it must drive the Blood farther, and through greater obstructions, it should be thicker and stronger.

938.  *Inf.*—Since the same Blood that goes through the lungs is to be thrown out, BOTH HEARTS SHOULD BEAT THE SAME NUMBER OF TIMES

---

932. What —?  933. What do statistics of heart show?  934. What if—?  935. Of what tissues —?  936. What —?  937. Why need —?  938. Should —?

EXACTLY, and therefore may act together, while their relative size and expansion must be in exact ratio to the activity in the lungs.

## *Arteries.*

FIG. 214.

FIG. 215.

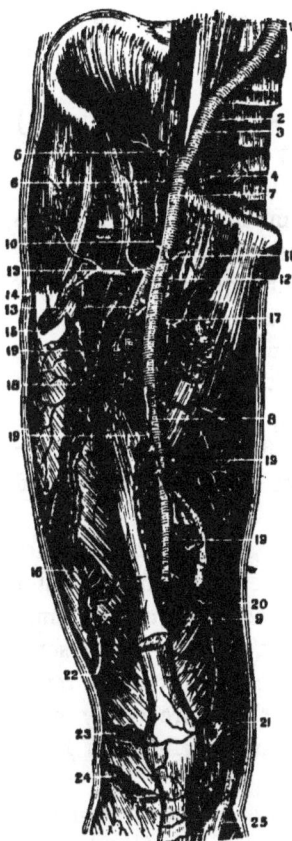

Fig. 214 represents the thigh dissected to show the large artery, 1, and numerous branches of it. Notice at 7 where it is over the bone, and near the surface, and easily compressed. Notice at 16 (anastomosis) where a branch from below unites with one from above. Notice that from 10 to 19 is the region for applying a tourniquet to compress the artery.

Fig. 215 represents the large artery, 9, 10, of the arm beneath the inner edge of the biceps, 2. Observe how it winds forward into the elbow-joint, as 1 does behind the knee. Notice the anastomosis of 16.

Describe Fig. 214. Describe Fig. 215. What does anastomosis mean? Point out the position of main arteries.

939. ARTERIES are very firm tubes that commence by a single trunk from each heart, and lead, in one case, into the lungs, and, in the other, into all parts of the body.

940. THE NUMBER AND CAPACITY OF THE ARTERIES is only about one third that of the veins in the system generally, and about one half of that in the lungs; but as the arteries receive the blood directly from the heart, it flows rapidly through them.

941. THE ARTERIES OF THE SYSTEM ARISE (Fig. 44) from the left, or systemic heart, by a single trunk (Aorta) that turns down with a beautiful arch, from which branches sweep out into the arms and lead directly up to the head. The main subdivisions above and below are noticeably few and large, throughout their length giving off very numerous small branches and twigs. It is also noticeable that all the branches lead off at acute angles to the current of the Blood. The large ones take a deep course in the limbs along the inner side of the bones, always winding into the flexures of the joints; all of which facts, and more, are admirably illustrated by Figs. 214, 215.

942. THE ARTERIES ARE CONSTRUCTED of an inner coat like that of the heart and the veins, and an outer coat (sometimes subdivided) of sinewy fibres, the inner part mostly of the elastic variety, together with some unstriated muscular fibrillæ. The outer coat is much thicker than that of the veins; these are easily compressed, while the arteries maintain a cylindrical form, unless forcibly compressed.

943. THE ELASTIC AND MUSCULAR STRUCTURE OF THE ARTERIES allows them to distend when the Heart forces the Blood into them, and, when its action intermits, that of the arteries, like that of the compressed air in fire-engines, continues to propel the Blood onward.

944. THE ARRANGEMENT AND CONSTRUCTION OF THE ARTERIES AFFORD every facility for the rapid flow of Blood in a concentrated volume, buried where its warm current will be least exposed to loss of heat, and least likely to be tapped by superficial injuries; and even when the arteries are severed their contraction diminish-. es their calibre very much, and if the injury is such as to cause the artery to be extended, as is usually the case, it is closed by that very action.

945. WHEN AN ARTERY, THOUGH NOT VERY LARGE, IS CUT, and not closed by the effect of the injury, its structure and use cause the Blood to flow in jets, or pulses, with serious rapidity, and life will be lost, unless the flow is speedily stopped by forcible compression.

946. *Remark.*—THE FACT MENTIONED IN ¶ 945 SHOWS the importance of having every person understand the position of the large arteries in the thigh and arm, and know how to compress them at that point, for that cuts off the flow below. Forcible pressure may be made with the thumb over the artery upon the hip-bone, or with the fingers over the artery in the arm; or a bandage, handkerchief, or cord, can be drawn tightly around the limb or twisted tightly by the aid of a stick-inserted beneath it, loosely tied around the limb; a knot may be made in the bandage and placed over the artery, or a smooth stone, a chip, half an inch or more thick, or a few pieces of coin, may be used to produce pressure directly on the artery. The best bandage is an elastic one, like an elastic suspender, or ladies' belt, wound tightly several times around, each turn of course increasing the pressure. *It is not best to stop the entire flow of Blood; it should only be well checked, since Blood is essential to the vitality of the parts immediately around the wounded part, and a little blood had better be lost, than not to have any received below the point of pressure upon the artery.* (See Ap. M.)

947. THE ARTERIES OPEN INTO the capillaries; their basement membrane, like that of the veins, is continuous with the walls of the capillaries.

948. THE USE OF THE ARTERIES is not only to lead the Blood from the Heart to the capillaries, but to assist in forcing it along.

---

944. What does —? 945. What effect —? 946. What does —? Should blood be entirely checked? 947. — what? 948. What is —?

949. *Remark.*—In the same manner as the hearts may be said to be a modification of a portion of the veins, so may the arteries be said to be an extension of the heart into every part of the body.

## Lymphatics.

Fig. 216.

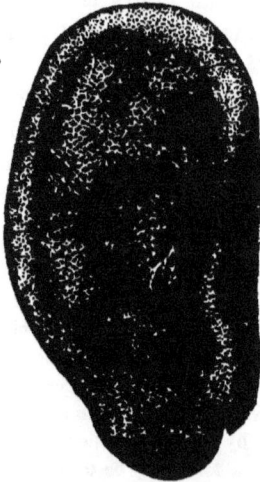

Fig. 216.

950. LYMPHATICS GENERALLY COMMENCE as a network of capillaries, larger than those of the red Blood, as represented in the adjoining cut of those of the skin of the ear. (See also Fig. 91.) In the papillæ of the second stomach they commence, as shown by Fig. 88, as club-shaped tubes. The form of the network differs in different kinds of organs.

951. LYMPHATIC CAPILLARIES ARE CONSTRUCTED of walls more delicate than even those of the red blood capillaries.

952. THE LYMPHATIC TUBES OR VESSELS COMMENCE from the network, as the veins do from theirs, the membrane of the capillary walls being continuous with the basement membrane of the tubes, in which it is lined with cells, and surrounded with sinewy fibres, not, however, so thickly as that of the veins.

Fig. 217.

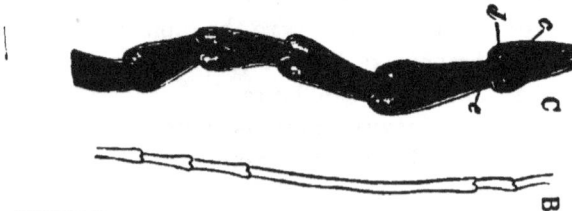

949. What may arteries be said to be? 950. How do —? 951. How are —? 952. How do —? Describe Fig. 217.

953. THE LYMPHATIC TUBES APPEAR beaded or with numerous enlargements (B, Fig. 217), owing to the valves within, as shown by *C*, a lymphatic laid open and enlarged.

954. THE LYMPHATICS EXTEND nearly parallel (See Pl. 5*), and are very uniform in size, opening together without forming a larger trunk, several leading into a gland, from which a smaller number issue, and all at last opening into veins.

955. THE CONTENTS OF THE LYMPHATICS, the lymph, appears like the watery or serous part of the Blood, and seems to be that portion gathered from the various tissues, into which it has passed from the red capillaries, and by which it has not been used.

956. THE USE OF THE LYMPHATICS SEEMS TO BE to gather the lymph and pour it into the glands, which appear to have the office of originating the Blood-cells, or the cells from the nuclei of which the Blood-cells are produced.

957. *Remark.*—THE LYMPHATICS ABOUND in all parts of the body except the brain and nerves, in the substances of which they have not been detected; and this fact may assist in determining additional facts in regard to both the brain and lymphatics.

## SECTION VII.

### *Respiratory Organs.*

958. THE RESPIRATORY ORGANS INCLUDE air-passages to the lungs, the lungs, and the walls of the trunk, including the diaphragm.

959. THE PASSAGES LEADING TO THE LUNGS INCLUDE the nose (and incidentally the mouth), the pharynx, the larynx, the trachea, and the commencement of the bronchii.

960. THE NOSTRILS ARE INTENDED FOR the passage

---

953. How do —? 954. How do —? Describe Figs. 92, 93. 955. What said of —? 956. — what? 957. Where do —? 958. What do —? 959. What do —? 960. — what?

and warming of the air breathed, and should be preserved in such a condition that the air can pass freely.
They open into the front upper part of the pharynx.

961.  THE PHARYNX is an irregular, cylindrical cavity, extending from beneath the cranium down back of
the nostrils and mouth, and upper part of the windpipe,
where it is narrowed into and becomes continuous with
the œsophagus leading into the stomach.

962.  THE OPENINGS OF THE PHARYNX are seven : two
at its upper central forward part, into the nostrils; two,
one on each side, at the upper back part, into each ear;
one at its front middle part, into the mouth; one at its
front lower part, into the larynx; and one directly downward, into the œsophagus.   (See Fig. 2, Pl. 30.)

963.  To the upper part and sides of the opening
into the mouth A CURTAIN IS ATTACHED, that can either
be drawn perpendicularly down against the back of the
tongue and close the opening into the mouth, or can be
drawn horizontally across the pharynx against its back
part, dividing it into an upper third and lower two
thirds, and cutting off communication between the parts
below it and the nose and ears.

964.  THE CURTAIN (soft palate) SHOULD BE DRAWN
UP when food or drink is swallowed, and hang or be
drawn down when the air is inhaled through the nose,
and at various other times, during sneezing, etc.

965.  UNDER THE BACK OF THE TONGUE a cartilage is
situated, against which the upper portion of the larynx ·
being raised, the opening into it is closed, and the food
and drink allowed to pass safely on into the œsophagus,
the opening into which is relaxed at such times. (Pl.30.)

966.  THE FRAME OF THE PHARYNX is the base of the
cranium, the cervical spinal column, and the interior
part of the facium.

967.  FROM VARIOUS PARTS OF THE FRAMEWORK mus-

961. What is —?  962. Describe —.  963. Where is —?  964. When should —?
965. What found —?  966. What is —?  967. What extend —?

cles extend in various directions, forming the chief part
of the immediate walls of the pharynx, as illustrated by
Figs. 5, 7, Pl. 19.

968.   THE PHARYNX IS LINED with mucous membrane,
continuous with that of the nostrils, of the middle ear,
of the mouth, of the windpipe, and of the œsophagus;
it partakes of the nature of all those linings, and is very
apt to partake, more or less, of their condition in disease, while its diseases are liable to extend to them.

969.   *Inf.*—The physician will frequently look into the throat to
enable him to judge of the state of the parts with which its lining is
continuous.

970.   THE LARYNX is a small part of the windpipe,
that has no other relation to breathing than has been
indicated by the opening and closure of its upper orifice,
and, as an organ of speech, can best be described after
the breathing apparatus has been constructed.

971.   THE TRACHEA is a tube commencing from the
lower portion of the larynx, that is a complete ring, and
extends down into the chest, where it divides into two
branches, or Bronchii.

972.   THE TRACHEA is constructed of a framework
of C-shaped cartilages, connected to each other by white
sinewy fibres, closed at the back part by elastic and some
muscular fibres, and finished by a lining of mucous membrane, in which there are a great number of minute
mucous glands.

973.   THE LUNGS are organs for bringing the air
and blood within the compass of the action of each
other; they are therefore situated on each side of the
heart, and at the lower extremity of the air-passages.

974.   THE LUNGS ARE CONSTRUCTED (see Fig. 218) of
divisions of the bronchii, 4, 5; of air-cells; of capillaries; of arteries, 13, 14; of veins, 17, 18, 19, 20; of
sinewy tissue, white and elastic (Fig. 43), binding all

Fig. 218.

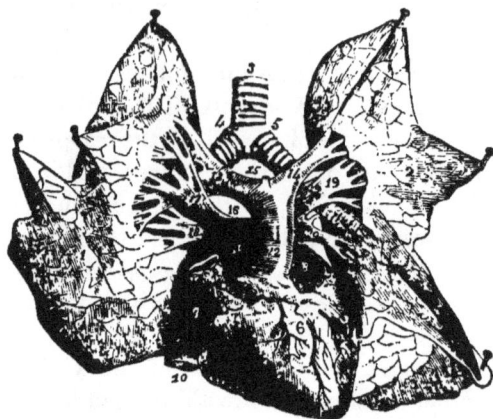

Fig. 218. 1, 2, lungs, drawn aside from the hearts; 3, the trachea; 4, 5, the bronchii; 6, ventricle of front, right, pulmonary, or respiratory heart; 7, its auricle; 8, left, back, or systemic heart; 9, 10, vena cava, opening into 7; 11, aorta; 12, pulmonary or respiratory artery; 13, 14, its branches; 15, a cord; 16, upper edge of 8; 17, 18, 19, 20, pulmonary or respiratory veins, opening into 16 or 8.

parts together; and of an external covering (pleura pulmonalis). (See Fig. 96.)

975. THE BRONCHII ARE CONSTRUCTED like the trachea, only in the smaller branches, instead of one C-shaped cartilage, several pieces complete the ring, in which also there is more muscular tissue. The mucous membrane also grows thinner as the air-cells are approached.

Fig. 219.

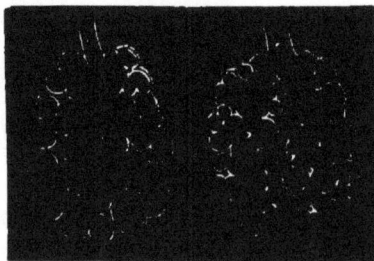

976. AIR-CELLS are an expansion of the lining of the minute bronchii, its fibrous portion being diminished almost to nothing, and its lining cells also being modified. The sides of the cells are saculated for the purpose of presenting a greater extent of surface. (See much magnified cells in adjoining Fig.)

977. MILLIONS OF SUCH CELLS cluster about the extremities of the divisions of the bronchii, presenting an almost incomputable extent of surface to favor the action of the air and blood upon each other.

978. *Inf.*—With such an extent of surface for the action of the air, it can hardly be in the Lungs so short a time as not to have its action complete, and therefore it should be at once thrown out, and a fresh portion taken in.

979. THE PULMONARY OR RESPIRATORY CAPILLARIES are an exquisitely beautiful network of the minutest vessels, the meshes between which are less than the diameter of the vessels, existing in the walls of the air-cells, and in part forming them. The appropriate contents of the cells and of the capillaries pass through the separating membrane, as freely as if it were not there; indeed, it appears to assist the action.

980. THE ARTERIES (pulmonary or respiratory) extend into the lungs by the side of the bronchii, and divide with them, till they reach the air-cell or pulmonary capillaries, in which they terminate.

981. THE NUTRIENT (bronchial or systemic) ARTERY OF THE LUNG is another small artery that branches from the aorta, and extending into the lung by the side of the bronchus, divides and subdivides *in all parts* of the lung, supplying its nutritive wants.

982. *Inf.*—There is a double circulation in each lung, one from the right heart to the left, very voluminous, and the other small, from the left to the right heart.

983. THE PULMONARY OR RESPIRATORY VEINS commence from the pulmonary capillaries, and lead back, two in number, by the side of each division of the bronchii, and pour their contents into the left or systemic heart.

984. *Remark.*—It is noticeable that while the systemic arteries carry bright red, the pulmonary carry dark red, blood; and while the systemic veins carry dark, the pulmonary carry bright red, blood. The

---

977. What said of —? 978. How long should air remain in a lung? 979. What are —? 980. What said of —? 981. What is —? 983. What are —?

expressions "arterial" and "venous" Blood are not proper, for the character of the vessel has nothing to do with its contents; but if they are received with great force the vessel must be an artery; if with light force, a vein is sufficient. The vessel receiving blood from a heart is, therefore, always an artery.

985. THE INTERSPACES OF THE BRONCHII, BLOOD-TUBES AND CELLS OF THE LUNGS are filled with the sinewy tissue necessary in working them, and upon the elasticity of which the expulsion of the air from the lungs is chiefly dependent.

986. The sinewy tissue near the surface of the lungs is wrought into the form of a membrane, and covered with basement membrane, that is covered with serous cells, thus producing a serous coat, called pleura, and in this position PULMONARY PLEURA.

987. THE LUNGS thus constructed MUST BE PLACED in an air-tight box, the dimensions of which can be enlarged and diminished. How this is done is shown by Fig. 220 and Fig. 221.

FIG. 220.

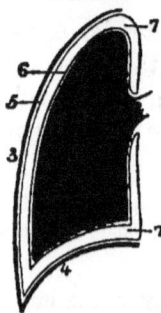

Fig. 220. 1, section of a lung; 2, the root, composed of branches and blood-tubes; 6, the external covering or pleura, at the root turning up, extending out, passing down at 5, to turn up over 4 (diaphragm), and again up to 2, where it becomes continuous with 6. At 5 it is called pleura costalis, because against the ribs. The two pleuræ are shown to be merely one continuous serous membrane. The lung is also shown to be attached only at its root, and elsewhere has a free surface; not hanging from its root, but, filled by air, is sustained by the action of the external walls of the chest.

988. *Remark.*—There is no space between the box in which the lung is placed and its surface. THE REPRESENTATION AT 7 IS NOT CORRECT, therefore, nor is that at 18, if the chest is entire; but as soon as it is open, the external air acts on the surface of the lung, balancing the internal pressure, and the elasticity of the lungs causes them to contract. The upper part of the lung will then fall down, and it will hang by its root.

---

985. What said of —? 986. Describe —. 987. Where must —? Describe Fig. 220. 988. Is —?

FIG. 221.

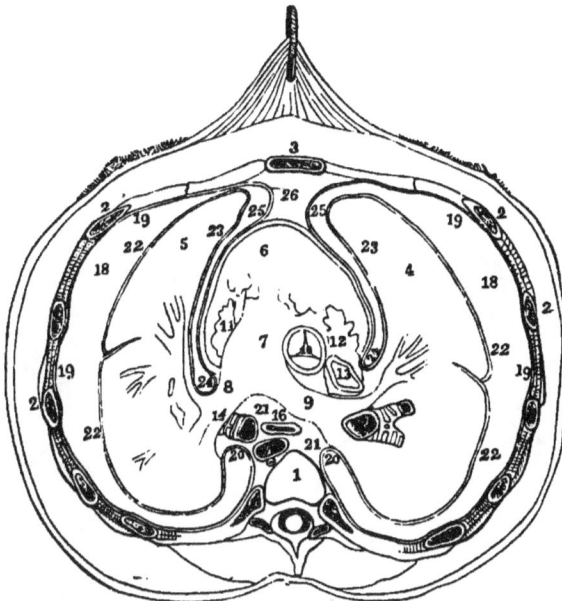

Fig. 221 represents a cross section of the chest and contents, just above the level of 10, semilunar valves of the heart. 1, Body of vertebra; 2, section of ribs; 3, breast-bone; 4, 5, lungs; 6, heart; 7, P. artery; 8, left, 9, right branch; 11, part of left, 12, part of right auricle; 13, vena cava; 14, left, 15, right bronchus; 16, œsophagus; 17, aorta; 18, cavity between the pleura, owing to the contraction or collapse of lungs, when the cavity is opened and air admitted; 19, pleura (5, Fig. 220), applied to walls of chest; 20, where 19 is continuous with 22; 23, the pleura pulmonalis; 21, space called posterior mediastinum; 24 corresponds to 20. From 24 to 25 the pleura and heart-case adhere; 26, a space unduly large, called anterior mediastinum. Notice also the two layers of muscles between the ribs.

989. THE PLEURA COSTALIS AND PULMONALIS are, in health, constantly moistened with serous fluid, so that they move against each other without any friction, and, from the elastic structure of the lung, only touch in the gentlest manner imaginable (except the lung is closed); and, as the former is attached to the inner surface of the ribs, forming, indeed, a lining to the chest,

Describe Fig. 221. Are the lungs represented as full or empty? In the sound chest how are they? 989. What are —?

and to the upper surface of the diaphragm, all the movements of these parts must be accompanied by a conforming action of the lungs.

990. *Inf.*—IF THE WALLS OF THE CHEST AND THE DIAPHRAGM ARE CAPACIOUS, and extensively movable, the lungs must be roomy, and cause the action of a great quantity of air upon the Blood, and vice versa.

FIG. 222.            FIG. 223.

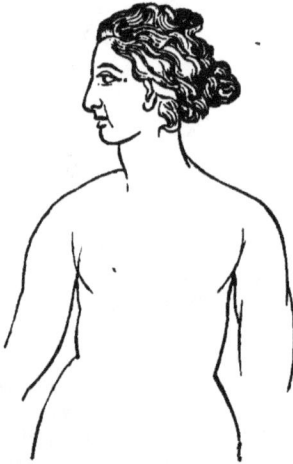 

991. *Illus.*—Fig. 222 represents a chest that is movable, and can supply all the blood its body requires, while Fig. 223 represents a chest that cannot supply the requisite air to the body to which it belongs; and all the inconveniences of a proper want of air must be experienced, in ill health, imperfect complexion, deficient expression, and all the ills that life is heir to.

992. *Remark.*—It is not necessary under this head to farther refer to the walls of the trunk as respiratory organs.

### Digestory Organs.

993. THE DIGESTORY ORGANS ARE CONSTRUCTED of a digestory canal and its appendages.

---

990. What —? 991. Describe Figs. 222, 223. Can the clothing be tight, and a person not suffer? 993. How are —?

994. THE DIGESTORY CANAL IS DIVISIBLE into the Mouth, Pharynx, Œsophagus, Stomach, Second Stomach, and Colon.

## The Mouth.

995. THE MOUTH IS CONSTRUCTED of a framework of the upper and lower jaws and the palate-bone, of the teeth, of the lips and cheeks, of the tongue, of the salivary glands, of the Amygdaloid glands, of the mucous glands, and of the veil of the palate (soft palate).

996. *Remark.*—THE FORM, SIZE, AND POSITION OF THE JAWS have been sufficiently well shown, and may be reviewed by looking at the illustrative figures. The position of muscles moving the jaw may also be re-examined.

997. IN THE LOWER EDGE OF THE UPPER JAW, AND IN THE UPPER EDGE OF THE LOWER, sockets are wrought, the sides of which are called the alveolar processes, that are built up around the roots of the teeth, and are absorbed when the permanent teeth are removed, diminishing the width of the jaws by so much, and bringing the chin and mouth nearer together, as seen in old people.

998. The TEETH IN THE FIRST SET are 20 in number (Fig. 224), succeeded by thirty-two permanent teeth: 8 incisors, 4 canine, 8 bicuspids (two pointed, pre, or small molars), and 12 molar or grinding teeth. A front view of one of each kind is shown by Fig. 225, and a side view by Fig. 226.

999. THE SECOND SET BEGIN TO GROW immediately after the first teeth begin to be formed, the fangs of which usually melt away to give place to the permanent ones.

1000. *Remark.*—IF THE SECOND TOOTH BEGINS TO PUSH OUT BY THE SIDE OF THE FIRST, the latter should be extracted to make room for the former. The mouths of children should be often examined when the second set are to be expected.

---

994. How is —? 995. How is —? 996. What said of—? 997. What is wrought —? 998. How many —? 999. When do —? 1000. What —?

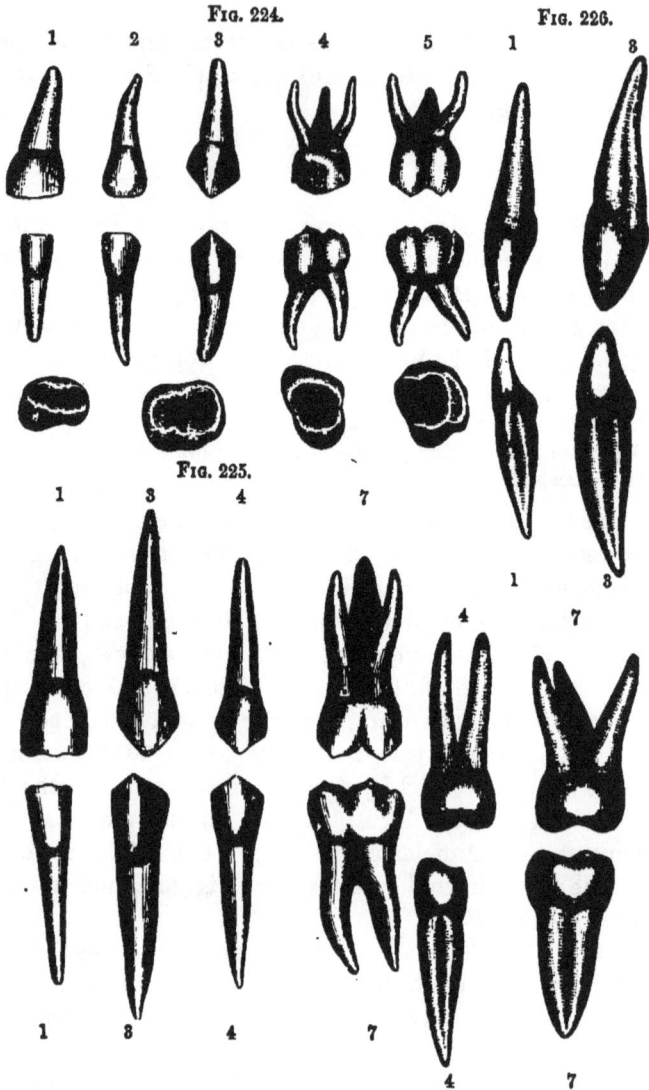

Fig. 224.

Fig. 226.

Fig. 225.

1001. THE SECOND SET REQUIRE MORE ROOM THAN THE FIRST, both because larger and more numerous.

Describe Fig. 224. Describe Fig. 225. Describe Fig. 226. Why are the back molars called the wisdom teeth? 1001. Do —? How correct irregularity of teeth?

*Remark.*—The jaw does not always expand as rapidly as teeth appear, whence they must be irregular and crowd upon each other. This defect can be corrected by pressure on the teeth when age has expanded the jaw; or, if it does not expand the jaw sufficiently, a tooth can be extracted and the others regulated.

FIG. 227.

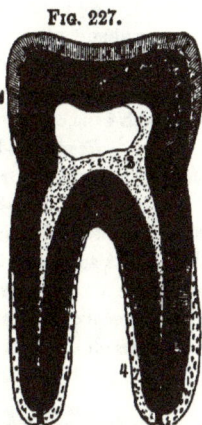

1002. A TOOTH IS DIVISIBLE into the Crown, or exposed part, the neck, or part covered by gum, and the fang, or part buried in the socket.

1003. A SECTION OF A TOOTH (Fig. 227) SHOWS that it is constructed of the ivory or *dentine* body, 2, covered with the hard, polished *enamel*, 1, and the softer *cement*, 4. The body contains a considerable quantity of pulp, in which are numerous blood-vessels and nerves, that enter by the minute channels in the fang. (See Fig. 50.)

1004. TEETH DECAY because constructed of improper material, that will not bear necessary exposure, or because the enamel is cracked or worn through, exposing the ivory, or because tartar is allowed to collect about the gums and become deposited between them and the teeth.

1005. *Remark.*—AS THE TEETH ARE FORMED IN INFANCY and youth, it is of exceeding importance that wholesome food should be taken and general health maintained during that period.

1006. *Illus.*—IF A CHILD HAS A FEVER DURING CONSIDERABLE TIME, the teeth growing at that time will manifest a ridged appearance.

1007. *Remark.*—FROM THE KNOWN EFFECTS OF HEAT AND COLD, successively applied to similar substances, it would be expected that anything hot or cold, acting on the enamel, would have a tendency to crack it; and sometimes it looks like a glaze-cracked piece of earthenware.

1008. *Illus.*—HOT WATER AFTER COLD, OR COLD WATER AFTER HOT, poured into a tumbler, will crack it.

---

1002. How is —? 1003. What does —? 1004. Why do —? 1005. — what follows? 1006. What effect —? 1007. What inference —? 1008. What effect of —?

1009. *Inf.*—HOT OR COLD SUBSTANCES should not be brought in contact with the teeth. Children should be cautioned against eating snow, ice, etc.

1010. *Remark.*—BITING HARD SUBSTANCES, cracking nuts, biting off thread, untying knots with the teeth, endanger the enamel.

1011. *Remark.*—In THE SALIVA OF SOME PERSONS' MOUTHS the tartar is so abundant that very frequent brushing is necessary to prevent its deposit: the general health of the person should be improved, as it also should be if the gums are spongy, for by proper attention to diet, etc , both evils can be remedied.

1012. *Remark.*—IF A TOOTH MANIFESTS DECAY, the decayed part should be thoroughly removed and its place supplied with gold.

1013. *Remark.*—SOMETIMES, WHEN THE TEETH BECOME VERY TROUBLESOME, loosened from their sockets, or are so extensively decayed, and the sockets so diseased, that they are extremely painful and almost unendurable, perchance exciting neuralgia, dyspepsia, &c., it will be advisable to remove them altogether, and have a set of artificial ones inserted, the result of which will be in some cases a most extraordinary improvement in health and the enjoyment of life.*

1014. THE USE OF THE TEETH is to masticate the food; a most important preparation to its after-digestion.

1015. *Illus.*—It is very difficult to fatten animals that cannot masticate; and old people, who have lost their teeth, are very much subject to derangement of the digestory canal.

1016. *Remark.*—When the teeth become so few as to be useless, or when gone entirely, their place should be supplied by artificial ones, that can be made to serve the purpose almost as well as natural ones. Many people would very much promote their comfort and health, and of course prolong their lives, if they would adopt artificial teeth as soon as they cannot masticate well with their natural ones.

1017. THE LIPS AND CHEEKS ARE CONSTRUCTED of muscles covered with the skin, and lined with the mu-

---

* The author believes, from what he has seen, that sets of teeth upon vulcanized rubber, though very much less expensive, are preferable to those upon gold. He has had the occasion, and been interested to notice, several such made by Dr. Fuller, a dentist in his neighborhood, and they serve their purpose most admirably, as it respects appearance, comfort in wearing, and durability; and if he had occasion for artificial teeth, he certainly should have them set in this manner, even if they cost the more.

---

cous membrane, that blends with the skin at the edge of the lips.

FIG. 228.

1018. THE TONGUE chiefly constructed of different muscles, as shown by Fig. 7, Pl. 19, and Fig. 228, is by some of them attached to the lower jaw, the hyoid bone, and the styloid process of the temporal bone.

1019. THE MUSCLES FORMING THE TONGUE ARE COVERED by mucous membrane, wrought into numerous papillæ of three classes, and excavated into mucous follicles along the side of the base.

1020. THE SALIVARY GLANDS are three in number, upon each side, called parotid, sub-maxilary, and sublingual.

1021. THE PAROTID GLAND is the largest of the three, situated and to be felt at the angle of the jaw, just in front of the ear. (See Fig. 49.) Its outlet tube, that may be felt like a cord beneath the skin, leads over the masseter muscle, and opens at a little eminence perceptible at the centre of the inside of the cheek.

1022. THE PAROTID IS CONSTRUCTED, like any racemose gland, of an immense number of minute follicles, clustered about small tubes, that all unite at last in one duct (Steno's). The inner surface of the pouches and tubes is composed of secretory tissue, the whole interwoven with nerves and capillaries, and bound together with sinewy tissue.

*Remark.*—The duct (Steno's) might be described as commencing at the inside of the cheek, and leading back beneath the outer skin to the angle of the jaw, where it divides into many branches, about the twigs of which many follicles are clustered, thus forming the parotid.

1023. **The Saliva formed by the parotid** is very watery, and apparently of chief use in moistening the mouth and dissolving the masticated food. It flows most freely when the jaws are active, in speaking, and particularly in chewing, etc., the pressure of the muscles being so many hearts to move Blood through the glands.

1024. *Remark.*—The parotid gland, in a most remarkable manner, shows the influence of the mind upon secretion, since the mere thinking of desired food makes the mouth water, and the relish of food increases the flow of the saliva, useful in digesting it.

1025. *Inf.*—A good appetite, desired and well-chewed food, are great promoters of digestion, by reason of the increased flow of saliva they cause.

1026. **The sub-maxillary gland** (*a*, Fig. 50) is a small racemose organ beneath the side of the tongue, constructed much like the parotid on a small scale, yet differing from it somewhat. A small duct from it opens beneath the bridle of the tongue.

1027. **The fluid of the sub-maxillary** is more viscid than that of the parotid, and flows chiefly while eating, seeming to be of especial use in digesting food.

1028. **The sub-lingual** is the smallest of the three, lies under the side of the front part of the tongue, pours out its fluid through several minute tubes, and seems to merely another sub-maxillary. (*b*, Fig. 50).

1029. **The Amygdaloid glands**, or tonsils, are two almond-shaped bodies, situated in each side of the back part of the mouth. They exhibit many pits, in the bottom and sides of which mucous glands open. Thus the Amygdaloid are merely clusters of mucous glands. (Pl. 30.)

1030. **The use of the tonsils** is to pour out mucus and lubricate the food forced between them as it is being swallowed.

1023. What is —? 1024. What does —? 1026. What is —? 1027. What said of —? 1028. What is —? 1029. What are —? 1030. What is —?

1031. *Remark.*—The tonsils are very apt to swell and be trouble-some, being sore to the touch externally, and partially closing the passages within. In the latter case, if the trouble becomes permanent, a portion of them can be removed by a very simple operation.

1032. THE MUCOUS GLANDS OR FOLLICLES of the mouth are very numerous, and scattered throughout its lining; they are also clustered as described above.

1033. THE VEIL OF THE PALATE, or soft palate, is an extension from the hard palate, constructed mostly of muscles, lined below and above, or before and behind, by the lining of the mouth and nose, and in the centre exhibiting a dependent point, called the uvula.

1034. THE USE OF THE SOFT PALATE is to close the communication between the upper and the middle part of the pharynx, or between the mouth and the pharynx.

## Pharynx.

1035. *Remark.*—The pharynx has already been described. (¶ 961.)

## Œsophagus.

1036. THE ŒSOPHAGUS is merely a distensible, con-tractile tube, extending from the pharynx downward in front of the spinal column nearly to the diaphragm, when it turns slightly forward and to the left through the œsophageal opening in the diaphragm (13, Fig. 181), and connects with the stomach. (See Fig. 70.)

1037. THE ŒSOPHAGUS IS CONSTRUCTED of annular and longitudinal muscular fasciculi (see Fig. 79), and lined with an extension of the mucous membrane of the pharynx and stomach.

1038. *Remark.*—The muscular rings at the lower part of the œso-phagus are somewhat stronger than those above.

1039. THE USE OF THE ŒSOPHAGUS IS to receive food from the pharynx and force it into the stomach. For

---

1031. If tonsils are troublesome? 1032. What said of —? 1033. What is —? 1034. What is —? 1035. Describe pharynx. 1036. What is —? 1037. How is —?

this purpose the rings in front of, and embracing, substances swallowed, relax, and those behind them contract.

1040. *Remark.*—It is noticeable that substances do not fall from the mouth into the pharynx, nor from it into the stomach.

1041. *Illus.*—When a horse drinks water it may be noticed rising along his neck.

1042. *Remark.*—Sometimes the muscular rings contract upon the food passing through them, and detain it; this is painful, and particularly unfortunate when it takes place at the upper part; for at this instant the windpipe is raised, and its orifice into the pharynx is closed, admitting no air to the lungs. The food must be removed before the passage of air will be allowed; hence continued choking soon causes death. Push the food down if it cannot be withdrawn. Action must be speedy: if consciousness is lost, proceed as in drowning. (See Ap. O.)

## Stomach.

1043. THE STOMACH is one of the simplest organs in the body, being merely a distension of the œsophagus as soon as it has descended through the diaphragm, having a few modifications easily comprehended. (See Fig. 70.)

FIG. 229.

Fig. 229 represents, 3, 3, the upper inner surface of the distended stomach, a section of it being made through the smaller extremity. 1, under surface of liver; 2, gall-bladder; 4, œsophageal orifice; 5, pyloric orifice; 6, the thick muscle around 5; 7, Second Stomach.

1044. THE LINING OR MUCOUS MEMBRANE OF THE ŒSOPHAGUS becomes thicker as it enters the stomach, and is wrought into minute papillæ, like the piles of velvet, giving the inner surface of the stomach a very delicate velvety appearance.

---

1040. What noticeable? 1041. Illus. 1042. What causes choking? 1043. What is —? Describe Fig. 229. 1044. What said of —?

1045. THE CAPILLARIES of the membrane here are very much increased in number, giving the beautiful appearance of the blush of the peach to the inner surface of the stomach when digestion is taking place.

1046. BETWEEN THE PAPILLÆ the minute mouths of two kinds of glands ARE FOUND, one the mucous gland of the stomach, the other the gastric glands.

Fig. 230.

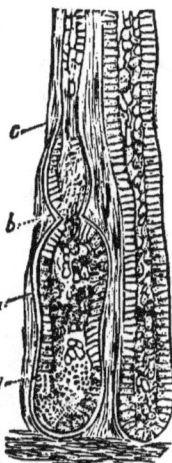

1047. THE GASTRIC GLANDS, almost infinite in number, are exceedingly small, straight tubes of basement membrane, lined with cells, as represented in Fig. 230, the right-hand tube being healthy and the other deranged.

1048. THE CELLS OF THE GASTRIC GLANDS are very rapidly produced at the commencement of the digestive process, and, charged with their important contents, move out into the stomach, dissolving there, or on their way, and yielding the important digestory fluid to the food in the stomach.

1049. THE FLUID SECRETED BY THE CELLS is called gastric juice, its most important component having the name of pepsin.

1050. UPON THE HEALTHFUL CHARACTERISTICS AND ABUNDANT FLOW OF THE GASTRIC JUICE depends, in a great degree, the first process of digestion.

1051. THE GASTRIC JUICE does not FLOW constantly, but when required food is taken, or even thought of, if very necessary, the juice starts and continues to flow from a few minutes to half an hour, according to the necessities of the body and the quantity of food taken.

1052. *Remark.*—The gastric juice gushes freely, if much food is required; but if only a little food is taken, the juice soon ceases to flow, while, if unnecessary food is eaten, the juice does not flow at all; or, if

---

1045. Describe —. 1046. What —? 1047. What are —? 1048. What said of —? 1049. What of —? 1050. What depends —? 1051. When does —?

more food is taken than is needed, only the quantity of juice requisite to digest a proper quantity of food will appear.

1053.   THIS FLUID, most extraordinary in its modes of appearing, and in its actions upon food, has a very simple, watery appearance, without anything very peculiar in its taste or smell, yet it defies the arts of the most subtle chemistry to produce its like; it must be secreted.

1054.   To SECRETE HEALTHY GASTRIC JUICE, (1) a healthy condition of the secretory tissue of the gastric glands, (2) good blood, in sufficient quantity, and (3) an efficient nervous influence, ARE REQUIRED.

1055.   To MINGLE THE JUICE WITH THE FOOD, the alternate action of the muscular fasciculi (see Fig. 80) of the stomach is required.

1056.   WHEN REQUIRED FOOD IS EATEN, three things will be at once noticed in the passive stomach : it will be gradually distended; the juice will begin to start; the color of the inner surface will redden, and the large extremity begin to slowly contract, after which it will relax, and the smaller extremity begin to contract.   The three actions will become more and more conspicuous, till a sufficient quantity of juice is secreted, when it will no longer flow; the color will fade, but the motions will become more rapid.

1057.   AS THE FOOD IS MOVED ABOUT IN THE STOMACH, little by little, it is thoroughly acted on by the gastric juice and dissolved, a part vanishing from the stomach by passing into its blood-vessels, and the remainder going on into the second stomach, to undergo further process there, the stomach gradually diminishing in size till its contents are all evacuated, when it becomes again passive.

1058.   THE TIME REQUIRED FOR FOOD TO BE DIGEST-ED in the Stomach varies with the kind of food, its method of preparation, its mastication, the requirement for food, the quantity taken, the health of the stomach, the blood, the nervous system, and the state of the mind.

1059. SOME KINDS OF FOOD are much more easily digested than others, and some kinds are digested by one person that are not by another. The accompanying table, from Dr. Beaumont, illustrates the first-mentioned fact.

## TABLE,

EXHIBITING THE AVERAGE TIME OF DIGESTION OF CERTAIN ARTICLES OF DIET.

| Articles. | Preparation. | Time | Articles. | Preparation. | Time |
|---|---|---|---|---|---|
| | | h. m. | | | h. m. |
| Pigs' feet, soused | Boiled, | 1 | Pork | Stewed, | 3 |
| Rice | Boiled, | 1 | Soup, chicken | Boiled, | 3 |
| Tripe, soused | Boiled, | 1 | Soup, bean | Boiled, | 3 |
| Apples, sweet | Raw, | 1 30 | Beef, with mustard | Boiled, | 3 10 |
| Soup, barley | Boiled, | 1 30 | Pork steak | Broiled, | 3 15 |
| Trout, salmon, fresh | Boiled, | 1 30 | Pork, recently salted | Broiled, | 3 15 |
| | Fried, | 1 30 | Oysters, fresh | Roasted, | 3 15 |
| Venison steak | Broiled, | 1 35 | Mutton, fresh | Roasted, | 3 15 |
| Sago | Boiled, | 1 45 | Bread, corn | Baked, | 3 15 |
| Apples, sour, mellow | Raw, | 2 | Carrot, orange | Boiled, | 3 15 |
| Cabbage and vinegar | Raw, | 2 | Sausage | Broiled, | 3 20 |
| Codfish, cured, dry | Boiled, | 2 | Beef, fresh, lean, dry | Roasted, | 3 30 |
| Eggs, fresh | Raw, | 2 | Bread, wheat, fresh | Baked, | 3 30 |
| Liver, beef's, fresh | Broiled, | 2 | Butter | Melted, | 3 30 |
| Milk | Boiled, | 2 | Catfish | Fried, | 3 30 |
| Tapioca | Boiled, | 2 | Cheese, old, strong | Raw, | 3 30 |
| Milk | Raw, | 2 15 | Eggs, fresh | Boiled hard, | 3 30 |
| Eggs | Roasted, | 2 15 | | Fried, | 3 30 |
| Turkey, wild | Roasted, | 2 18 | | | |
| ———, domesticated | Boiled, | 2 25 | Flounders, fresh | Fried, | 3 30 |
| ——— | Roasted, | 2 30 | Oysters, fresh | Stewed, | 3 30 |
| Potatoes, Irish | Baked, | 2 30 | Potatoes, Irish | Boiled, | 3 30 |
| Pig | Roasted, | 2 30 | Soup, mutton | Boiled, | 3 30 |
| Parsnips | Boiled, | 2 30 | Soup, oyster | Boiled, | 3 30 |
| Meat, hashed with vegetables | Warmed, | 2 30 | Turnips, flat | Boiled, | 3 30 |
| Lamb, fresh | Broiled, | 2 30 | Beef, fresh, lean, with salt only | Boiled, | 3 36 |
| Goose | Roasted, | 2 30 | Corn, green, and beans | Boiled, | 3 45 |
| Cake, sponge | Baked, | 2 30 | Beets | Boiled, | 3 45 |
| Cabbage head | Raw, | 2 30 | Beef, fresh, lean | Fried, | 4 |
| Beans, pod | Boiled, | 2 30 | Ducks, domesticated | Roasted, | 4 |
| Chicken, full-grown | Fricas'd, | 2 45 | Fowl, domestic | Boiled, | 4 |
| Custard | Baked, | 2 45 | | Roasted, | 4 |
| Apples, sour, hard | Raw, | 2 50 | Salmon, salted | Boiled, | 4 |
| Oysters, fresh | Raw, | 2 55 | Soup, beef, vegetables and bread | Boiled, | 4 |
| Bass, striped, fresh | Broiled, | 3 | Veal, fresh | Broiled, | 4 |
| Beef, fresh, lean, rare | Roasted, | 3 | Pork, recently salted | Fried, | 4 15 |
| ———, steak | Broiled, | 3 | Beef, old, hard, salted | Boiled, | 4 15 |
| Corn cake | Baked, | 3 | Cabbage | Boiled, | 4 30 |
| Dumpling, apple | Boiled, | 3 | Ducks, wild | Roasted, | 4 30 |
| Eggs, fresh | Boiled soft, | 3 | Suet, mutton | Boiled, | 4 30 |
| Mutton, fresh | Broiled, | 3 | Veal, fresh | Fried, | 4 30 |
| ——— | Boiled, | 3 | Pork, fat and lean | Roasted, | 5 15 |
| Pork, recently salted | Raw, | 3 | Suet, beef, fresh | Boiled, | 5 30 |

1059. What said of —? Mention the kinds of food digested most easily; with most difficulty, etc.

1060. SOME METHODS OF COOKING hasten and others delay or altogether prevent digestion. Frying meats is not as wholesome a mode of cooking as broiling. (See Dr. Beaumont's Table.)

1061. FOOD NOT PROPERLY MASTICATED cannot be readily digested, because it is not sufficiently comminuted nor mixed with sufficient gastric juice.

1062. IF FOOD IS EATEN WHEN NOT REQUIRED, it does not remain harmless in the stomach until required, but begins to ferment and provoke disorder. Food should not be eaten in advance of appetite.

1063. IF MORE FOOD IS EATEN THAN IS REQUIRED, it retards the digestion of that which is needed. Food should not be eaten beyond the point of satisfying the appetite. When all parts say enough, and there is a complete tranquillity of the body produced, eating should be stayed.

1064. Many times THE HEALTH OF THE GASTRIC GLANDS is such that but a small portion of food can be digested, or, perhaps, none at all, when it is really needed by the body. Appetite will not at such times usually exist; if it does it must not be gratified, since rest is essential to allow the gastric glands to recover their healthy condition. To force digestion is only to make a bad matter worse.

1065. *Remark.*—When digestion cannot take place the body must be so adjusted, as far as possible, as not to require the results of digestion; viz., it must rest and be kept warm.

1066. IF THE BLOOD IS BAD, good gastric juice cannot of course be formed; hence it is important to digestion that the respiratory organs should be moveable and well supplied with pure air. Tight dressing and bad ventilation are always productive of deranged digestion.

FIG. 231.     FIG. 232.

Fig. 231 shows how, by improper position, the digestory organs are directly compressed, and indirectly suffer for want of pure air that cannot be freely inhaled; while Fig. 232 shows a proper position. The first will be soon broken down, while the second will improve in health by a reasonable amount of such exercise. When the work or reading cannot be seen on account of near-sightedness, glasses should be worn, to obviate the necessity of curving over the work; or it should be raised to a convenient seeing distance.

1067. It is also evident that the BLOOD MUST BE SUPPLIED TO THE STOMACH IN LARGE QUANTITIES during the formation of the gastric juice; hence, no violent or engrossing exercise, muscular or nervous, ought to be engaged in immediately after eating.

1068. THE EXHAUSTION OF THE NERVOUS SYSTEM by severe labor or mental application prevents the formation of gastric juice, and rest must be taken before eating in such cases.

1069. *Remark.*—ON THE OTHER HAND IT WILL BE SOMETIMES FOUND that nervous exhaustion arises from want of food of the right kind: in fact, this is frequently the case with those who are called upon to make powerful or continued mental effort, and they require a more generous diet, or one well adapted to nourish the nervous tissues. It is not every man who can digest out of ordinary food the amount of nutrition necessary to supply an active brain. In this case he must select brainial food.

Describe Fig. 231. Describe Fig. 232. 1067. Why must —? 1068. What said of —? 1069. What —?

1070.  Nothing is more easily observable than that the state of THE MIND INFLUENCES DIGESTION; pleasant emotions, a relish of food, and an undistracted, unburdened mind facilitate digestion, while gloom, depression, anger, nervousness, and anxiety heavily impede it.

1071.  *Remark.*—It MAY BE THOUGHT IMPOSSIBLE to know what takes place in the stomach during digestion.  In the year 1822, in the State of Michigan, a young man named St. Martin, serving in our army, was injured by the accidental discharge of a gun, the muzzle of which was about a yard from his body.  The buckshot tore open his side, lacerating his lung and stomach.  He fell into the hands of Dr. Beaumont, by whose aid he recovered, leaving, however, an irregular opening into his stomach, about an inch and a half in diameter, that adhered to his side in such a manner that food could be put into or taken out of the stomach, and the organ examined under various circumstances.  He lived with Dr. B. two years, whose account of the case is deeply interesting and profitable.  St. Martin, however, fearing the experiments would affect his health, plunged into the depths of Canada, and was lost sight of till the railroads brought him to light a few years since, when he was alive and well, and went to Europe to exhibit himself, the aperture never having closed.

1072.  *Remark.*—DR. BEAUMONT MADE only a series of physical observations and experiments, yet he conclusively determined many important facts : that different substances, and different methods of cooking them, require different times for digestion; that partially masticated food is tedious in digesting, and often irritates and inflames the stomach; that, indeed, *thorough mastication* is one of the most important steps toward health ; that food not needed is not digested, and soon deranges the stomach; that a hearty appetite and moderate distension of the stomach, combined, favor digestion ; that moderate exercise. facilitates digestion, while active exercise of either muscles or brain as surely retards it; that sleep immediately after eating is not advantageous; that ill-temper or other ill dispositions check the entire digestory action, which, after a time, goes on again, but not perfectly; that too much liquid with food retards digestion ; that the frequent use of alcoholics, wine, cider, etc., *invariably inflames the stomach ;* that the stomach is not sensitive when touched, and may even be so diseased as to bleed without exciting pain !  Other experiments and observations upon man and animals, since those of Dr. B., have also still further and more clearly shown the true nature and use of the first process of digestion.

---

1070. Does —?  1071. What —?  Describe the case of St. Martin.  1072. — what ?  How many points did he determine ?

1073. THE USE OF THE PROCESS TAKING PLACE IN THE FIRST STOMACH is to dissolve the food and allow the fat and starch to pass on into the Second Stomach; to set at liberty any sugar and gum the food contains, either to pass into the blood-vessels of the stomach or to go on into the second stomach; to dissolve the gelatine, and change it in some manner not yet understood; to liberate the albuminous portions of food, and change them into albuminose, ready for nutritious purposes.

1074. The distension of the Stomach by food, and its contraction into a small space when empty, as well as its motions, require that THE STOMACH SHOULD HAVE AN EXTERNAL FREE SURFACE moving without friction against surrounding parts.

1075. THE EXTERNAL SURFACE OF THE STOMACH is serous membrane, pearl-colored, smooth, and moistened with the glairy serum.

1076. THE STOMACH IS THEREFORE SAID TO BE CONSTRUCTED of three coats, mucous, muscular, and serous; or it may be described as constructed of sinewy tissue, differently wrought in different parts of its thickness, with muscular fasciculi in the centre and blood-vessels, lymphatics, and nerves throughout, lined with basement membrane and mucous and gastric cells, and covered with basement membrane and serous cells.

1077. THE SEROUS MEMBRANE OF THE STOMACH extends upward from each side of its smaller curvature toward the liver, forming a ligament to sustain the stomach, as seen at 4, Fig. 233, while at the larger curvature the serous membrane from each side (5) extends down like an apron (Omentum, caul) in front of the organs below, in which omentum an abundance of fat is netted most beautifully (see Fig. 234). It is used by butchers to improve the appearance of their meat, that taken from a fat creature being spread upon the quarters of a lean one.

---

1073. What is —? Write the uses in tabular form. 1074. Why should —? 1075. Describe —. 1076. How is —? 1077. Trace —.

FIG. 233.

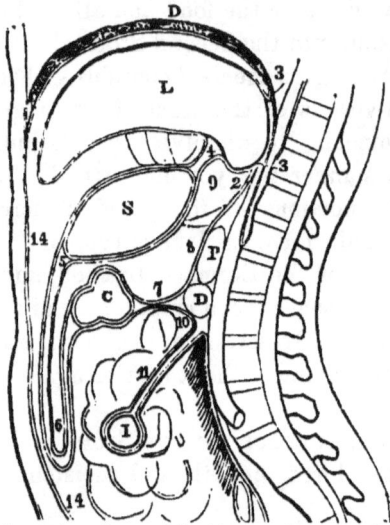

Fig. 233. THE SEROUS MEMBRANE OF THE STOMACH CAN BE TRACED up at 4, then forward under the liver, L, up over it to 3, where it passes to the under surface of the diaphragm, D´, from the front part of which it extends down, forming the inner surface of the walls of the abdomen, and is there called peritoneum. From 5, the serous membrane of the front part of the stomach extends down, as seen, turns up, passes to the under part of C, colon, the surface of which it forms, then ex·tends to the duodenum, D, turning down at 10 from the spinal column, to ex-tend to I, small intestine, the external coat of which it forms, and extends up back, adhering throughout to 11, thus forming what is called a mesentery, in the midst of which blood-vessels, lacteals, and nerves are found connecting with the second stomach. Again, at 5, the serous membrane from the back of the stomach extends down to 6, adhering all the way to the layer from the front of the stomach; it turns up, dividing at C, the colon, to form its front and upper surface, a part of its mesentery, at 7, a small portion of D, binding down the pancreas P, and so on around to L, and down to S, the starting-point.

## Second Stomach.

1078. The food, having been dissolved, and its several varieties set at liberty in the first stomach, THE PASTY MASS, CALLED CHYME, IS READY TO RECEIVE the influence of different fluids adapted to act upon each variety and prepare it to pass into the Blood, while a great extent of surface is necessary in order that every particle of useful substance may be abstracted.

Describe Fig. 233. Are the contents of the abdomen attached to its front walls? Do the contents in reality perfectly fill the walls? 1078. What is — ?

Fig. 234. 1, flaps of front walls of abdomen; letters and straight lines indicate regions arbitrarily made for convenience in describing positions of organs. *a*, epigastric, *b*, hypochondriac, *c*, umbilical, *d*, lumbar, *e*, hypogastric, *f*, iliac regions. 2, 3, lobes of liver; 4, gallbladder; 5, round ligament; 6, part of suspensory ligament; 7, 8, stomach; 9, duodenum; 10, spleen; 11, great omentum; 12, small intestine; 13, coecum; 14, vermiform appendage; 15, ascending, 16, transverse, 17, descending, 18, sigmoid colon; 19, epiploic appendages; 21, points to lower edge of diaphragm.

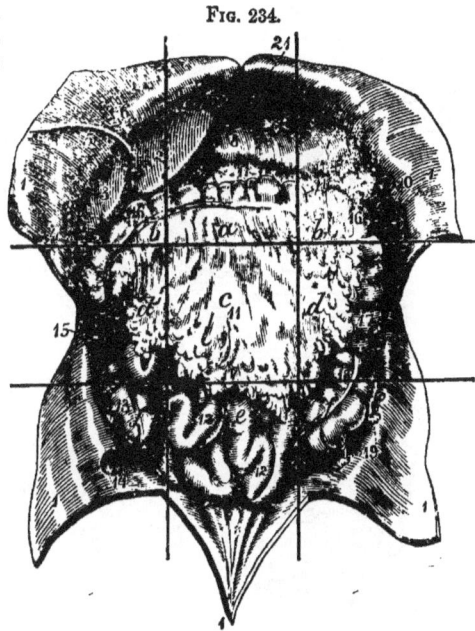

FIG. 234.

1079. THE SECOND STOMACH AND ITS APPENDAGES ARE PERFECTLY ADAPTED to the farther preparation of the food, and the taking from it of everything valuable.

1080. THE SECOND STOMACH is a narrowed or tubular extension of the Stomach, averaging twenty-five feet in length, (sometimes as short as five, and as long as thirty-four,) and an inch to an inch and a half in diameter.

1081. THE SECOND STOMACH IS CONSTRUCTED, like the Stomach, of three coats, except that the serous is wanting at points (see D, Fig. 233). The mucous coat being modified into folds (valvulæ conniventes), as seen in plate 29, its internal surface is immensely increased.

1082. THE SECOND STOMACH IS DIVIDED nominally

Describe Fig. 234. Locate the organs in each region. 1079. To what are —? 1080. What is —? 1081. How is —? 1082. How is —?

FIG. 235.

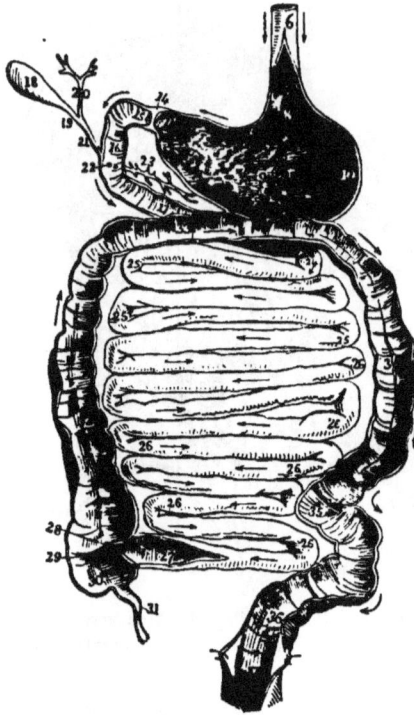

Fig. 235, plan of distended digestory canal, from œsophagus, 6, 7, to rectum, 36; 8, opening into distended stomach; 9, back inner surface; 10, left or large extremity; 11, small extremity; 12, larger, 13, lesser curvature; 14, pylorus contracted; 15, ascending, 16, descending, 17, transverse, duodenum; 18, gall-bladder; 19, duct; 20, duct from liver; 21, duct in common; 22, opening of 21; and 23, duct of pancreas; 24, curvature forward of duodenum; 25, jejunum; 26, ileum; 27, ileum slitted, and opening, at 28, into colon; 29, fold-like valve; 30, coecum; 31, vermiform appendage; 32, ascending, 33, transverse, 34, descending, 35, sigmoid, 36, rectal colon. The jejunum and ileum do not pass directly nor so many times from side to side, as shown. (Fig. 100.)

into the *duodenum*, about twelve finger-breadths long, as its name signifies; the *jejunum*, or empty, about ten feet, and the *ileum* about fifteen feet, long.

1083. THE DUODENUM extends upward and backward toward the liver, turns down, and then, deeply situated (see Fig. 233), crosses to the left side, near which it extends forward.

1084. THE APPENDAGES OF THE DUODENUM are the pancreas, Brunner's, Lieberkühn's, and Peyer's glands, the liver, and gall-bladder.

Describe Fig. 235.   Describe Fig. 100.   1083. Describe the course of —. 1084. What are —?

FIG. 236.

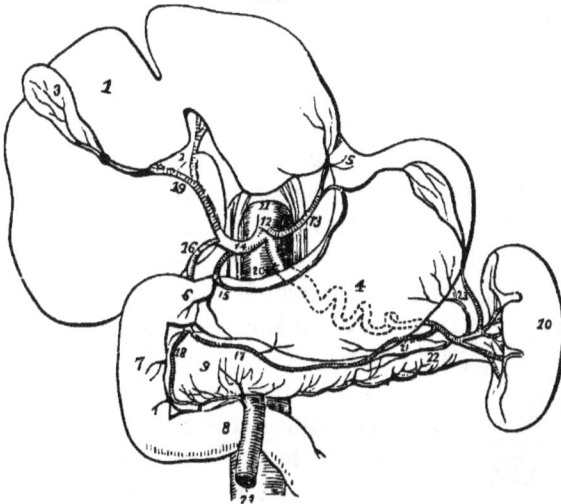

Fig. 236. 1, liver, turned up; 2, its fissure; 3, gall-bladder; 4, stomach; 5, œsophagus; 6, pylorus; 7, descending, 8, transverse duodenum; 9, pancreas; 10, spleen; 11 to 24, arteries.

### Pancreatic, Brunner's, Lieberkühn's, and Peyer's Glands.

1085. THE PANCREAS is an oblong racemose gland (Figs. 101, 102), constructed of sinewy, secretory, and nervous tissue, situated above the transverse duodenum, and opening into it.

1086. THE USE OF THE PANCREAS is to secrete a fluid called pancreatic juice, having the general appearance of saliva.

1087. THE USE OF THE PANCREATIC JUICE is to form an emulsion with fat, and thereby enable it to enter and mingle with the blood, which alone it could not do.

1088. BRUNNER'S GLANDS are minute racemose glands in the walls of the duodenum, apparently of the same character as the pancreas, and secreting the same kind of fluid.

Describe Fig. 236. Do the arteries branch off singly or double? 1085. What is —? 1096. What is —? 1087. What is —? 1088. What are —?

1089. LIEBERKÜHN'S GLANDS, called also the tubular glands, are minute straight tubes, in immense numbers, throughout the Second Stomach, lined with cells (see Plate 29) that secrete and pour into the intestinal canal a fluid called intestinal juice.

1090. THE USE OF THE INTESTINAL JUICE appears to be to change starch into sugar, in which condition this calorific element can pass into the Blood.

FIG. 237.

1091. STARCH KERNELS (Fig. 237, magnified) are too large to pass into the Blood, and if they did would delay its circulation in the capillaries: hence, starch must undergo a transformation; if in the body, vital power will be required. Any process of cooking, therefore, or any other preparation that assists in the transformation, is an advantage. The continued application of heat is one of the best and cheapest means for effecting the desired end. All kinds of grain or breadstuffs, cooked for a long time, with a slow fire, are improved. One of the best ways of preparing wheat is to take it in the kernel and cook it for hours, in water nearly at the boiling point. Nurses frequently tie up flour in a cloth and boil it for several hours, and grate up the inside for the use of delicate children. The starch is thus changed to a great extent. One of the best things for laxity of the second stomach is parched corn; it was carried by the Indians as a corrective of the effect of bad water.

1092. PEYER'S GLANDS are minute enclosed sacs (g, Fig. 238), containing granular matter, just beneath the mucous membrane, in which there will be a depression, f, over the sac, owing to the absence of tubular or Lieberkühn's glands.

1093. THE MODE OF ACTION OR USE OF PEYER'S

GLANDS is not known.
They are found singly,
and called solitary, or
in clusters, when they
are called agminated.

FIG. 238.

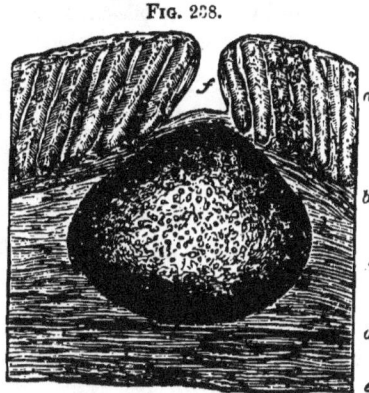

Fig. 238, magnified view
of Peyer's gland, *a*, tubular
glands. *f*, depression of membrane, owing to absence of
*a*: *b*, *d*, muscular tissue; *c*,
sinewy tissue; *e*, serous surface.

*Remark.*—The second stomach and colon are also, of course, supplied with mucous glands.

## Liver and Gall-Bladder.

1094. THE LIVER is the largest gland in the body,
and, like all compound glands, is merely an aggregate
of minute parts, all of which are similar to each other,
and in the liver called lobules.

1095. THE LOBULES OF THE LIVER CAN BE SEEN by
tearing it, when it will exhibit a fine granular structure, each granule being a lobule, and the whole bound
together by a small amount of sinewy fibres; they are
interwoven with an immense number of blood-tubes and
some nerves, while the whole are enclosed by sinewy
tissue, finished with cells, forming a serous membrane.

1096. EACH LOBULE IS CONSTRUCTED of supplying
blood-vessels which encircle its surface, and a vein spring-
ing from its centres; between the former and the latter
an exquisitely delicate network of capillaries exists, in the
meshes of which are the secretory cells of the lobules,
secreting the bile from the blood of the capillaries, and
pouring it into the bile-ducts between the lobules.

1097. *Remark.*—However curious it may seem, PHYSIOLOGISTS ARE

Describe Fig. 238. 1094. What is —? 1095. How can —? 1096. How is —?
1097. In what are —?

NOT YET AGREED how the bile-ducts commence, nor how they obtain the bile from the cells in which it is secreted.

1098. THE LIVER IS FURNISHED with only a small artery, supposed to supply the nutrient blood; but that from which the bile is secreted is supplied through the branches of a vein (portal), that is composed of five, one from the stomach, spleen, pancreas, and two from the second stomach: these unite to form one, and through it pour their contents into the liver.

1099. *Inf.*—It is obvious that THE BLOOD FROM THE FIVE ORGANS MENTIONED does not FLOW back directly to the heart, but ere that organ is reached it is operated upon by the liver; a peculiarity of the circulation worthy of notice and being remembered. ☞

1100. IT IS ALSO ASCERTAINED that some of the cells of the liver secrete from the Blood, and yield back to it, a sweet substance, called liver-sugar; for the Blood that flows from the liver is sweeter than that which flows into it.

1101. *Inf.*—THE LIVER MAY BE CALLED A DOUBLE GLAND, having a double office, to secrete bile, and remove it from the Blood, and to transform some of its elements into sugar.

1102. *Remark.*—BY WHAT PROCESS, change, or from what source, THE SUGAR IS PRODUCED IN THE LIVER, is not yet understood. It may be that the secretion of the bile produces such a condition in the Blood that sugar is formed as a necessary result, and without any further secretion.

1103. THE LIVER, BY PRODUCING SUGAR, BECOMES an assistant heat-producing organ.

1104. *Remark.*—Hence, if it is deranged, and the other organs can deposit fat, it may be expected they will do so, that the heat produced may be preserved thereby. Accordingly it not unfrequently happens that derangement of the liver, as effected by the use of ale, lager beer, etc., will be followed by deposits of fat. On the other hand, if the liver is deranged, and the other organs cannot deposit fat, as is generally the case when distilled alcoholics are used, leanness and pallor follow.

---

1098. How —? 1099. How does —? 1100. What is —? 1101. Why may —? 1102. By —? 1103. — what? 1104. How does the liver effect deposit of fat?

1105. THE BILE is a very complex fluid, of an orange-green color, rather sweetish than bitter, copiously secreted in nearly all animals; yet its complete and precise uses have not been satisfactorily determined.

1106. THE GALL-BLADDER or cyst is a conical bag or pouch of one to three ounces' capacity, beneath the liver, and by a duct opening into its duct at a point near where it leaves the liver. The cyst is constructed of a strong layer of sinewy tissue, lined with mucous membrane.

1107. THE USE OF THE CYST, as usually described, is to receive and retain the bile till the proper time for it to pass into the duodenum. But the contents of the cyst are intensely bitter, and this character must be the result of a secretion there, even if the whole of the gall it contains is not secreted in the cyst.

1108. THE JEJUNUM AND ILEUM are merely extensions of the duodenum, with slight modifications, the valvulae conniventes (Pl. 29) becoming more numerous, and the villi on their surfaces increasing to an almost infinite number. In the villi lacteals take their rise, and numerous absorbing capillaries also exist. (Pl. 29.)

1109. THE CHYME FROM THE STOMACH, passing little by little through the pylorus (porter), is mingled with the pancreatic, hepatic, and intestinal juices, and such changes are wrought in it as gradually adapt all the useable parts to pass into the circulation; by the gentle vermicular or worm-like action of the

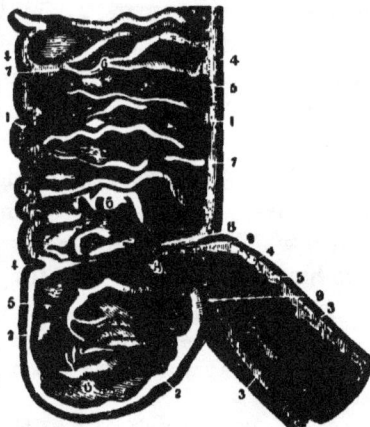

FIG. 239.

1105. What is —? 1106. What is —? 1107. What is —? 1108. What are —? 1109. What said of —? How is changed chyme spread over intestinal surface?

16

canal the semi-fluid mass is pressed slowly over the delicate surface, and each particle ultimately brought in contact with it, and removed either into the lacteals or into the capillaries.

1110. THE RESIDUAL MASS, constituting but a small part, is urged onward into the colon, from which it cannot readily return on account of the valve-like arrangement shown at 8, Fig. 239. Another portion excreted by the glands and cells of the second stomach is also passed on into the colon.

### Colon.

1111. THE COLON is a larger canal than the second stomach or small intestine. Its mucous coat is formed into pouches by the contraction of its muscular bands, and its surface is studded with punctures, the mouths of tubular glands, in appearance like those of the second stomach; but their office seems to be different, it being to eliminate from the Blood useless material, which forms the chief part of the contents usually to be found in the colon.

1112. THE LONGITUDINAL MUSCULAR FIBRES OF THE COLON are gathered into three bundles, shorter than the mucous or serous coats, giving the colon a pouched appearance, and producing a condition by which the contents of the colon may be detained.

1113. THE LENGTH OF THE COLON is about five feet, commencing within the right hip (Fig. 234), where it is fastened down quite closely (Fig. 240). It then ascends to the liver, turns across the abdomen to the left side, down which it follows, curving over the inner surface of the hip, and then becomes straight, or the rectum.

1114. *Inf.*—It MUST BE EVIDENT that tight clothing or bandages just above the hips will tend to obstruct the passage of the contents of the colon, both upward into the transverse, and from it downward into the descending.

1115. *Remark.*—HUNDREDS OF LIVES ARE YEARLY SACRIFICED to

---

1110. What said of —? 1111. What is —? 1112. Describe —. 1113. What is —?
1114. What —? What figures show the effect of tight clothing on colon?

constriction of the colon, producing accumulations in the first or cœcal and the transverse part of it, and consequent inflammation. Clothing should not be fastened upon the hips, but should hang suspended upon the shoulders. (See Ap. N.)

FIG. 240.

Fig. 240, cross section of abdomen, to show 10, the ascending colon, bound down by 28, 29, the lining (peritoneum) of the abdomen; and 14, the descending colon, confined by 18, 20, while 12, ileum, is held in place by the portion of the mesentery formed by 23 and 25. The same membrane can be followed quite around. 7, kidneys.

1116.　THE USE OF THE COLON is to receive whatever is thrown into it from the ileum, to eliminate waste matter from the Blood, and to serve as a portable reservoir for those substances.

1117.　ONE OF THE MOST IMPORTANT LAWS OF HEALTH is to have a regular daily evacuation of the accumulations in the colon.

### The Lacteals.

1118.　THE LACTEALS are a variety of the lymphatics, commencing in the papillæ of the second stomach as club-shaped roots (Fig. 88), and, gradually uniting together at glands in the mesentery (Figs. 93, 104) form at last a small trunk, the thoracic duct, 1, 2, Fig. 241, that extends up into the neck, turns down, and empties into the veins.

Describe Fig. 240.　1115. How are —?　1116. What is —?　1117. What is —?
1118. What are —?

Fig. 241.

1119. THE LACTEAL ROOTS ARE CONSTRUCTED of membrane only, but to this are added a lining of cells and a covering of sinewy tissue to form the tubes.

1120. THE LACTEALS TAKE UP, among other substances, the fat, that gives their contents the color of milk; hence the name, lacteals. Their contents are called chyle.

1121. THE CHYLE TAKEN UP BY THE LACTEALS IS MODIFIED by their action and that of the glands, so that by the time it reaches the veins it has a pinkish cast, but its light color can be traced in the Blood, into the lungs, after which no farther trace is perceptible to the eye, though it is abundantly evident if properly sought.

SECTION IX.

ELIMINATORY ORGANS.

*Kidneys.*

1122. THE KIDNEYS are two similar glands, of a dark brown color, about four and a half inches in length by two in breadth and one in thickness, of the form of kidney-beans, with a smooth surface, dense to the touch, and SITUATED one on each side of the spinal column, corresponding to two dorsal and lumbar vertebræ, the right, however, being a little lower; it is in contact with the liver, the descending duodenum and ascending colon, the left being in contact with the spleen, pancreas, stomach, and descending colon.

1123. *Remark.*—Sometimes there is only one very large kidney, of an irregular shape, and extending across the spinal column, like the pancreas; sometimes the two are connected by a band, the whole having the form of a horseshoe. Sometimes when disease, accident, or experiment destroys one kidney, the other increases so as to supply the use of both.

1124. THE KIDNEYS ARE BURIED in a large quantity of Adipose tissue, the precise purport of which is not determined, further than that it serves to preserve a uniform temperature in the kidneys.

1125. A SECTION OF A KID-NEY SHOWS that it is constructed of an external coat or tunic (2, Fig. 241) of sinewy fibres, woven very densely, within which is a granular layer, 3, with a tubular portion, 4, surrounding a cavity, 7, from which a tube or duct, 8, 9, leads to the vesicle or reservoir.

1126. AGAIN, THE KIDNEY MAY BE DESCRIBED as constructed of an immense number of minute tubes that radiate from the pelvis, 7, outward to the granular-appearing cortical part, where it will be found that each granule is a group of capillaries beautifully enveloped by the extremity of the tube.

FIG. 241.

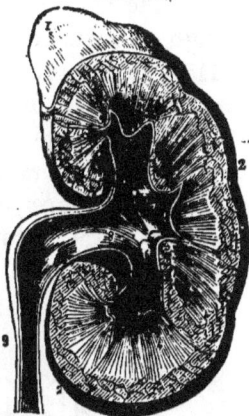

1127. THE STRAIGHT PART OF THE TUBE IS CONSTRUCT-

ED of basement membrane lined with cells, while the outer extremity is constructed of the membrane only, and the water of the Blood circulating through the group of capillaries there, can exude very freely when occasion requires, washing out the eliminations from the cellular part of the tube, where they are excreted. Veins arise from the group of capillaries, and lead to capillaries around the tubes. There are, therefore, two classes of capillaries in the kidneys, and a peculiar circulation.

1128. THE ARTERY LEADING INTO THE KIDNEY is larger, in proportion to the size of the gland, than the artery leading into any other part except the brain. The artery divides and subdivides among the tubules, until microscopic twigs reach the capillaries of the granules or Malphigian bodies, and the veins smaller in capacity than the arteries lead back first to the capillaries of the tubules, and again others arising from them, lead out of the kidney by the side of the arteries. Nerves also extend through all parts of the kidney.

1129. THE STRUCTURE OF THE KIDNEY INDICATES that it is adapted to a double use, which is the removal of water and the elimination of waste matter, produced by the action of the brain in particular, much of which is in a crystalline form, or readily become such. They are illustrated in the following figures of urates and phosphates.

FIG. 242.                          FIG. 243.

FIG. 244.    FIG. 245.    FIG. 246.

1130. THE ELIMINATION OF A CER-
TAIN QUANTITY OF WATER would be
necessary to dissolve and float away the solid eliminations, but the removal of water in large quantities would
rapidly diminish the entire quantity of Blood circulating in the Body, though not the heat-producing part;
that would, of course, pass through its rounds, and produce heat so much the more rapidly. The kidneys,
therefore, assist in regulating the heat of the body.

1131. *Remark.*—THE SHORT TIME IN WHICH SOME SUBSTANCES taken
into the stomach will MANIFEST THEIR ACTION THROUGH THE KIDNEYS, has
often caused the question, Is there not a direct communication between
the two? There is none: the ordinary circulation soon subjects all the
Blood to the action of these organs, and it shows the rapidity of the
circulation.

## The Perspiratory Glands, etc.

1132. THE PERSPIRATORY GLANDS are minute tubes,
some of them straight and some coiled at the inner ex-

FIG. 247.

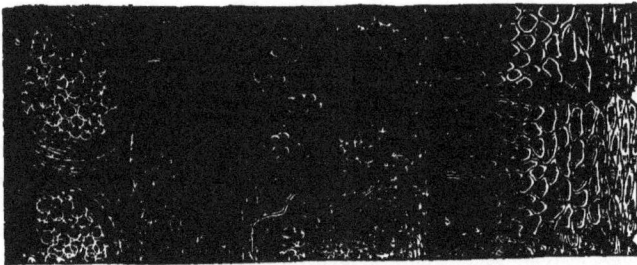

tremity, situated in all parts of the skin, yet most numerous in some parts.   (See Fig. 109.)

1133. THE COILED PART OF THE PERSPIRATORY GLANDS IS CONSTRUCTED of secretory membrane only: the part between the coil and the surface is lined with cells.

1134. THE STRUCTURE OF THESE GLANDS INDICATES their office to be twofold, as is the fact: to remove water, and to eliminate a viscid substance composed of materials no longer fit for use in the body.

1135. THESE GLANDS REMOVE WATER in small quantity to assist in removing the viscid substance, but in large quantities when the heat of the body is excessive, in order that by evaporating from the surface, the skin and consequently the body may be cooled.

1136. THE ELIMINATION OF THE VISCID SUBSTANCE varies in quantity, independently of the watery, but in accordance with the necessities of the Blood, a diurnal increase taking place about midnight, commencing earlier and continuing longer in infancy.

1137. *Remark.*—THIS PERIOD CORRESPONDS to the night-sweats of some kinds of disease, that are merely extraordinary conditions of a healthy, increased action.

1138. *Illus.*—A ROOM IN WHICH A PERSON SPENDS THE HOURS OF NIGHT will be charged with a greater quantity of, and much more unpleasant, exhalations, than will be found under like circumstances in the day-time, though if a person remains awake the increase of perspiration will not be as considerable as if he sleeps; and the discomfort following a sleepless night is partly owing to this fact.

1139. *Inf.*—THE MORNING IS THE APPROPRIATE TIME to cleanse the skin; the clothing worn next the skin during the night should not be worn during the day, but aired; the bedding should be well aired every morning, and often thoroughly cleaned; sleeping rooms should be well ventilated during the night, and thoroughly aired every morning.

1140. *Remark.*—It IS NOT STATED that thorough washing of the skin is absolutely necessary every morning; it may be sufficient to do this two or three times per week: but to rub the skin with a towel every morning will not only do no harm, but will do much good.

---

1133. How is —? 1134. What does —? 1135. Why do —? 1136. What said of —? 1137. To what does —? 1138. What said of —? 1139. For what is —? 1140. What—?

1141. *Remark.*—THE BEST METHOD OF CLEANSING THE SKIN will depend upon the constitution and health of a person. If he can afford to lose heat, cool water may be used; but if he suffers from this, warm water will never be harmful if care is taken to prevent evaporation, by rapidly washing, drying, and covering one part before exposing another. The vapor-bath is the most delicious and thoroughly cleansing of all baths, and if taken during a very brief period, only till the perspiration begins to start on the face, no other than good results can follow; and in case of incipient colds, and in rheumatic and asthmatic affections, the most desirable results may be anticipated. The idea that colds are more likely to be taken after them is erroneous, for heat being by them added to the body it can afford to lose it without detriment. Hence, the Russians pass from their vapor-baths into the ice-cold river without a shock, but with pleasure.

1142. THE USES OF THE PERSPIRATORY GLANDS are among the most extensive and important in the body; and this is the reason why they are so numerous, and one reason why the skin should be supplied with an extraordinary proportion of Blood. It must also be circulated through the skin to keep it warm and in a sensitive condition.

1143. *Illus.*—WILSON HAS COMPUTED that the amount of tubeing in the perspiratory glands of the human body is about twenty-eight miles; but this computation is certainly much too small. The amount thrown out by the skin is never, in health, less than one pound in the twenty-four hours, running up to five pounds in the hour. I have observed, in one case, a man to lose four pounds in one hour and thirty-five minutes: this included that lost by the lungs; he neither ate nor drank anything meantime; most of this was, of course, water. Sanctorius weighed himself and all he ate and drank for thirty years, thus giving all the necessary data on this point.

1144. THE CERUMINOUS AND ODORIFEROUS GLANDS are coiled tubes, like the perspiratory, the former in the ear-tube, and secreting the ear-wax, the latter being clustered with the sweat-glands in the armpit.

1145. *Remark.*—In some persons the odors from these glands are very unpleasant. Cleanliness will prevent them from accumulating; but only strong perfumes, like musk, will overpower them, and are the only resources at present known.

---

1141. What is —? 1142. What are —? 1143. What has —? What amount lost by skin? 1144. What said of —?

1146. THE OIL OR SEBACEOUS GLANDS are purse-shaped pouches of basement membrane, lined with cells, opening usually into the hair-follicle, and pouring their oily secretions out by the side of the hair (Fig. 248), and in other cases on the surface of the skin.

FIG. 248.

1147. THE USE OF THE OIL-FOL-LICLES is to supply oil to promote the softness and flexibility of the skin and hair.

1148. The oil being formed from the Blood, a free circulation of it is ESSENTIAL TO THE FORMATION OF THE OIL.

1149. *Inf.*—RUBBING AND BRUSHING THE SKIN, and keeping it warm, will promote the formation of oil.

1150. *Remark.*—THE TENDENCY OF COLD WEATHER, and much washing of the skin with alcalies, is to prevent the formation of oil, and to remove it faster than it can be re-formed, especially if the alcalies are strong and applied so long that the oil is drawn out of the glands as well as removed from the surface. Then the skin cracks or chaps, and the hair becomes dry. For the skin, in such cases, one of the best things is an ointment of honey, lard, and sweet oil, one third each; and for the hair, the best artificial application is lard, sweet oil, sweet almond oil, and castor oil, mixed with a little borax water; each article should be pure; the proportions vary with the season, being such as will make a cream. The very best oil is that furnished by the head itself.

1151. THE HAIR-FOLLICLES are short tubes lined with cells, the lower ones of which grow up into the hairs themselves, becoming scales at the surface of the hairs, and remaining more pulpy within.

1152. THE USE OF THE HAIRS is in some cases orna-mental, in some protective, as well as eliminatory, and in others it appears to be only the latter.

1153. THE HAIRS GROW from the Blood; and only such causes as supply it with appropriate material and

---

1146. What said of —? 1147. What is —? 1148. What is —? 1149. What said —? 1150. What is —? 1151. What said of —? 1152. What is —? 1153. From what —?

circulate it rapidly around the roots of the hair can promote its growth.

1154. *Remark.*—The various lotions and nostrums that are sold to promote the growth of the hair can have no other effect than to increase the circulation of Blood, and this the brushing alone would do equally as well. Some are said to prevent dandruff; that is only the external layer of scales, and ought to come away.

1155. THE NAILS are only the cells of the external layer of the skin, with an increased quantity of horny matter deposited in them, and modified in a few other particulars.

FIG. 249.      FIG. 250.      FIG. 251.

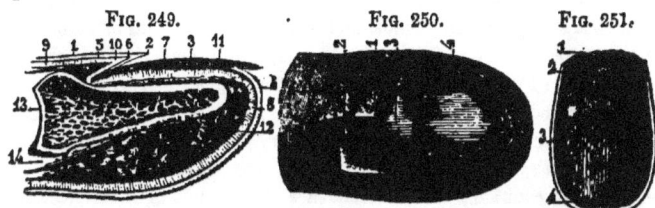

Fig. 249, vertical section of finger-end. 1, cuticle; 2, where it turns back to become continuous with the nail, 3; 4, cuticle leaving nail; 5, 6, 7, 8, surface of the dermis or first layer of cuticular cells; 9, 10, 11, 12, dermis or true skin.

Fig. 250 represents the surface (matrix) on which the nail grows; 1, being the skin overlaying the back edge or root of the nail; 2, a portion of 1 turned back; 3, shows less vascular part, and 4, the more vascular.

Fig. 251 represents the under surface of a nail, grooved to fit the matrix.

1156. THE NAILS GROW forward from the root at the same time that additional cells are applied from below, increasing their thickness toward the front part.

## SECTION X.

### MODIFICATORY ORGANS.

### *Blood-Cells.*

1157. *Remark.*—BLOOD-CELLS cannot, properly speaking, BE CALLED ORGANS, since an organ is constructed of several tissues, and is under nervous influence; while Blood-cells are constructed of a single secretory

1154. What said of nostrums, etc.? 1155. What are —? Describe Fig. 249. Fig. 250. Fig. 251. 1156. How do —? 1157. Why cannot —? How considered?

tissue, and, as far as can be judged, float in the Blood independently of any nervous influence. If, however, they were massed as a gland, they would form the essential part of it, and if their action is no higher than that which exists in the plant, there would be no necessity for nervous influence. Hence they may be considered the simplest form of a gland, dissolved, so to speak, in the Blood; or they may be looked upon as floating plants, and, as their office is modificatory, they are correctly classed under the above head. If the cells need nervous influence, it may be received as they pass through the capillaries of the nervous centres.

1158. *Inf.*—Since the Blood in the body weighs not less than twenty pounds, IF THESE CELLS HAD BEEN MASSED, constituting when dry one tenth, and distended one half the bulk of the Blood, they alone would have weighed ten pounds, and with the other tissues of an organ, at least fifteen pounds. Where in the body could this have been conveniently located? Observe the beauty of the arrangement by which this immense gland is dissolved in the Blood. Also notice, that the blood-cells exceed in weight all the other secretory tissues collectively!

1159. *Remark.*—SINCE THE BLOOD-CELLS HAVE THE NATURE OF PLANTS, it IS PROBABLE that the influence of the sun's light and heat assists in perfecting them and their action, especially as it is certain that an active circulation of Blood through the skin, and a free exposure of it to the influence of the light, and heat, and chemical rays of the sun, is essential to perfection of health, and always conducive to its improvement, while the opposite conditions, viz., feeble circulation in the skin, cold, and darkness, is sure to deprave health, however good.

1160. *Inf.*—WINDOWS SHOULD BE large and numerous, especially in a school-house, and every room in a house should be so constructed as to receive directly the rays of the sun.

1161. *Remark.*—ALL WINDOWS SHOULD BE DOUBLE IN WINTER, that they may preserve the heat that flows in from the sun, as well as that produced within by artificial means; not with the object of keeping out the cold air, an abundance of which, for purposes of ventilation, should be properly introduced, and can be afforded, if windows are double, for rooms are chiefly cooled by heat passing through the glass itself. Double windows are a great economy.

## Lymphatic Glands.

1162. LYMPHATIC GLANDS are small bodies, about the size of grains of wheat, and exceedingly numerous in various parts of the body, particularly in the neck and groin, where, when they are swollen and hard, they are easily felt.

1163. THE LYMPHATIC GLANDS ARE CONSTRUCTED of an external coat of sinewy tissue, also woven through the gland so as to leave numerous passages, a kind of network, lined with basement membrane and cells, that grow, loosen, and float off to form the white blood-cells, from the nuclei of which the red ones are developed.

1164. THE PASSAGES OF THE GLANDS communicate with or open from or into lymphatic tubes, through which lymph is received and removed,—modified by cellular action.

1165. THE GLANDS ARE ALSO CONSTRUCTED of capillaries and nerves; by the former Blood is circulated through, and by the latter nervous influence exerted upon, the glands.

1166. *Remark.*—It is evident that the cellular action of the gland may be influenced by the Blood it receives; and it may be that a part of the lymph is taken up by the capillaries.

1167. *Remark.*—It is not possible to calculate the size of a gland that would be formed by the Lymphatic glands if amassed in one organ, but it would weigh several pounds. Hence these glands are distributed where there is room for them, where their office will be best performed, and without disfiguring any member.

## *Spleen.*

1168. THE SPLEEN is a purple-brown, mottled gland, four to five inches long by three to four broad and one to two thick, shaped like half an egg cut lengthwise, the inner surface, however, being slightly concave.

1169. THE POSITION OF THE SPLEEN is in the left side, the outer or convex surface, shown by Fig. 251, being in contact with the walls of the trunk, and the opposite concave surface, Fig. 112, being in contact with the stomach and pancreas. It is attached to the dia-

FIG. 252.

phragm above and to the large extremity of the stomach.

1170.  *Remark.*—Though THE SPLEEN is located in the vicinity of the digestory organs, and usually described under the same head, it does not appear to HAVE ANY DIGESTORY OFFICE.

1171.  THE SPLEEN IS CONSTRUCTED of an external serous coat, to prevent friction; the sinewy fibres of this intersect the spleen in every direction, forming a sponge-like structure, in the spaces of which a pulp, looking like clotted blood, is found.  This pulp is composed chiefly of blood-cells, unchanged; others in various states; a fine granular matter, etc.

1172.  THE DIVISIONS OF THE SPLENIC ARTERY are exceedingly numerous, and the capillaries and veins correspond.  Lymphatics exist in the spleen, and it is also supplied with nerves.  On the minute arteries small bodies (splenic corpuscles) are found; use unknown.

1173.  As the spleen has no inlet but an artery, and no outlet but a vein, THE USE OF THE SPLEEN MUST BE to modify the Blood in some manner; and every indication at present shows that its use has reference to the blood-cells, and is probably adapted to the change of the nuclei of the white into the red cells.

1174.  *Remark.*—The spleen has been removed from animals by experiment, and from man by accident, without affecting health; it is therefore evident that its office must be unimportant, or can be performed by some other part.  And since the spleen at times enlarges enormously, as when the passage of the Blood through any part is obstructed for a time, it appears probable that the cells that cannot cir-

1170. Does —?  1171. How is —?  1172. What said of —?  1173. What must —?
1174. Effect of removing spleen?

culate freely in Blood are amassed in the spleen, especially as they are always small in number in the Blood when the spleen is enlarged.

## Thyroid Gland.

1175. THE THYROID GLAND, Fig. 113, is a deep-red, small, saddle-shaped gland, situated upon the windpipe below the larynx, not usually conspicuous, yet can be felt.

1176. THE THYROID GLAND IS CONSTRUCTED of an external sinewy tissue, the fibres of which intersect the gland, forming small cavities lined with basement membrane, the inner surface being clothed with cells, and the central cavity filled with a viscid, amber-colored liquid containing granular matter. It is also constructed of blood-vessels, lymphatics, and nerves.

1177. THE USE OF THE THYROID GLAND can at present only be said to be, to modify the Blood, of which there is assurance, since its only inlet and outlet is blood-tubes, and its structure is in part secretory.

1178. *Remark.*—THE THYROID FREQUENTLY ENLARGES to a very uncomfortable size. This is more usual among females; in certain families also there is this tendency. But so far as the author has observed, in this country, the enlargement of the thyroid is associated with the use of hard water, especially that produced by magnesian limestone; with rare exceptions the enlargement can be prevented or diminished by the use of soft water. The surest course for a person or family thus affected is to move to a locality where the water is soft and the air pure.

## Thymus Gland; Suprarenal Capsules.

1179. THE THYMUS GLAND is located back of the upper part of the sternum; it is of noticeable size in infancy, but diminishes to a trace at maturity; use unknown.

1180. THE SUPRARENAL CAPSULES are small bodies, found at the summit of the kidneys; use not understood.

1181. THE THYMUS AND SUPRARENAL GLANDS ARE CONSTRUCTED of sinewy and secretory tissue, and supplied with blood-vessels, lymphatics, and nerves.

---

1175. What is —? 1176. How is —? 1177. What is—? 1178. Does—? In what cases? 1179. Describe —. 1180. Describe —. 1181. How are —?.

# CHAPTER IV.

## Synthesis of Organs into Apparatus, and of uses into Functions.

### Introductory.

1182. *Remark.*—A REVIEW OF THE ORGANS WILL SUGGEST that they act together in ten classes to accomplish ten results. Each class is called an apparatus, and each result a function.

1183. *Remark.*—THOUGH EACH ORGAN HAS BEEN FOUND TO BE CONSTRUCTED IN PART OF BLOOD-VESSELS, when the organs are classed as apparatus, the blood-vessels are considered as parts of the circulatory apparatus superadded to other apparatus.

### SECTION I.

### *Motion: Motory Apparatus.*

1184. THE PRODUCTION OF MOTION REQUIRES a ganglion to produce, a nerve to transmit, and a muscle to receive, nervous influence, and usually some parts of a skeleton for the attachment of muscles.

1185. All the muscles of the body, their connecting nerves and appropriate ganglia, with the entire skeleton, CONSTITUTE A GRAND MOTORY APPARATUS. (Pl. 1, 2, 3, and Figs. 1, 2, 3, Pl. 6.)

1186. THE MOTORY APPARATUS IS SUBDIVIDED into voluntary, or that over which the mind has control, and involuntary, or that which the mind does not control.

1187. *Inf.*—THE LATTER are included in the different kinds of Blood-making apparatus.

1188. *Remark.*—THE MUSCLES UNDER THE INFLUENCE OF THE MIND OUGHT TO ACT INVOLUNTARILY at various times, as when the action of one muscle or group of muscles is necessarily, or by habitual action, associated with that of another.

---

1182. What will —? 1183. — how considered? 1184. What does —? 1185. What necessary to —? Describe Pls. 1, 2, 3, and Figs. 1, 2, 3, of Pl. 6. 1186. How is —?

1189. *Illus.*—In walking, the action of muscles necessary to balance the body should be involuntary.   In playing a familiar tune, or in speaking, certain muscles should be involuntarily associated in action.

1190. THE VOLUNTARY MOTORY APPARATUS MAY BE SUBDIVIDED in accordance with the groups of muscles naturally associated in action.

1191. *Remark.*—THIS VIEW WILL ACCOUNT for the continuous ganglia of the spinal cord, the lower ones involuntarily associating and harmonizing the action of the muscles of the lower extremities, other parts of the cord doing the same in regard to muscles corresponding to them, till at the upper part of its motor tract the cerebellum is found crowning its involuntary office by harmonizing the action of all the muscles of the body, while the voluntary portion of the tract extends still higher, and reaches forward to receive the wilful action of the mind.

1192. *Inf.*—EXERCISE OF THE MOTORY APPARATUS is something more than exercise of muscles—it is also that of nerves, ganglia, and even the skeleton; and disease or derangement of the Motory apparatus may be of either ganglia, nerves, or muscles, or of all at once.   When motion cannot be produced, the question is, which part of the apparatus is at fault ?   Perfect action of the Motor apparatus requires exercise and perfection of the ganglia and nerves, and exercise of the motor ganglia is particularly essential.

1193. THE DECOMPOSITION ATTENDANT UPON EXERCISE OF THE MOTOR APPARATUS REQUIRES it to be also associated with centres that excite appetite, influence digestion, and especially with those influencing respiration.

1194. ONE OF THE MOST IMPORTANT ADVANTAGES OF THE MOTORY EXERCISES, is that they cause more air to be breathed ; in fact, if the chest is constrained, or the air impure, muscular exercise is harmful rather than beneficial.   Let the trunk-walls be free to move, cause pure air to be inhaled, then shall exercise invigorate the whole body, giving a hearty appetite, good digestion, a glowing complexion, sound sleep, and in every way a full measure of physical enjoyment, and of course mental profit.

1195. *Inf.*—Those exercises that cause free respiration, as SPEAKING, SINGING, READING ALOUD, ROWING, etc., must be the best.

---

## SECTION II.

### *Intellection: Intellectory Apparatus.*

1196.  The Intellectory Apparatus is constituted of ganglia only.

1197.  *Remark.*—This apparatus requires no less attention because composed of only one kind of organs.  It must be associated with centres that excite an appetite for food appropriate to satisfy its demands.  The regulation of its exercise is especially important, lest it be too great and cause serious harm; and it should be graduated so as to develop that large circulation of Blood needed by this apparatus, when in advanced years the mind requires its intense and continued activity, for which the apparatus must be prepared during a long course of training.  Time is an element quite as requisite, or even more so, in developing the apparatus as in the acquisition of knowledge.  Make haste slowly is here the true motto.

## SECTION III.

### *Emotion: Emotory Apparatus.*

1198.  The Emotory Apparatus is also constituted of ganglia, those only being needed.

1199.  *Remark.*—This apparatus has a remarkable relation to the respiratory, being most active early in life, and less so in advanced years, and always greatly influenced by respiration and the air respired.  It is also associated with the motory apparatus, the gestures of the emotions exciting them.  It is related to the secretory organs, as the flow of tears, saliva, gastric juice, etc., abundantly testify.  Indigestion depresses emotions.  But of all things pure air and an abundance of it excite emotions.  As soon as children burst loose from the close room, and the pure air strikes their Blood, it stirs their emotions; this makes them breathe still more and better, and they become more noisy.  Impure air, and little of it, will keep children very still; in fact, after a while, so still, they will not even breathe.

## SECTION IV.

### *Sensation: Sensatory Apparatus.*

1200.  *Remark.*—In one sense the whole body may be called a sensatory apparatus, since every part of it may be concerned in excit-

---

1196. How is —?  1197. What does —?  1198. How is —?  1199. What relations has Emotory apparatus?  1200. Why may —?

ing sensation, and, in fact, is constantly thus acting; for the very comfortable sensation of health is merely the aggregate result of the healthy action of all parts. Only a part of the organs of the body are voluntarily applied to receiving sensations, and these are what are usually included under the head of Sensatory Apparatus.

1201. THE VOLUNTARY SENSATORY APPARATUS INCLUDES organs of sense, nerves, ganglia, and portions of the skeleton.

1202. *Remark.*—Though the organs of sense are the most conspicuous parts of the Sensatory Apparatus, they are by no means the only essential parts. Indeed, EVERY VARIETY OF SENSATION MAY BE PRODUCED by ganglia alone, but cannot be by nerves or organs of sense alone, those organs being merely the means of exciting the ganglia to action.

1203. *Illus.*—When the organs of sense are removed, or nerves cut or diseased, the ganglia can still act and produce sensations. In dreams also the ganglia, being active, cause sensations of all kinds, objects being apparently seen and directions perceived as in real vision. In delirium tremens unreal sights, sounds, and other sensations are as vivid as if there was a cause outside the brain. The sensations are real, of course, but the apparent cause is not. In nervous headaches light appears to glimmer before the eyes, rumbling sounds are heard in the ears, etc. In these cases rest and a restored condition of the digestory organs will often remove the cause that is acting on the ganglia.

1204. *Inf.*—EXERCISE OF THE SENSATORY APPARATUS INCLUDES that of its ganglia and nerves, as well as that of the organs of sense.

1205. THE EXQUISITE SENSATIONS that result from the proper culture of the senses are CHIEFLY DUE to the proved condition of their ganglia and the mind.

1206. *Illus.*—IMPERFECT MELODY may be the fault of the ear alone, but the CHARMS OF HARMONY depend upon the relations between the ganglia and mind; melody may be appreciated, and harmony not be. Colors may be accurately distinguished, yet their most exquisite blending cause no especial delight, on account of defect in the ganglia.

1207. THE VOLUNTARY SENSATORY APPARATUS may be subdivided into as many kinds as there are organs of sense, viz., six; and the appropriate exercise of each develops its perfection, both in skill and in affording

pleasure, by securing a proper circulation of Blood
through it.

1208. *Remark.*—As the ganglia of the senses (SENSORIUM) are near
the other brainial ganglia, and in relation with them, it is not strange
that their activities mutually influence each other, since if activity of
one class of ganglia solicits Blood to itself it would easily be turned
from that channel to another near by. Hence, to address the senses,
and arouse the circulation toward their ganglia, is to advance one step
toward arousing the other ganglia; and as the senses are always open
to attack, or solicitation, one of the best ways of reaching the inatten-
tive mind with instruction is often through the senses, as they, especially that of hearing, have a very intimate relation to the emotions, and
these may be easily awakened through the portals of the ear. So also,
through the sensatory and emotory, the intellectory ganglia are aroused.

1209. *Inf.*—When children are listless, let them sing some lively
song; the tones heard, the air inspired, and the exercise enjoyed, will
quicken the Blood through all the ganglia. For this purpose, object-
teaching must, of course, be admirably adapted.

### SECTION V.

### *B. Circulation: B. Circulatory Apparatus.*

1210. The Capillaries, Veins, Hearts, Arteries, and incidentally the Lymphatics, are but parts
of a whole, adapted to keep the Blood in rapid motion
through every organ of the body, and very frequently
expose it to the action of the air in the lungs.

1211. The Circulatory Ap-
paratus is divisible into the
pulmonary or respiratory and the
systemic; each is constituted of
a heart, arteries, capillaries, and
veins, both sets being required
to complete a circuit, as shown
by adjoining table of initials.

1212. The systemic circu-
lation may be subdivided according to the ten kinds
of apparatus; for practical purposes, however, it may

better be divided into *digestory, brainial, muscular, bronchial, cutaneous,* and *renal.*

1213. *Remark.*—Of course, if one part of an apparatus requires more Blood than usual, all parts of it will, and it will be philosophical and correct to suppose that Blood will in such cases be supplied to all parts if to any, as otherwise action could not take place; yet THE LATTER DIVISION IS MOST PRACTICAL.

1214. THE DIGESTORY DIVISION OF THE SYSTEMIC CIRCULATION is peculiar in this, that the Blood flowing through it, and that from the spleen, does not pass directly to the lungs, but is first circulated through the liver.

1215. THE MUSCULAR AND CUTANEOUS DIVISIONS have the same general branches, which should be the case, since the activity of the muscles is attended with the production of heat, which should be lost through the skin; and therefore the exercise of the muscles brings Blood to themselves and the skin at the same time.

1216. THE BRONCHIAL DIVISION must be carefully distinguished from the pulmonary circulation.

1217. *Remark.*—It MUST BE REMEMBERED that the bronchial artery arises from the systemic aorta, and opens into the nutrient capillaries of all parts of the lungs, which may thus be overcharged with systemic Blood, limiting the space to be occupied by the air, as well as tending to excite disease in their inner surface membrane (bronchitis), in their outer surface membrane (pleurisy, pleuritis), or in the substance of the lung (pneumonitis), in accordance with the natural divisions of the bronchial artery and its capillaries.

1218. THE BRAINIAL DIVISION MAY BE SUBDIVIDED in accordance with the classes of ganglia that it supplies.

1219. *Inf.*—(1.) Increased circulation through one division only of the systemic circulation must be attended with diminished circulation in one or all of the others.  (2.) If Blood is required in large quantities in one it should not be directed to another.  (3.) If Blood is shut out from one it must overcharge some others.  (4.) The action of the heart must harmonize with the conditions of the other parts of the circulation.

---

1213. Why is —?  1214. What peculiar in —?  1215. What said of —?  1216. What said of —?  1217. What —?  1218. How —?  1219. Inf.

1220. *Illus.*—(1.) Digestion is frequently attended with chilliness; brainial activity, with cold feet. (2.) Persons should not bathe nor abstract the mind during digestion, etc. (3.) If the skin contracts, the Blood overcharges the brain, air or digestory passages, causing death, colds, or derangement of the bowels. (4.) The activity of any part quickens the pulse.

1221. *Remark.*—If it is remembered that THE SKIN SHOULD CONTAIN one third of the Blood, the effects of its exposure till its Blood is driven inward will not be surprising; they would be worse and more often seen were it not for the friendly office of the kidneys, that rapidly remove water from the Blood, reducing its quantity and increasing the amount of heat producible by it. But the waste substance the skin ought to remove, must in this case be eliminated by the air or digestory passages, overtasking them, or it must remain in the body, oppressing all parts, especially the brain; both evils result to a degree. Hence, checking the circulation through the skin is the greatest of evils, and to keep it warm is the most important of rules. It will also, if the facts of ¶ 1217 are considered, appear natural that when the Blood is driven in from the skin, it should engorge the lungs, as they can most readily allow it a place, the digestory canal being next in order, and the brain last. It will also be most likely to distend those parts first that are most active, as in winter the lungs, in summer the digestory canal, producing cough in one case, bowel affections in the other; and will also be most apt to distend parts diseased, they not being able to resist the pressure.

## SECTION VI.

### *N. Circulation: N. Circulatory Apparatus.*

1222. *Remark.*—IT MUST HAVE BEEN ABUNDANTLY EVIDENT DURING THE STUDY OF THE ORGANS, that they require different quantities of Blood at different times; and now it is clearly seen that while the Heart drives out the Blood through a single artery and its branches toward all organs, there is need that its flow be increased and diminished through the different branches at different times. It is also important that the tendency of the Blood to flow to or from parts should at times be resisted.

1223. *Inf.*—When the cold acts upon the skin, and drives the Blood inward, it should be forced outward again.

1224. There are good reasons for believing that THE SYMPATHETIC OR GANGLIONIC NERVOUS SYSTEM (Fig. 192)

IS CHIEFLY OF USE in regulating the flow of Blood in accordance with the need of parts.

1225. *Illus.*—When cold first acts on the skin of a healthy person, it is blanched for an instant, then flushed.

1226. THE DESIRED OBJECT CAN BE GAINED by enlargement of the capillaries of a part, and by a more rapid motion of the Blood through them.

1227. The degree of perfection, and THE ENDURANCE WITH WHICH THE CIRCULATION CAN BE REGULATED, WILL DEPEND upon the constitution and health of the person.

* 1228. *Illus.*—A person was observed in Vermont sitting still in very cold weather out doors, and perspiring freely though thinly clad. He said he never felt cold. A man, fatigued by labor, came out of the woods, was struck by a cold blast, and fell from his load dead before he had time to speak.

1229. *Remark.*—It is not THE CONTRAST OF COMING FROM A WARM ROOM, nor because a person is hot, that causes him to take cold when coming into cold fresh air; but the bad air of the room has exhausted him, and the cold is taken before he leaves the warm room. Let him clothe warmly and breathe the fresh air as speedily as possible.

1230. ONE OF THE MOST IMPORTANT LAWS OF HEALTH IS, preserve an equable circulation of Blood in the body, especially in the skin.

1231. *Remark.*—It has been said that if a person should keep his head cool, the skin warm, the bowels daily active, and a clear conscience, he would live forever. Nobody has disproved it.

1232. *Inf.*—IF A PERSON IS EXHAUSTED by fatigue, sickness, loss of sleep, or in any manner, it MUST BE EVIDENT that reaction cannot take place energetically nor for any considerable length of time, and that therefore he should not then expose the skin to the cold; that if he has been exercising vigorously and sits in a breeze, especially if his skin or clothing is moist with perspiration or otherwise, his health is endangered; that he should not bathe with cold water when fatigued, etc.

1233. TO SUSTAIN A LARGE CIRCULATION OF BLOOD THROUGH ANY PART, the temporary action of the stimulated nervous centres must not be relied upon; but by

---

1225. Effect of cold? 1226. How can —? 1227. On what will —? 1228. Illus. 1229. What said—? 1230. What—? 1232. What —? 1233. What necessary —?

gradually acquired habits the *channels for the Blood must be enlarged* and an EQUABLE FLOW of it thus secured.

1234. *Remark.*—IT IS OF ESPECIAL IMPORTANCE IN CASE OF THE GANGLIA OF INTELLECTION, that by frequent and regular use during the long *period of growth*, the paths of the Blood should be amplified, and the ganglia acquire an aptitude for the rapid action necessary in mathematical studies, for the more deliberate uses to which they must be applied in the study of the languages, and for the observational methods of the natural sciences.

## SECTION VII.

### *Respiration: Respiratory Apparatus.*

1235. THE ORGANS CONJOINED IN RESPIRATION partake of a threefold character; 1st, those that circulate Blood; 2d, those that receive the air; 3d, those that cause the air to pass into and out of its receptacles.

1236. Those that circulate Blood THROUGH THE LUNGS consist of a Heart, Arteries, Capillaries, and Veins; the three latter are parts of the lungs.

1237. THE LUNGS ARE ALSO COMPOSED of air passages, elastic tissue, and enclosing membrane (Pleura).

1238. *Remark.*—Though the LUNGS ARE USUALLY SPOKEN OF and considered as receptacles of air, they are equally such of the Blood, and the quantity of it that passes through the lungs is even more startling than is the amount of air breathed. More than a barrel of Blood per hour, night and day, is driven through the lungs; and if there is a failure of fresh air to act upon it for the brief space of a minute, it becomes so much loaded with poisonous carbonic acid that consciousness begins to waver, and in another minute is gone.

1239. THE MEANS CAUSING THE AIR TO PASS INTO THE LUNGS are the walls of the trunk, the contrary action of which, with the elasticity of the lungs, CAUSES THE AIR TO PASS OUT OF THEM.

1240. THE ELASTICITY OF THE CARTILAGES OF THE TRUNK, both of the ribs and of the spinal column, assist in inspiration and expiration; for when the chest is raised above the medium point the cartilages are stretched and tend to restore the chest, causing expiration, and when the chest is drawn below the medium point the natural action of the cartilages tends to restore the chest to the medium point, and inspire air.

1241. THE MUSCLES OF THE ENTIRE TRUNK-WALLS are at times engaged in respiration, and are divisible into two sets, those of inspiration and those of expiration. One set and the diaphragm enlarge the chest, the diaphragm at the same time pressing down all the organs beneath it, and of course distending the walls of the abdomen; then alternately these contract in harmony with another set of intercostals, and the chest is drawn down at the same time the organs below the diaphragm are pressed against it relaxed, carrying it up, the elasticity of the lungs expelling the air.

1242. THUS THE RESPIRATORY APPARATUS MAY BE SAID TO INCLUDE a receptacle of air and Blood, with a sub-apparatus on each side, one to drive the latter and the other the former through the receptacle.

1243. *Inf.*—Respiration affects all the digestory organs, and the more active it is the more pressure there is upon them.

1244. *Remark.*—Of course all these different parts must act together harmoniously, and all the muscles concerned in the ACTS OF INSPIRATION MUST BE DEPENDENT on the same centres for influence to govern their movements. (See pneumo-gastric nerve, Pl. 3*.)

1245. *Remark.*—RESPIRATION is the most important of the Blood-making functions; related to every organ; eliminating carbonic acid, that, retained, is an oppressive nervous poison; essential to heating the body; by its abdominal action assisting the activity of the digestory canal; most likely to be disturbed and to suffer from exposure of the skin. Is it not astonishing that it should be the most neglected and abused of all the functions? Many forcibly impede, or even constrict it; most supply it with air, especially during sleeping hours, that is extremely impure, or quite intolerable; while few so well appreciate its character as to fully realize the importance of its development, often

1241. What said of —? 1242. What may —? 1243. What said of respiration? 1244. — on what? 1245. What farther said of —?

1 7

supposing that muscular exercise is useful chiefly in reference to the muscles rather than by increasing respiration; while in fact the relations of air to the Blood are more constant, more extensive, and more effective upon health, than those of any other thing, even food.

## The Atmosphere.

1246. THE ATMOSPHERE, WHEN PURE, IS COMPOSED of nearly four-fifths nitrogen, one-fifth oxygen, a trace of carbonic acid, and a variable quantity of watery vapor, that should be near the point of saturation.

1247. THE TEMPERATURE OF THE ATMOSPHERE affects its action in the lungs in two ways. The colder the air, the more of it in a given space, and the more it expands in the lungs, filling its cells and acting more completely on the Blood circulating about them; the warmer the air, the less of it in a given space, and the less it expands in the lungs.

1248. *Inf.*—In winter, when more heat should be produced in the body, the air inhaled is adapted to that purpose.

1249. AIR THAT HAS BEEN BREATHED once, contains about one-fifth less oxygen than pure air; carbonic acid, equal to the loss of oxygen; a variable quantity of other impurities; and is saturated with moisture.

1250. THE AMOUNT OF OXYGEN LOST, AND CARBONIC ACID GAINED, by the breath, depends upon the purity and temperature of the air, the condition of the Blood, the activity of its circulation, the vigor of respiration, and the health of the body.

1251. *Inf.*—THE AIR IN WHICH A PERSON RESPIRES is, with every breath, becoming unhealthy, from diminished oxygen, and from increased carbonic acid, various other impurities, and moisture.

1252. *Remark.*—From the skin many of the most noxious exhalations are constantly thrown into the air. Indeed, from this source there is so much that is poisonous, that when large numbers of persons live compactly, disease is very apt to be developed, as on shipboard, in tenant-houses, and even in schools and colleges.

1246. How is —? 1247. What said of —? 1248. Relation of air to winter. 1249. What said of —? 1250. What said of —? 1251. What said of —? 1252. Remark.

1253. *Inf.—The evil can always be avoided* by sufficient attention to cleanliness and ventilation, especially of the bed-clothing.

1254. IF THE SAME BREATH IS AGAIN INHALED, the oxygen is diminished, and the carbonic acid and other impurities increased still further, but in smaller proportions, while the moisture remains the same.

1255. *Inf.*—THE SAME AIR SHOULD NEVER BE INHALED A SECOND TIME, since it can neither yield the appropriate oxygen to the Blood, nor extract from it the noxious carbonic acid.

1256. *Inf.*—WHATEVER DIMINISHES THE OXYGEN of the air, or increases its carbonic acid, makes it unhealthy.

1257. *Illus.*—Charcoal burning in a room has often destroyed life; so has a candle in a small room: gas, or anything burning in an unventilated apartment, diminishes the oxygen and increases the carbonic acid.

1258. *Remark.*—ALLOWANCE MUST ALWAYS BE MADE in ventilating apartments in which anything is burned. To heat rooms with gas without ventilation is very unhealthy.

1259. MOISTURE is always an essential part of the atmosphere, and should exist nearly to saturation.

1260. WHEN ROOMS ARE HEATED ARTIFICIALLY, moisture should be evaporated into the air, or it will be so dry as to exhaust too much moisture from the lungs.

1261. *Remark.*—DAMP AIR is often thought to be injurious to the lungs, and cold *dry* air is welcomed as wholesome. It is not, however, dry to the lungs; its frost melts and saturates the air ere it enters the lungs. Dampness, if cold, is not obnoxious to the lungs, but it is to a thinly-clad skin. The electric condition of the air has more to do with its uncomfortableness than its dampness. When the air is positively electric, though warm or cold, it will be bracing and refreshing, while if negatively electric, it will be oppressive and unhealthy. These influences are by some attributed to the presence or absence of ozone, a condition of oxygen. But is not that owing to electric action? One thing is certain; air that has swept over ice, and become positively electric, produces a more invigorating effect than can be accounted for by its purity or temperature. The northwest wind in summer comes out from the great polar ice-box bearing healing on its wings and strengthening the well. Rooms ventilated in summer with air passed over ice are not only refreshingly cool, but health-giving to a remarkable degree, while the same temperatures produced by the negatively electric northeast winds are raw, piercing, and unsalutary; the same is true of cellars and basements, which should never be occupied for any length of time, the upper rooms being the most healthy. On the other hand, air passed over nearly red-hot surfaces of iron, is made negatively electric, and will be wanting in ozone; the air will be called burnt, and is very unhealthy; yet furnaces or heaters constructed so as to allow air to pass over very hot surfaces, are used in many, yes, even in most educational rooms. Such apparatus ought to be discarded at once, unless disease is a blessing. Is it not possible to have people taught, by example as well as by precept, the true office, importance, and relations of air, the choicest gift of Divine Benevolence?

1262. VARIOUS IMPURITIES, ARISING FROM OUT-DOOR CAUSES, are found in the air at various times; sewers, marshes, etc., are such sources, while unknown causes of disease often infect limited or extensive localities.

1263. VENTILATION, OR PURITY OF AIR, HAS REFERENCE to various things: 1st, Locality, to be selected or made free from impurities; 2d, Houses, in all their parts, to be constructed so as to allow free passage of air through them without exposing any one to a draft; 3d, The air to be made, or preserved, positively electric; 4th, Air should be kept moist, and free from dust; 5th, Clothing, bedding, etc., should be often cleaned, and very often aired; 6th, The lungs themselves should be often thoroughly ventilated by deep inspiration and expiration. In brief, there must be thorough ventilation of outdoors, indoors, and the lungs.

1264. *Illus.*—ETHER INHALED ONE EVENING was perceived in the breath the next, after active exercise; showing that air may remain a long time in the outskirts of the lungs during quiet breathing.

1265. *Inf.*—It is evidently important that THE AIR SHOULD BE VERY PURE AND QUITE COOL during the quiet breathing of sleep, in order to effect as thorough a ventilation as possible in the lungs, and that deep expiration and inspiration should be made when a person first awakes.

1266. *Remark.*—There need be no fear of A COLD, or other harm, BEING TAKEN DURING THE NIGHT, from cold or damp air, provided it does not draft over a person, and the skin is properly clad. It is only in the case of infants, the aged, or the sick, when heat enough cannot be produced in the lungs to equal that which cold air would take from them, that they cannot with safety be allowed to receive cold air. It should then be warmed, but always pure. Nothing can be worse than impure air. Its influence is illustrated by drowning. (See Ap. O.)

## *Vocal Apparatus.*

1267. THAT THE AIR EXPELLED FROM THE LUNGS MAY BE AUDIBLE, it must be thrown into vibrations, and for this purpose must issue in a rapid, impulsive current.

1268. VIBRATION CAN ONLY BE PRODUCED by considerable pressure upon the air emitted, to effect which the air-passages must be closed hastily, and suddenly opened.

1269. BY RAISING THE WINDPIPE, closing it, and at the same time compressing the chest, the air in the lungs is forced upward against the epiglottis, the raising of which, or the dropping of the windpipe, will allow the air to gush out with a forcible current; one process.

FIG. 250.

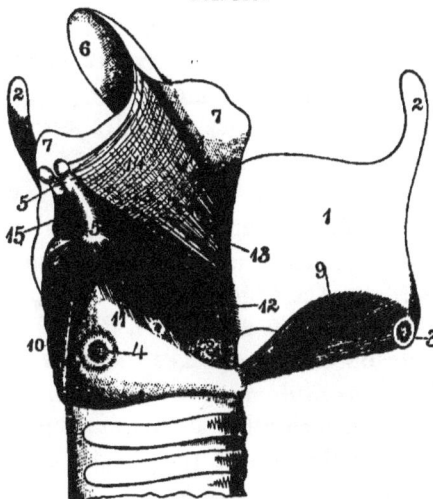

1270. THE SECOND PROCESS CONSISTS in throwing the jet of air into vibrations, which is accomplished in the larynx.

1271. THE LARYNX IS CONSTRUCTED of framework, muscles, nerves, and a lining of blood-tubes.

1272. THE FRAMEWORK IS CONSTRUCTED of

1267. What necessary —? 1268. How can —? 1269. What done —? 1270. In what does —? 1271. How is —? 1272. How is —?

five cartilages, including the epiglottis. The lower, called the cricoid, is shaped in front like a ring of the windpipe, but instead of being open behind it is closed and broad, rising up as shown by 4, Fig. 250. At the

FIG. 251.

upper edge two small triangular carti-lages, called Arytenoid, 5, are jointed in such manner that they can move forward and backward, and toward and from each other. A fourth cartilage, called the thyroïd, is large, forming the prominence of the throat, and overrides the cricoid. In Fig. 250 one half is represented as if turned forward, 1 being the inner sur-face, and 3, the point where it is joint-ed to 4; 2 is its upper horn; 6, epiglottis, 7, turned up to show 14.

Fig. 251. Side view of larynx. 8, trachea; 7, crycoid; 6, thyroid; 5, membrane; 1, 2, 3, 4, hyoid.

1273. THE MUSCLES OF THE LARYNX are very well shown by 9, 10, 11, 12, 13, 14, 15, Fig. 250, their use be-ing to move the arytenoid and thyroid upon the crycoid, that is, the immovable part of the larynx; 12 is not well represented; it stretches across more nearly hori-zontal than it appears to, leaving but a small aperture, like a button-hole, between it and its fellow. .

1274. THE MEMBRANE LINING THE LARYNX is con-tinuous with that of the pharynx and that of the wind-pipe, being somewhat thickened where it covers 12, thus forming the vocal cords, as the edges of the narrow pas-sage are called. (See Pls. 16, 19, 22, and 29, Pl. 30.)

1275. BY THE ACTION OF THE MUSCLES OF THE LAR-YNX the vocal cords can be made more or less tense, and more or less removed from each other at their back part, enlarging or diminishing the size of the slender trian-gular aperture between them.

1273. What said of —? 1274. What said of —? Can a tremulous motion be felt in the larynx when speaking? 1275. What effected —?

1276. The air gushing between the vocal cords is thrown into vibrations, and all THE VOWEL-TONES OF THE VOICE ARE THUS PRODUCED.

Fig. 252.

Fig. 252. Upper view of and through the larynx. 1, 2, thyroid; 3, vocal cords; 4, glottis, or opening between 3; 5, arytenoides; 6, muscle.

1277. THE THIRD PROCESS CONSISTS in the modulations of the tones and in articulations of the breath, which are produced by the muscles of the back and front mouth, these being entirely under mental control.

1278. THE APPARATUS OF SPEECH MAY BE CLASSED under three heads, *Detonatory, Intonatory,* and *Articulatory,* each of which requires distinct attention and exercise.

1279. THE FORMER TWO ARE ASSOCIATED in breathing, and should therefore be, as they are, associated with the same involuntary nervous centres (see Fig. 192); hence they work easily together.

1280. It IS TO BE PARTICULARLY NOTICED that the latter is not in any involuntary action associated with the former two, but that there is a converse relation.

1281. *Illus.*—When the mouth is in use, it is naturally associated with the pharynx and œsophagus, and the larynx is to be closed while what is swallowed passes over it.

1282. *Inf.*—Articulate speech is not therefore natural, but is an invention of man, and can exist only where there is a mind. Animals do not speak articulately, not because they have not articulating organs, out because they have not mind: all the articulating details must be attended to by the mind.

1283. *Remark.*—NATURAL SPEECH, that of the emotions, is merely that of detonation and intonation (naturally associated), except so far as acts of singing are concèrned, or the mere opening of the mouth, as in lowing, bleating, etc. Hence it will be noticed that stammerers can sing unhesitatingly, because the mouth is open in natural harmony with the action of the intonatory and detonatory organs. It is also noticeable that the best singers often do not articulate well.

---

1276. How are —? 1277. — in what? 1278. How may —? 1279. How are —? 1280. What —? 1281. Illus. 1282. Inf. 1283. What said of —?

1284. *Inf.*—THE CHIEF DIFFICULTY IN PERFECTING SPEECH will be in exercising the articulatory organs, and harmonizing their action with that of the other two classes, as it must be done by the control of the will unassociated with the assistance of any involuntary centres.

1285. *Remark.*—As the detonatory and intonatory organs are controlled by involuntary centres, associated intimately with the emotions, these must be active in order THAT SPEECH MAY BE FORCIBLE.

1286. *Inf.*—To read or speak with effect, a person must feel the sentiment he utters.

1287. *Remark.*—ANY PERSON CAN ACQUIRE the power of calling up at will the emotions he desires to have; the gestures of emotions will excite them; thinking of emotional subjects will also excite them, etc., provided the circumstances are favorable: when a thunder-storm is approaching, some persons will be very much depressed, and cannot overcome their feelings; other electrical conditions of the atmosphere exhilarate the emotions. Again, they are depressed by indigestion, derangement of the liver, etc., while they are exalted by health and pure air; which facts show, that the emotions on the one hand are subject to the control of the mind, and on the other are dependent upon their appropriate ganglia, which are acted upon by all the other ganglia of the brain, and of course directly or indirectly by all parts of the body, and the influences acting upon or through them. Therefore the mind can control the emotions by controlling the agencies that affect them.

1288. IF SPEECH IS DEFECTIVE, the precise cause is to be observed and the particular difficulty removed or overcome by exercise.

1289. *Illus.*—If a person stammers, he must be induced to speak with the mouth open; if he lisps, let him read and speak frequently with his teeth closed, until the lisping habit is broken.

1290. *Remark.*—THE ADVANTAGE OF EXERCISING THE VOCAL APPARATUS by speaking, reading aloud, and especially singing, does not depend upon its improvement merely; singing, in particular, causes active respiration, favorable movements of the digestory canal, excites pleasant emotions, drives the Blood to the surface, and withdraws it from the intellectory ganglia. Hence it is an admirable preparation to sleep. Also observe, that children like a good, vocal, laughing romp just before retiring, exciting perspiration, good humor, and full respiration. Laughing is excellent for breathing and good humor; it fattens. Let pupils have a good hearty laughing exercise occasionally; sing often, and particularly in the evening.

SECTION VIII.

## Digestion : Digestory Apparatus.

1291.  THE OBJECT AND NATURE OF DIGESTION SHOW that its organs must act, and of course be arranged, consecutively, since their collective use or function is to dissolve the food into its various elements, and to transform some of its varieties and pass them into the Blood; three steps of one process, the last having several branches.

1292.  TO FORM A DIGESTORY APPARATUS, nothing therefore remains to be done but to associate the organs with common nervous centres, related also to all parts of the body, and of course to the mind.

1293.  ANTECEDENT, WHOLESOME COOKING, is for the purpose of facilitating digestion by grinding the food; dissolving it, or preparing it to readily receive the digestory fluids; pleasing the palate, that excites the more profuse flow of those fluids; and changing starch into sugar.

1294.  *Remark.*—HEAT AND MOISTURE COMBINED are the chief agents in facilitating solution.  At a high temperature, after considerable time, they change the tough, sinewy tissue of meat, into gelatine, and partially or wholly dissolve it, freeing the elements it binds together.  At a higher temperature, in a closed vessel (Papin's digestor), they will dissolve bones into wholesome food.  Heat, by swelling the kernels of starch, cracks them, and, continuously applied, changes part of it into dextrine, a sweetish substance, on the way to become sugar.  Heat affects some articles prejudicially, eggs, cabbage, etc., since it does not facilitate, but retards their solution, and cannot produce favorable changes in them, as there are none to be made, but does produce unfavorable changes.  SCORES OF WHOLESOME DISHES, the fundamental elements of which are alike, may, by various combinations, and the addition of delicious flavors, be provided to please the palate, without in the least diminishing, but rather increasing, their digestibility and valuable characteristics.  In this art of combining food and flavors, and applying heat, the whole of good cookery consists.  By it the strictest economy and the "best living" are compatible; indeed, they are only attainable together. It is a valuable, a *manly* art.  Man only cooks.  Mind only could in-

1291. What do —?  1292. What is necessary —?  1293. What the purpose of —?
1294. Effect of —?  What said of —?

vent cooking. Man only of all creatures has the mind to make the greatest of inventions, fire, or its applications, in the highest degree humanizing; not the least of which is its application to cooking. Let it be always wholesome, always palatable. Economical science says, let nothing be wasted: the brown is the best part of wheat; sour, skimmed, and buttermilk are for warm weather, and being animalized, should not be fed to brutes, etc. Physiology says food should be enjoyed. Good cooking perfectly harmonizes the two.

## SECTION IX.

### Elimination : Eliminatory Apparatus.

1295. THE DIFFERENT ORGANS OF THE ELIMINATORY APPARATUS do not act together to accomplish the same precise works; therefore, in one sense, they should not be conjoined in one apparatus.

1296. Yet, WHEN ANY ONE ELIMINATORY ORGAN CANNOT ACCOMPLISH ITS OFFICE, it will be attempted by one or several of the others; thus showing that they are associated indirectly with the same nervous centres.

1297. THERE IS, THEREFORE, GREAT PRACTICAL BENEFIT in grouping the perspiratory glands and the kidneys as special, and the lungs, liver, and digestory canal as additional, organs of an eliminatory apparatus.

## SECTION X.

### Modification : Modificatory Apparatus.

1298. THE MODIFICATORY ORGANS can only be consistently grouped together as an apparatus, by supposing that the spleen and lymphatic glands are concerned in the production or perfection of Blood-cells. That is probably the case. But what is the office of the Thyroid? Yet to class together various organs will serve to keep them in the memory, and whatever their special office, they certainly in some way modify the Blood.

1295. What said of —? 1296. What occurs —? 1297. — in what? 1298. What said of — ?

# CHAPTER V.

*Mentory (Right and Left) Sanguificatory.*

1299. A REVIEW OF THE FUNCTIONS WILL SUGGEST that, though they are very dissimilar in some respects, several have a similar purpose, and may therefore be grouped together. Four are for the purpose of perfecting Blood; four for perfecting mind; and two for the purpose of circulating Blood through the other two apparatus.

FIG. 253.

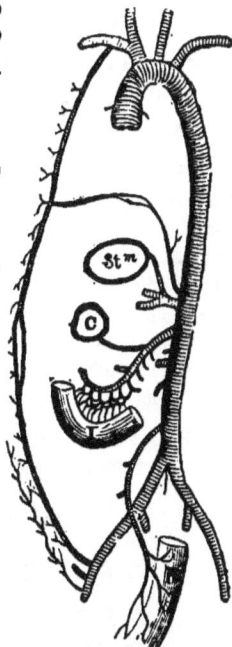

1300. *Inf.*—IT IS EVIDENT, THEREFORE, that the functions exhibit all that either the Mind or Blood can be.

1301. THE STRUCTURE OF THE APPARATUS equally exhibits the grouping. The organs of four kinds, Sanguificatory or Blood-making, are grouped about or appended to the Blood-tubes, being *cavities* of various forms, in the walls of which the Blood-vessels are distributed, as shown by St^m, C, I, and R, of the adjoining figure, while the organs of the mentory apparatus are *solid* and the blood-tubes extend through them in every direction.

---

1299. What will —? 1300. What is —? Describe Fig. 253. 1301. — exhibits what?

1302. *Inf.*—This ARRANGEMENT shows which is the dependent group, for it is evident that the Sanguificatory is appended to support the mentory, which is not merely for the purpose of superintending the supply of the former with proper material; that is only a secondary purpose.

1303. *Remark.*—THE WHOLE OF THE BLOOD-PERFECTING APPARATUS are grouped in the enclosing trunk-walls, where they are suspended to the spinal column and attached to the circulatory circuit of the Blood by appropriate connections, as illustrated—with the exception of the perspiratory tubes that are located in the skin.

1304. A REVIEW OF THE APPARATUS WILL ALSO SHOW that two of each kind of the mentory must be constructed complete, while of the Blood-perfecting one of each suffices.

1305. *Remark.*—EXCEPTIONS to the preceding remark SEEM TO EX-IST in the lungs, kidneys, skeleton, and skin. Of the lungs and kidneys it may be said, that they are each a half of the substance necessary to accomplish their functions; yet they are not—one is larger. The spinal column and the breast-bone or sternum appear to be composed of right and left halves, neither of which would be complete without the other, yet they are perfectly symmetrical; the lungs and kidneys are not. The skin appears to be a continuous surface across the middle line, yet at various places it shows the fact of its intrinsic double character.

1306. THE TWO APPARATUS OF SIGHT exhibit a remarkable deviation from the rule, exceedingly interesting, as explaining a phenomenon, and demonstrating the unity of Mind, and that it is not, though its apparatus is, dual.

Fig. 254.

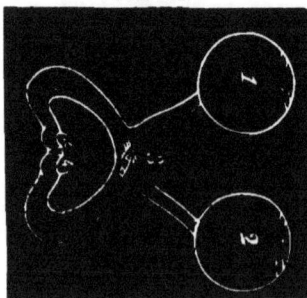

1307. THE OPTIC NERVES EXTEND back from the eyes, 1, 2, Fig. 254, toward each other, till at 3 they meet, and that part of the fibres that commences in the left part of the right eye extends across to the left, and those fibres that commence in the right half of the left eye extend across to the right, interweaving with each other at 3, the commissure.

1308. THE NERVES, THEREFORE, COMMENCING IN THE RIGHT HALF OF EACH EYE, connect with the right ganglion, 4, and those commencing in the left half of each eye, connect with the left ganglion, 4.

1309. *Remark.*—Objects on the right affecting the nerves in the left half of the eyes, cause sensations through the left ganglion, and objects to the left of the eye excite sensations through the right ganglion. But if an object is coming from the right, and must be warded off by the right hand, how is it to be done?

1310. THERE ARE ALSO NERVES communicating BETWEEN the right and left ganglia, and extending from the right motory ganglia across into the left spinal cord, and from the left ganglia into the right side.

1311. *Remark.*—Thus, if an object threatens from the right, the light from it affects the left side of the eyes, and a sensation is caused through the left ganglion, that at the same time causes an action across to the nerves influencing the muscles of the right side, and thus WARDS OFF THE DANGER.

1312. *Illus.*—A MAN MAY HAVE PARALYSIS of some parts upon one side, and at the same time of parts on the other influenced by the same nervous centres. A person may, by disease, as is frequently the case in "blind headache," be unable to see more than half an object, if only one ganglion is affected; if the right half is dark, the left ganglion is affected. There is a more remote cause, however, than disease of the ganglion, such as indigestion or nervous exhaustion, or both together: generally, rest is the only remedy. This condition sometimes seems to be produced by the weather—perhaps its electrical conditions.

1313. It IS THEREFORE LITERALLY TRUE for the most part, and ideally wholly so, that the Mind is served by double sets of apparatus, while in every sense the Blood is made by single sets of apparatus. It is also certain that the double sets serving the Mind are so intimately related to each other, and so associated, that in health there is a perfect unity of mental action produced by and through them.

# CHAPTER VI.

## SYNTHESIS OF GROUPS INTO MEMBERS.—BODY.

### PURPOSES APPLIED. MIND.

1314. WHEN THE TWO GROUPS OF APPARATUS ARE BROUGHT TOGETHER, they will be found perfectly adapted to each other; the trunk-walls that are for the primary purpose of sustaining the head are exactly adapted to enclose the subsidiary group, and can perform all their acts the better for its being there.

1315. THE TWO GROUPS BROUGHT TOGETHER INSTANTLY EXHIBIT five kinds of members, HEAD, Neck, Trunk, lower extremities and upper extremities, to which a little consideration will add a sixth, the larynx —and thus the *Body* stands complete.

1316. Thus, from the primary *Chemical Elements*, have we come up through *Organic Elements, Tissues* and *Fluids, Organs, Apparatus, Mechanisms, Members,* to the fully developed *Body*, "most fearfully and wonderfully made,"—fit instrument for the attainment of the highest purposes. But how shall its latent powers be developed? What shall attune its lips to eloquence? What shall gather wisdom through its activities and adaptation? What shall provide for its wants under all the varied circumstances of life? To the *Body* let the *Mind* be added, and MAN stands forth complete, potential to the development of all his powers, and to the invention of whatever is necessary to aid therein.

---

Those who desire to read upon these subjects in greater detail, are referred to Leidy's splendid work on Anatomy, also to Harrison's, and to the admirable Physiological works of Draper and Dalton. They may be had of the publishers of this work, at a discount to Teachers.

| | SPECIAL | ANATOMY, | PHYSI- | |
|---|---|---|---|---|

$Body =$  $=$  *Six Members Applications.*  $=$  *Two Groups or Mechanisms Purposes.*  $=$  $=$ 

†*  ‡

The *Right+Left* M, Mental, Mentory, or Prime $G'$; $=$

and

The B, Blood-making $G''$, Sanguificatory or Secondary. $=$

$$R + L$$

$$H + N + T\text{-}w + l\,ex + u\,ex + l \qquad =$$

C of T - w alls . . . .
T runk
ontents $=$

$$H + N + T + l\,ex + u\,ex + l\,arynx$$

tremities pper
tremities ower
runk { Trunk-walls + Contents do.
eck
HEAD { HEAD-walls + Contents do.

$$\text{MAN} = \text{Body} = H + N + T + l\,ex + u\,ex + l$$

WEALTH =
{ Time, Nature, *Mind*, Body, *Labor*, Riches.

{ Mind
Body =

☞ See next two leaves. Cut out near their back, trim near reading. Cut top of this, the next, bottom of next two, outside stars; lap next on this, match daggers, paste fast. Ditto second on next. Behold from left analysis; from right, synthesis!

# APPENDIX.

## A.

*a.* As man is composed of Mind and Body, Anthropology must bridge across the dividing line between the branches of science that treat upon material and those that treat upon immaterial existences. It is the connecting link between them.

*b.* Noology, from *noos*, mind, treats upon mind in general, the mind of the Deity, Theology; the relations of mind to mind in social life, Sociology; as well as of the human mind, Psychology; to wit:

$$\text{NOOLOGY} \quad\ldots\ldots\ldots\ldots\quad \left\{ \begin{array}{l} \text{Theology.} \\ \text{Sociology.} \\ \text{Psychology.} \end{array} \right.$$

*c.* Psychology is embraced under Anthropology; indeed, it takes hold upon Sociology, while on the other hand it embraces all of Human Biology, though usually restricted to the popular part of it. For better illustrating the whole subject, let the whole field of the sciences be presented in a brief view.

*d.* Ontology, from *onta*, being, and *logos*, is the name given to that science that includes all the sub-departments of science, treating upon the nature of everything that exists, or has *ta*, *a*, *on*, being. It corresponds to Pantology, meaning all sciences.

*e.* Whatever has being must be material or immaterial; so Ontology must be divisible into Cosmology, from *cosmos*, matter, that treats upon all material existences, and Noology, that treats as said above.

*f.* All material existences must either exhibit life (organic), or be devoid of it (inorganic); therefore Cosmology is divisible into Biology and Physics.

*g.* As life is exhibited by Vegetables, Animals, and Man, Biology is naturally subdivided into Human, Animal, and Vegetable; certain distinctions between man and animals exalting him above them.

*h.* Inorganic matter has certain properties in common with organic; for instance, it exhibits Gravity, *Astronomy*, heat, light, and electricity, *Natural Philosophy*, and affinities, *Chemistry;* so that Physics is also naturally subdivided into three departments.

*i.* A single table shows the subordinations and relations, as follows: [See page 402.]

OLOGY, AND HYGIENE.

++ *Ten Apparatus (ory) Functions (ion,).*

*Thirty-nine kinds of Organs* — Uses or Actions.

**MENTORY ORGANS.**

| | | ORGANS OF SENSE | | |
|---|---|---|---|---|
| Sensat- | ion = | *b* GANGLIA + SIX KINDS OF SENSATORY NERVES + | Ears . | N + + |
| Emot- | ory = | M *r* GANGLIA | Eyes . | O + + |
| Intellect- | ory / ion = | IN *a* GANGLIA | Noses . | T + + |
| Mot- | ory / ion = | D *i* GANGLIA | Mouths . | |
| | | *n* GANGLIA + MOTORY NERVES | Skins . | E + |
| | | | Muscles + | L + + |

**CIRCULATORY ORGANS.**

| | | | | |
|---|---|---|---|---|
| N. Circulat- | ion = | *Ganglia + Nerves, (Sympathetic)* | | E + |
| B. Circulat- | ory / iorr = | *Hearts + Blood-tubes, (Arteries, Veins, Capillaries, Lymphatics).* | | K + + |

**SANGUIFICATORY ORGANS.**

| | | | | |
|---|---|---|---|---|
| Respirat- | ion = | (Nose + Pharynx) + Larynx + Trachea + Lungs + (Diaphragm + T-w) . | | R + . |
| Digest- | ory = | (Mouth), Salivary-gl. + (Pharynx) + Œsophagus + Stomach, + Gastric-gl., 2d Stomach, (Duodenum, Jejunum, Ilium) + Pancreas + Liver, Gall-bl. + Brunner's, Lieberkühn's, Peyer's gl. + Colon + Lacteals. | | |
| Eliminat- | ory / ion = | Kidneys + Perspiratory gl. + Hair and Sebaceous gl. | | K + + |
| Modificat- | ory = | Spleen + Blood-cells + Thyroid, Thymus, Supra-renal, and Lymphatic gl. . | | S + + |

DEVELOPING MIND. — CIRCULATING BLOOD. — DEVELOPING BLOOD.

R.&L. — G'. — G'.

```
                                  { Theology.
                    Noology. . . . . . { Sociology.
                                  { Psychology.
ONTOLOGY . . . {
                                  { Biology { Human.
                                  {         { Animal.
                                  {         { Vegetable.
                    Cosmology {
                                  { Physics { Chemistry.
                                  {         { Natural Philosophy.
                                  {         { Astronomy.
```

ELEMENTARY . . . . Observation, Language, Mathematics, Drawing.

*j.* Observation, Language, Mathematics, and Drawing, have been placed below the table as Elementary, only to be so regarded to a limited extent, for many of the ideas embraced under those heads arise from the study of the highest departments.

*k.* All the studies of the table, as well as the Elementary, have a Primary, Academic, Collegiate, Philosophical, and Professional aspect.

*l.* What is common Geography but Pictorial Ontology? Is not the germ, a superficial outline or taste, of all the sciences, except perhaps Psychology, found in the most primary Geography? In one sense Psychology is not there, because its subject, the mind, cannot be pictured, and in another sense it is found; but Theology is illustrated in the modes of worship; Sociology is shown in the pictures of capitols, &c.; Human Biology is found in the representation of the races, &c.

*m.* In our commonest works, therefore, we have the elements of all that it is desirable for a person to study; yet objections are sometimes made to introducing so many "ologies" into schools, when in fact they are only advanced expansions of those in use.

*n.* Again. From the departments, as seen above, spring all the Professions: from Theology the *Theological;* from Sociology the *Legal;* from Psychology the *Teachers* or *Professorial;* from Human Biology the *Medical;* from Animal Biology the *Veterinary Surgeons* and *Farriers;* from Vegetable Biology the *Horticulturists;* from Chemistry the *Architects, Engineers, Dyers,* &c.; from Natural Philosophy the *Machinists;* from Astronomy the practical *Astronomers,* the *Navigators, Surveyors,* &c.

*o.* It will be seen, that the order of dependence is that in which the topics are mentioned. To understand well any one, a knowledge of those mentioned below it is essential. Thus a Farmer, for the best prosecution of his labors, requires a knowledge of the constituents of his soil (chemistry), what promotes vegetation, and how to care for his creatures (vegetable and animal Biology); and as himself is to be taken into the account, to be sure he should understand Popular Human Biology. Indeed, any department is much illuminated by light reflected from above itself.

*p.* Human Biology cannot, even for popular use, be com-

[See page 404.]

GENERAL ANATOMY, PHYSIOLOGY, AND HYGIENE.

*Six kinds of Tissues* Properties. + *Fifteen Liquids + Gases* Properties. = *Organic and Chemical Elements.* Properties.

PRACTICAL HINT-WORDS.

*Educate ;*
*Exercise ;*
*Arrange ;*

*Rub ;*
*Clean ;*
*Clothe ;*

*Air ;*
*Water ;*
*Food ;*

*Repose ;*
*Sleep ;*
*Habits.*

= Elements { Organic or Proximate. Compound Chemical. Simple Chemical. }

General {
Blood,
Lymph,
Flesh-Juice,
Serum,
Mucous,
Oil, } Surface.

Special {
Tear-fluid,
Salivas, (two)
Gastric-juice,
Bile,
Gall,
Pancreatic
Brunner's
Lieb'k'ns
Carbonic Acid,
Oxygen, } Juices,

Passive { Bony (Hard), Gristly (Firm & Elastic), Sinewy (Tough), }

Active { Secretory (to Secrete), Muscular (to Contract), Nervous (to Excite), }

ORGANS.

pletely studied without some knowledge of Psychology, because
the Mind and Body are so intimately associated that whatever
influences one necessarily affects the other.

*q.* It will also be perceived that a thorough acquaintance
with all the Departments of the Table, fully developed, secures
all knowledge; for outside of them is nothing, and under these
heads all things will necessarily, naturally, and easily fall. Geo-
logy is but Historic, and Physical Geography but Comparative,
Cosmology; History falls under Sociology; Meteorology and
Moral Philosophy are Comparative Physics and Noology, etc.

*r.* Anthropology, as has been seen, is not a new department,
but is made up of two of them, as follows:

|  |  |
|---|---|
| NOOLOGY . . . . . . . . . . . . | { Theology.<br>{ Sociology.<br>{ Psychology. |
| ANTHROPOLOGY . . . . . . . . . . | |
| BIOLOGY . . . . . . . . . . . . | { Human.<br>{ Animal.<br>{ Vegetable. |

and while the consideration of Psychology, in treating upon the
science of Human Biology is very brief, the relations of the mind
are of such importance, that the little that is introduced is of the
utmost value, and the true name of the science upon which this
work treats would be, Popular Anthropology, except that there
is not so much said upon mind as that name given would neces-
sitate. The better name is that given in ¶ 24.

# B.

| BIOLOGY . . . | { Human,<br>{ Animal,<br>{ Vegetable, } | { PHYSIOLOGY,<br>{ ANATOMY,<br>{ HYGIENE,<br>{ *Pathology,*<br>{ *Therapeutics,*<br>{ *Materia-Medica,*<br>{ *Surgery.* | } POPULAR.<br><br>*Professional.* |

All the divisions of Biology are equally divisible into the
Popular and the Professional. In Vegetable Biology there is
no person who devotes himself exclusively to the cure of diseases of plants, but the intelligent Horticulturist or gardener
may well be styled a professional man.

The Veterinary Surgeon is one who devotes himself to the cure
of the diseases of domestic animals. In case of curing diseases of
man, animals, or vegetables, the questions that will arise will be,
the condition or Pathology of the diseased parts; the methods
or Therapeutics of curing it; the medicines or Materia-Medica
that will be useful; and the manipulations or Surgery that will
be necessary; all of which, to be done skilfully, presuppose
a thorough knowledge of the condition of the parts in health.

# C.

A singular, striking impression, even if not correct, is sometimes admirably adapted to impress the idea it is used to illustrate. " John barked at a dog " will excite the attention of a child, and impress its mind with the grammatical idea it is used to illustrate, much better than the correct form. So will "Little John barked at the big dog," or "the big blue dog." Count up the ten kinds of Apparatus on the fingers, " to keep them at the fingers' ends," or " to show why there are ten digits."

A figure upon a blackboard, if a little grotesque and laughable, will be often serviceable.

Thus break up monotony; make sure the mind is active to receive, and that ideas are so presented that they are sure to enter and be welcomed.

# D.

## Meaning of Train, Educate, Exercise.

A trained man=educated mind+exercised body. These words are not always used in their proper sense. Education cannot properly be applied to the body, nor to an animal; trained may be. Exercise can better be applied to mental activity, but not with the greatest propriety. It should be restricted to the activities of the body. Nor should it be at all limited to activity of the muscles, but should be applied to the activity of all parts: the exercise of the brain, or of the stomach, or of the lungs, being as proper expressions as exercise of the muscles. Muscle culture is but a small part of the physical duties that require repeated and regular exercise—brain culture, lung culture, eye culture, and that of some other parts, being even of greater consequence.

## External World.

But, in addition to the training of man, and for the purpose of training him, the importance of arranging properly the external world must be taken into consideration. In one sense, the training of man will provide for this arrangement, but in another respect it will be better to make a distinction, as it will fix upon the mind precisely what is to be done.

Mind;     Body;     External World.

The position of these words exhibits the relations in which they stand. The Mind exerts a double influence, upon the Body, and through it upon the External World. So, also, does the External World exert a double influence, upon the Body, and through the Body upon the Mind.

Educated;     Exercised;     Arranged.

The above three words express what should be done in order that the influence of Mind, Body, and External World may be

favorable upon each other, and tend to promote the welfare of
man.   Again,

<table>
<tr><td>Mind;</td><td>Body;</td><td>External World,</td></tr>
</table>

correspond to the three grand divisions of science,

<table>
<tr><td>Noology, or Mental Science;</td><td>Biology;</td><td>Physics,</td></tr>
</table>

and

| Mind<br>Educated; | Body<br>Exercised; | External World<br>Arranged, |
|---|---|---|

correspond to

| NOOLOGY<br>Developed; | BIOLOGY<br>Practical; | PHYSICS<br>Applied. |
|---|---|---|

Another view will still better illustrate all the influences
that conduce to Man's

### Welfare or Wealth.

The ancients fabled that Time, personified as Saturn, was the
Father of all things, or Nature.

Man, composed of Mind and Body, being introduced, not
finding the world adapted to his liking, at once applied himself
to appropriate and improve Nature by his Labor, thus producing
Riches with which to assist in satisfying both his Mental and
Physical wants.

Thus, two by two, one material, the other immaterial, there
are six

$$
\begin{matrix}
& \text{Elements} \\
& \text{of} \\
& \text{W E A L T H} \\
& \text{or} \\
& \text{Human Welfare,}
\end{matrix}
\left\{
\begin{matrix}
\text{TIME,} \\
\textit{Nature.} \\
\text{MIND,} \\
\textit{Body.} \\
\text{LABOR,} \\
\textit{Riches.}
\end{matrix}
\right.
$$

### Meaning of Riches.

Riches is not here used to signify an abundance merely of
material possessions, but all, the merest trifle as well as the most
extensive domains that man can obtain by appropriation or by
the improvement of Nature.

### Distinguished from Wealth.

Wealth is not used in the same sense as Riches.  A prominent and important means of enjoyment, they are far from being the only element, or even the chief one, since Nature and
Mental and Physical activity, without any reference to increase
of Riches, are directly the most considerable sources of happiness; as Solon says,

> " The man who boasts of golden stores,
> Of grain that loads his bending floors,
> Of fields with fresh'ning herbage green,
> Where bounding steeds and herds are seen,
> I call not happier than the swain
> Whose limbs are sound, whose food is plain,
> Whose joys a blooming wife endears,
> Whose hours a smiling offspring cheers."

Indeed, it is evident that proper occupation of Time, acquaintance with Nature, education of Mind, exercise of Body, skilful Labor, and adequate Riches, are each and all in themselves, and collectively, essential to our welfare, and he who is best in each of these respects is most wealthy. In a tabular view,

Essential to our welfare are
{
Occupation of      Time.
Acquaintance with  Nature.
Education of        Mind.
Exercise of         Body.
Skilful             Labor.
Adequate            Riches.
}

### Increase of Wealth.

This increase of wealth is not only to arise from the increase of riches, but, and chiefly, from the increased enjoyments afforded by Nature; for while each breath yields a rich enjoyment to mere animal health, and while the gorgeous splendors of the sky, the sublime beauty of the rainbow, the roar of the ocean, or even the pensive quiet of solitude, may awaken a measure of delight in the latent soul of the untutored savage, it is only to the cultivated intellect, the fully developed mind through a properly exercised body, that Nature becomes fully potential; it is only such who can fully realize, with the poet,

> " There is a pleasure in the pathless woods,
> There is a rapture on the lonely shore,
> There is society where none intrudes,
> By the deep sea, and music in its roar.
> I love not man the less, but nature more;
> For these our interviews, in which I steal
> From all I may be or have been before,
> To mingle with the universe and feel
> What I can ne'er express, yet cannot all conceal."

For there is not only all the high delight experienced by the influences of nature pouring in upon a finely cultivated mind, but also the reaction of the awakened mind searching with its philosophy into the hidden mysteries that underlie the marvels that so perplex or affright the ignorant.

> " The flow of riches, though desired,
> Life's real goods, if well acquired,
> Unjustly let me never gain,
> Lest vengeance follow in their train;
> For never, sure, shall Solon change
> His truth for wealth's most easy range,
> For vice, though plenty fills her horn,
> And virtue sinks in want and scorn;
> Since virtue lives and truth shall stand,
> While pelf eludes the grasping hand."

### Elements of Wealth, Bestowed, Obtained.

It is noticeable that some of the elements of Wealth are bestowed, and some the result of effort; accordingly the six can be arranged in two groups, in each of which there will be four, which is curious.

Time and Nature, equally bestowed on all, which we can

neither increase nor diminish, only occupy and enjoy, or waste and misemploy, belong entirely to the first group.

Riches and Labor, wholly the result of effort, as evidently belong wholly to the second group.

Educated Mind and Exercised Body belong to both groups. Mind and Body undeveloped belong to the first group, for they are bestowed; developed by Education and Exercise, they belong to the second group, for they are then the result of effort. In a tabular view the groups of elements are thus presented:

$$\text{ELEMENTS OF WEALTH} \begin{cases} \text{Bestowed} \\ \text{Obtained} \end{cases} \begin{cases} \text{Time.} \\ \text{Nature.} \\ \text{Mind Educated.} \\ \text{Body Exercised.} \\ \text{Labor.} \\ \text{Riches.} \end{cases}$$

## Attainable Wealth Unequal.

As Mind and Body are bestowed unequally, it follows that the measure of Wealth attainable by different persons will be unequal.

The practical fact, however, is, that most of the inequality is owing not to the inequality of bestowments, but to the non or improper use of them.

No one thing makes a greater difference in the Wealth of men than the manner in which time is used. To one man, from morning till night, each moment drags its slow length along; while to another the dawn never comes too early, and twilight fails to close his busy enjoyments. Nearly the same may be said of Nature. To one man it is lessonless; while to another, the first blush of morning, the crimson glories of evening, the brightness of noonday, and the majestic darkness of midnight, are alike full of meaning. He finds " sermons in stones, books in brooks, and good in everything."

Let no one be discouraged because genius has not been bestowed upon him ; let him use that he has, and his wealth shall be abundant.

## Educated Mind, Exercised Body, Essential to Wealth.

Since, for a proper occupation of Time, and a practical acquaintance with Nature, an Educated Mind and Exercised Body are necessary, and also for the application of Skilful Labor, the acquisition and enjoyment of Riches, and especially to counterbalance any inequality in the bestowment of mind and body, it is evident that Educated Mind and Exercised Body are the two most important of all the elements of Wealth, those two being, in fact, the means by which the other four are obtained and enjoyed, and they might therefore be considered as the primary and essential basis of Wealth.

How shall this proper Education and Exercise be accomplished ? is therefore a momentous question.

A partial answer is obtained from a very interesting fact in regard to the three material elements.

*Double Relation of Material Elements.*

|  |  |  |
|---|---|---|
| ELEMENTS OF WEALTH { | IMMATERIAL | TIME, *Nature.* MIND, *Body.* |
|  | *Material* | LABOR, *Riches.* |

Two relations in the body. The brain is directly associated with the MIND, the stomach ministers to the *blood.* The eye has been beautifully called the window of the SOUL, the lungs admit the vivifying air to the *blood.* Thus, from head to foot, each part of the body may be classed according to its relations Riches and Nature can - be correspondingly classed. Food nourishes the *blood;* music delights the MIND; silk, by its texture, preserves the warmth of the *blood,* while its glossiness and brilliant colors please the MIND. The rugged mountain and the trembling ocean excite emotions of sublimity in the MIND, while bracing air and cool water depurate and refresh the *blood.*

Through the six channels of hearing, seeing, smelling, tasting, touching, and the muscular sense, the mind is acted upon, and under the six aspects of food, water, air, clothing, warmth, and shelter, the *blood* is acted upon. In tabular view as follows:

RICHES AND NATURE ACT ON THE {

Mind through the channels of {
HEARING.
SEEING.
SMELLING.
TASTING.
TOUCHING.
MUSCULAR SENSE.

*Blood* under the aspects of {
Food.
Water.
Air.
Clothing.
Warmth.
Shelter.

Each and all the three material elements of Wealth can therefore and should be arranged in two corresponding classes, one adapted to develop and improve the MIND, and one adapted to develop and improve the *blood.*

Education is of course also twofold. The world at large says that wheat is practical, beef is practical, the pantry is practical; they are so only as every organ, when healthily active, conduces to happiness: a rose is practical as well as a cabbage, a flower-yard as well as a kitchen-garden, a parlor as well as a dining-room, a library as well as a larder, the labors of Praxiteles and Raphael as well as a reaper and a steam-plow. Nor is the sun, that shuts in our knowledge to the narrow limits of daylight, any more practical, though he melts the icebergs and makes the corn grow, than is the night, which lets out our minds from this restricted boundary to the vastness beyond our own planetary system, making us acquainted with the boundlessness of space, and the immensity of God's power, leading us to think, that if day shuts out so much glory which is revealed by night, why may not life shut out a corresponding glory, which shall be revealed by death.

18

## E.

It was the custom, but a few years ago, to write and lecture upon medical subjects in Latin, or a jargon so called. This has left an incubus upon these studies, both in terms and style, which the French teachers a long time since discarded, to the great advantage of their pupils. Prof. Leidy, one of the leading anatomists in America, in his Anatomical Treatise for the use of Medical Students, says: "Much of the difficulty in the acquisition and retention of anatomical knowledge arises from an excessive, and in some respects objectionable, nomenclature . . . founded upon no particular system, &c. In some measure to avoid the difficulty . . . a single name will be used for each part," &c.

If unnecessary difficulty is found by the professional student, who needs many technical terms, how much greater is the trouble given to the popular reader, who has no occasion for professional language?

Many persons favor the use of Technical terms, because they suppose them to be scientific. Now, technical and professional are not synonymous, nor are scientific and technical. Scientific refers to the order, fulness, and clearness of presentation, and may be used in connection with popular as well as with professional. Indeed, that which would be scientific addressed to a popular audience, would not be, if addressed to a professional one, and *vice versa*.

By some these technical expressions are liked because they seem to them to give an air of learning; they think the display a class makes in repeating them will be striking, and commended, and that a scholar will be satisfied he is wise, if he can mouth words that others do not understand. But, by all means, let the reality be first obtained, as thence comes the profit.

## F.

*Man is a twofold being, composed of Mind and Body. Though in many respects they may be, and in some must be, treated as distinct, yet the condition of either affects that of the other.*

This fact is the keynote of all practical Physiology. Bravery and courage, fear and cowardice, levity and cheerfulness, depression and gloom, depend on the condition of the body as much as on that of the mind; while dyspepsia, consumption, heart and other diseases, have their rise in mental anxieties as often as in purely physical causes.

Do not merely hopeful thoughts of delicious fruits make the "mouth water," while fearful emotions prevent the formation of juices? The rice-test, in India, detects the guilty servant on this principle.

The blush of modesty or shame, the flush of anger, and the pallor of fear, suggest the same truth.

On the other hand, military drills and decorum, requiring manly attitudes and regular muscular exercises, tend to reproduce in the mind manliness of character and regular habits of thought.

# G.

Some will be ready to ask if the Mind is diffused through the brain or located at some point. It is not known.

Again, some will be ready to ask, if all these questions cannot be answered, how is it certain that there is any mind distinct from the Brain; indeed, Gall said that mind was the name of the phenomena manifested by the activities of the Brain. But that whole collection of parts undergoes more rapid changes than any other portion of the body, so that the brain present in the head when any transaction takes place is in a short time entirely gone, while the memory of the act remains vivid for years. A person is also conscious, through his memory, of a continued individual existence from year to year, for scores of years, or during a long life, while he is served during that time by many Brains, even by many Bodies. Mind is free, hence intangible, and imperceptible to either of the senses, and therefore not manifested, lest it be coerced by any despotism, whether of opinion or of force.

# H.

True, the child uses its larynx to cry for food ; its upper extremities take everything direct to its mouth ; nothing solicits the action of its lower extremities more strongly than something to eat. The neck and the entire head also seem to serve, with the most watchful alacrity, the wants of the contents of the Trunk-Walls.

But from infancy up to manhood, as the growth of the body is becoming more and more complete, and it becomes more capable of its mental duties, so does the mind take control of it, and permits less and less attention and time to be bestowed upon the stomach.

# I.

The Sensational and Motional operations of the Mind will not be disputed, and the other two, Emotional and Intellectional, belong rather to the field of Psychology to discuss. It suits the purposes of Physiology to consider all the operations of the Mind not sensational nor motional as being Emotions or Intellections. Whether this classification is right or wrong, and in any case, the other mental operations, except Sensational and Motional, take place through the use of parts of the Brain only ; and of the mode of action, no one has any correct conception. Consequently to Physiology it is a matter of indifference how the Emotions and Intellections are classed, speaking of them abstractly, while practically there are some advantages in the method chosen.

# J.

If the two be placed on the thumbs' ends, the B on the left one, and the N on the right one, and the four of each of the others on the fingers, commencing with Motory on the fore finger of the right hand, and Respiratory on that of the left, and there will be a remarkable correspondence of the functions.

# K.

The proper activity of all the organs of Sensation is productive of much enjoyment. Their power of producing it is partly natural, and partly the result of culture or habit.

There is a skilful or scientific arrangement of objects to be made, such that, through perfect Sensatory Apparatus properly cultivated, the most exquisite sensations will be produced. Nor is this accidental or the result of habit. There is a science of music, colors, forms and sizes of objects, perfumery, savors, etc.

# L.

The following picture is introduced because it admirably contrasts the past and present, and is prophetic of the future. Labor is to be relieved of its drudgery, and turned over to brutes and to the inanimate forces of Nature. We copy by the press; travel by steam; post messages by telegraph; and the farmer mows, hoes, reaps, rakes, and even harrows, pleasantly riding over the ground his fathers tilled by the most arduous labor, yet not gaining half his results. Still more delightful to the philanthropic physiologist are the central views: the cramped, tedious, midnight toil of industrious poverty is abolished—pleasantly performed while the day is yet in the zenith, and repaid by comfort and deserved luxuries. Nor are these views comparatively too conspicuous; for certainly one of the greatest blessings ever bestowed upon mankind was the invention of the lifelike sewing machine;* exquisitely neat and compact; its use easily learned; in one hour doing the work of ten by hand, and of the most uninspiring kind; and instead of exhausting, improving the health of the operator, unless working too long at a time. Is there any need for that? Is not enough done by it even in four hours to command pay for a full day's work?

Mind, labor, and capital have each a right to share in the benefits conferred by inventive genius: the first by increased culture; the second by diminished hours and increased pay; and the last by increased percentages. It has been too illiberal; the former two are beginning to perceive and require their due; an average of six to eight hours' labor will accomplish now twice what twelve hours would a few years ago. Physiology says that six is more than the welfare of mind and body can allow, and Science declares that a full development of all her resources will render four hours more than will be necessary.

Ought not women to demand their share? Will it not be wisdom as well as justice, economy as well as propriety, to allow women more time for leisure and mental culture? Have not the worth, the wearing nature of their labors, and the time occupied

* Being frequently asked by teachers which is the best, the reply is, In our family we use Wheeler & Wilson's, and after several years' experience and observation think ⸺ ⸺ should be taught to all advanced classes of girls.

by them, been very much undervalued? Some are too indolent
for their health; but do not the most work too long, and more
in proportion than men? Do not their labors require much
versatility? Their household occupations are numerous, per-
plexing, and some of them very unpleasant, at best. Cooking
must be done with heat, and in hot weather cannot be made
very enticing. Yet what adds more to home-happiness than
good housekeeping, particularly cooking? Ought not, then,
woman's work to be honored, and made as easy, speedy, and
pleasant as possible?

Kitchens should never be in basements, unhealthy not only,
but uncheerful. Everything in them should be handy—fuel,
hot and cold water, etc. In particular, cooking utensils should

be simple, and easily and effectively used. Fire was the greatest invention of man; its most humanizing application was to
cooking; and a stove instead of a fire-place one of the greatest blessings to woman. The modern range-form, either set or
portable, is a great advance over the high stoves yet in very
common use. Being lower, the lifting on and off of articles gives
an aggregate, in the course of a day, of great relief to muscular

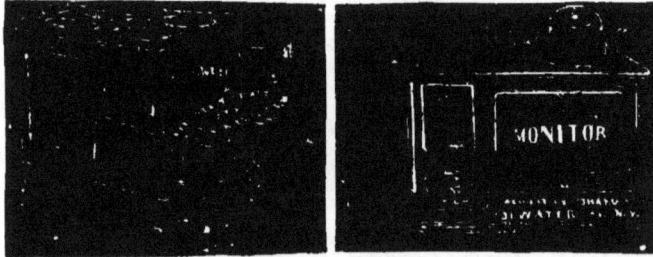

exertion. There is greater working surface; six holes are more
convenient than four; and an internal arrangement is allowable
that insures perfect baking at all times with coal, and with a
great economy of fuel, apparently otherwise unattainable; and
bad baking is a great vexation to woman, as well as a great waste.
The range-form is also made at less cost.* A wooden handle on
a kettle is a small matter, but will save many steps and motions.
All such little things must be observed. for woman's toils are
numerous, and she should receive corresponding facilities.

## M.

The following figures illustrate a neat form of Tourniquet
being applied. The lettered Fig. is a section of a limb at the
point of application. E indicates an elastic band, a yard or more
in length, that drawn tightly and repeatedly passed around the
arms of the Tourniquet, will surely compress the artery A,
yet not so as to altogether prevent the action of the neighbor-

* The author takes especial pride in having done something to improve cook-
ing-apparatus. The Monitor range, figured above, in external form does not differ
from others, but it has internal arrangements that render it the best for cooking,
particularly baking, and with much the greatest economy in fuel of anything yet
invented. He is also preparing a Treatise on Ventilation, Heat, and its household
applications, presenting economical and interesting suggestions.

ing muscles, or the flow of blood through some of the veins. Every Tourniquet should be elastic.*

# N.

Neither pantaloons, nor drawers, should be girded about the loins: especially in case of overalls worn by hard-laboring men an elastic band should form some part of the waist, allowing the most extensive motion of the trunk-walls, without injuriously constricting their contents ; many very serious evils would thus be prevented.

Skirts also should always be fully supported from the shoulders, and their bands should never gird or constrict the wearer. The invention of the hoop-skirt, though so much ridiculed, was a great blessing to ladies, enabling them to dress fashionably without the burden previously so oppressive. Their shadow may with propriety grow less, but it is physiologically desirable that parts of them should always be worn.†

# O.

## *Drowning, Choking, Suffocation, etc.*

The air-passages opening into the pharynx should be closed when food or drink is swallowed, and it will then be noticed that the tongue, in the very act of swallowing, presses against the roof of the mouth, firmly holding the epiglottis (30, Pl. 30) against which the larynx is involuntarily raised—as it also should be, and is, when any powerful substances, smoke, etc., enter as far as to the larynx.

If a person is plunged under water, none of it will enter his lungs, for his larynx will be involuntarily closed; neither will he swallow any considerable quantity of it.

The method of restoring a drowned person is suggested by reflecting upon his condition. He has no internal means of sustaining his heat; none therefore should be lost that can be saved, and it should be artificially increased. Remove wet clothing, and put warm articles about his chest in particular. Cloths dipped in water as hot as the hand can bear, often replaced, are best; bricks, stones, stove-covers, anything warm, but not burning, may be used. Keep him out of any draft of air. The next

* It will be of direct interest to a class to see the instrument applied, and its effects, and of practical value to each one to apply it. Its cost is small, and with full description will be supplied to Teachers at a discount by the Publishers.

† It may be of use to call attention to a very valuable improvement in this article of apparel invented by Mr. Bradley (the Duplex), since it is the lightest and best yet made ; it should, however, be suspended upon the shoulders.

thing necessary is to relieve the Blood of carbonic acid. If he have been insensible but a few moments, a smart slap on the shoulders, a dash of water in the face, or tickling the throat, may excite a gasp, and if he gives one he will give another in time, and nothing more need be done; he will soon be restored. The first step in respiration is to place him on the back, with the shoulders, head, and hips a little raised above the loins. The second step is to secure a free passage for air through the larynx, by seizing the tongue between the fingers, covered with cloth to hold it firmly, and extending it, thus raising the epiglottis, at the same time pressing down the Adam's apple, or larynx, away from the epiglottis. If this does not prove sufficient, let an incision admitting the little finger be made in the windpipe just below the larynx (it cannot do harm), and kept open in any convenient way. The third step is to expel air from the lungs by raising the loins as high as possible without lifting the hips, and then allowing them to sink, which will cause inspiration.

Let these motions for expiration and inspiration be repeated as often as a person would naturally breathe. Continue the processes of warming and artificial respiration for not less than three hours, unless signs of animation are previously shown. A case is known where two and one half hours elapsed, and the person now lives; and another, has been observed where, two hours after falling into the water, a person first gave signs of life.

In case of suffocation by smoke, gases in a well, etc., the same state exists as in drowning, and the same course is needed.

If a house is on fire, there will be a strata of pure air near the floor, that may be breathed; and if a person must go through smoke, he should close the nose that he may not strangle, and if he must breathe before the smoke is passed, let him put his face to the floor before freeing his nose. Smoke will not pass through a wet handkerchief or other cloth of several thicknesses, but air will. Therefore, if convenient, use it.

If a person is choked, the food should be drawn out from, or pushed down, the throat, and if a person does not then gasp, he is to be treated as drowned.

Hard substances swallowed are very apt to be refused by the œsophagus, naturally adapted only to receive soft substances, and in its effort to reject them they are sometimes crowded forward, under the epiglottis, and into the larynx, strangling by detention there, or, carried through it, obstruct the pipe below. In the former case, the windpipe must be opened; then proceed as in case of the drowned: in the latter case, the substance must be drawn out, a thing usually difficult to do. Young children should not have, nor older persons hold, in the mouth small hard substances.

In ulcerations of the throat, or when the ulcer breaks, or in case of croup in the larynx, opening the windpipe is sometimes advisable, and the only thing that can save life—can do no harm, and should always be assented to and encouraged when advised by the physician.

DESCRIPTION OF PLATES.

PLATE 1. Skeleton complete, with ligaments on right side; *a, a*, position of large arteries of arm and leg.

PLATE 2. General view of muscles, the superficial ones of the neck, chest, and abdomen, being removed.

PLATE 3. Fig 1, back view of brains and spinal cord, portions of the skull and the right side of the body being dissected to exhibit them. Fig.2, a front view of the spinal cord, roots of nerves, and under surface of the brain covered with thin membrane.

PLATE 3* or 26. Fig. 1, a beautiful view of the interior of the chest, dissected, and its contents drawn forward to show *d*, its left sympathetic ganglia, and the division of *b*, the left pneumo-gastric (8th) nerve. It will be observed that branches of it extend to the larynx. Fig. 2, parts dissected to show the front part of the spinal cord and the roots of its nerves. Fig. 4, white cords illustrate the nerves magnified in a finger.

PLATE 4. Fig. 1, Ear; see page 309. Fig. 2, see page 136. Fig. 3, see page 135. Fig. 4, see page 136. Fig. 5, a section of the nose, with a portion of the turbinated bone removed, to show 12, the lower extremity of the lachrymal, a tear-duct leading from the eye. Fig. 22, Eustachian tube. Fig. 6, Tongue. Fig. 7 represents touch. Fig. 8 speaks of the muscular sense. The Plate illustrates in part each of the six senses.

PLATE 5. Fig. 1. The dark lines upon the right side represent veins, the light ones on the right side arteries dissected into view. Fig. 2, plan of hearts, and large arteries, and veins.

PLATE 5* or 28. Fig. 1, general view of lymphatics; and in 2 a view of the lacteals is added. Fig. 3, lymphatics in the skin of a thumb. Fig. 4, the dotted lines show a network, or capillaries of lymphatics in a frog's foot; *A, V,* arteries and veins with capillaries between them.

PLATE 6. An admirable synoptical view of all the mentory organs. Fig. 1, skeleton; 2, muscles; 3, brain and nerves; 4, skin; 5, organs of sense, ear, eye, nose, mouth, skin, muscle; 6, Blood-tubes of mentory organs. This Plate explains at a glance the structure of all the members except the trunk, showing that there are only five kinds of organs in the arm, leg, trunk-walls, neck, and only five more (10) in the HEAD! This single Plate is a volume of Anatomy.

PLATE 7. Fig. 2, front walls of trunk dissected to give front view of its contents (see Fig. 54). Fig. 1, view of diaphragm, 4 (and 3, Fig. 2, D, Fig. 4). Fig. 3, pulmonary capillaries much magnified. Fig. 4, plan of perpendicular section of lungs with heart between them; the white lines and enlargements are divisions of the windpipe and air-cells, and the dark lines by the side of them blood-vessels. Fig. 5, perpendicular section of lungs in chest; *T,* trachea, leading into them; *H,* place of the heart. Fig. 6, cross section of the lungs and hearts, *H,* and of the chest-walls of one side. Fig. 7, same, on the level with the divisions of the trachea. Figs. 8, 9, 10, 11, views of the circulation. Fig. 12, 1, spleen; 2, pancreas; 3, duodenum; 4, liver; 5, gall-bladder; 6, 7, 8, gastric and mesenteric veins. Fig. 13, lacteals, lymphatic glands, and thoracic duct. Fig. 14, kidneys. This Plate illustrates the contents of Trunk-walls, being a counterpart to Pl. 6.

PL. 7* or 31. Fig. 1 (see Fig. 90). Fig. 2 (Fig. 100). Fig. 3 (Fig. 109).

PLATE 8. The Printer, Publishers, and Author may justly feel proud of this page, produced in four colors from electrotyped wood-cuts. It is a very difficult thing to register or adjust four colors as accurately as the necessities of these illustrations require, and as perfectly as has been done here. Better work of this kind has never been accomplished by any printer, yet it has not been possible to have every copy exact. The entire constitution of light, and the physical action of the eye upon it, is perfectly and beautifully illustrated by the several figures.

Fig. 1 represents light shining from one point of each of three candle flames through an opening in the cornea of an eye without any refracting media. Thus, the light from the three points is spread over a great extent of surface in the back part of the eye, and some from one point falls upon the same surface as that from another point, preventing the possibility of distinctness of vision.

Fig. 2 represents the light as in 1, acted upon by the media of the eye and refracted upon corresponding points or foci, r, y, b, by which sensations may be distinct and numerous, and perceptions of directions accurate.

Fig. 3 illustrates long sight, the media partially refracting the light.

Fig. 4 illustrates short sight, the light being refracted too much. It will be noticed that the effects upon the nerves in the back part of the eye are the same in 3 and 4. The nerves do not terminate as represented in these figures; they, however, induce a correct idea of the method of seeing.

Fig. 5 represents two rays of light acting on the same nerve, b d, in which case a single sensation (of purple, when red and blue act) will be caused.

Fig. 6 represents two similar rays acting on two nerves, a c, d b; in which case two sensations will be caused.

Fig. 7 illustrates white light, w, shining through a small orifice into a dark room, and passing through a prism, p, by which it is separated into three kinds, red, yellow, blue. The separation does not take place thus in the prism, as red and yellow will be blended at their margins, producing orange, and yellow and blue producing green, etc., exhibiting all the colors of the rainbow. The idea impressed by this figure, that there are three kinds of light, is correct; and

Fig. 8 more perfectly illustrates the proportions of the natural colors that acting together produce white, and, blended or acting consecutively, most highly please, namely, three parts red, five of yellow, and eight of blue, making white—sixteen. The figures 3, 5, 8, 16, may represent those colors, when 13 will stand for green, 11 for purple, and 8 for orange, etc., the proportions in which they must be combined to please most perfectly.

Fig. 9 represents an eye, 2, looking through a hole in eye 1. The light from l passes upward into eye 1, and comes downward into eye 2, upon the lower part of which it will act, and the direction from which it has come will appear to one eye to be the reverse of what it will seem to be to the other. This figure explains all the peculiarities of inverted images.

Figs. 10 to 17 represent by the black lines so many different objects with white light falling upon them, and absorbed, as in 10, causing the object to appear black, or reflected as in 11, causing it to appear white; or reflected in part, Figs. 12 to 17, causing the objects to appear red, green, etc.

PLATE 9. Fig. 1, skull. Fig. 2, upper jaw. Fig. 3, lower jaw. Fig. 4, palate-bone. Fig. 5, internal, Fig. 6, external surface of the occipital bone. Fig. 7, external surface of parietal bone.

PLATE 10. Fig. 1, internal surface of parietal bone. Fig. 2, upper, Fig. 3, lower surfaces of spenoid bones. Fig. 4, inner, Fig. 5, outer surface of frontal bone. Fig. 6, side view of ethmoid bone.

PLATE 11. Figs. 1, 2, 3, inner, side, and outer view of temporal bone. Fig. 4, posterior view of ethmoid bone. Fig. 5, under surface of skull. Fig. 6, inner surface of lower half of cranium.

PLATE 12. Fig. 1, side view of spinal column; 1, atlas (Fig. 2, upper, Fig. 3, lower view of same); 2, dentatus (Fig. 4, same); 3, cervical vertebræ (Fig. 5, one of same); 4 to 6, dorsal vertebræ (Fig. 6, one of same); 5 to 7, lumbar vertebræ (Fig. 7, one of same); 8, sacrum; 9, coxcyx. Fig. 8, side view of frame of trunk-walls. Fig. 10, section through the hip-joint; 6, the thigh-bone; 3, the hip. Fig. 9, hyoid bone.

PLATE 13. Fig. 1, front, 2, back view of spinal column; 3, inner view of ribs; 4, frame of chest; 5, front, 6, back view of sternum; 7, side view of pelvis.

PLATE 14. Fig. 1, front view of pelvis; 2, sacrum; 3, back, 4, front view of femur (thigh); 5, front, 6, back view of tibia (shin); 7, front and back view of fibula; 8, tibia and fibula, front view; 9 and 10, front and back view of patella (knee-pan).

PLATE 15. Figs. 1, 2, 3, upper, under, and side view of foot; 4, 5, upper and lower surfaces of clavicle (collar-bone); 6, back view of scapula (shoulder-blade); 7, humerus (upper arm-bone); 7, ulna; 8, radius.

PLATE 16. Fig. 1, lower portion of ulna and radius, and back view of hand; 2, front view of hand; 3, back view of opening into larynx. Fig. 4, view of ligament; 5, binding cranium, 4, to dentatus, 3. Fig. 5, showing ligament, 7, extending across behind the process, 5. Fig. 6, ligaments of spinal column. Figs. 7, 8, ligaments binding ribs in place.

PLATE 17. Figs. 1, 2, ligaments of hip; 3, 4, 5, those of knee; 6, section of knee, showing its lining membrane, the patella, 4, and some ligaments; 7, side, 8, back view of ankle and its ligaments.

PLATE 18. Fig. 1, ligaments beneath foot; 2, of the sternum, and ribs, and clavicle; 3, clavicle, scapula, and humerus (shoulder-joint); 4, of the elbow, and 10, between the ulna and radius. Fig. 5 shows the beautiful ligament, 4, that binds the head of the radius by the side of the ulna. Fig. 6, ligaments of wrist and hand; 7, a few of the same; 3, 5, 10, 17, 18, membranes lining the joints.

PLATE 19. Fig. 1, cranium and facial muscles. Fig. 2, facium and internal muscles of the jaw; 3, superficial muscles of the neck; 4, deep muscles of neck; 5, muscles of the side of the face and throat; 6, muscles of tongue; 7, back view of pharynx; 8, pharynx removed, and the larynx brought into view.

PLATE 20. Front view of superficial muscles of right side of head, neck, trunk, and lower extremity, and the deeper ones of the left side; 2, deep muscles of the throat; 3, deepest muscles of front part of the neck; 4, pharynx laid open to exhibit the orifice from it into the larynx, 12, the mouth, 10, and the nostrils, 6.

PLATE 21. Fig. 1, muscles of the side of the neck and under the shoulder-blade; 2, of the front walls of the abdomen; 3, deep muscles of the back of the neck; 4, deepest muscles of the back.

PLATE 22. Fig. 1, view of superficial muscles of the right side of the back and of the deeper ones of the left side. Fig. 2, superficial muscles of the right, and deeper ones of the left lower extremity. Fig. 3, section of larynx. Fig. 4, muscles of side view of larynx.

PLATE 23. Fig. 1, back view of hip-muscles; 2, front view of muscles of thigh; 3, back view of lower leg; 4, front view of lower leg; 5, superficial muscles in sole of foot.

PLATE 24. Fig. 1, deeper muscles in sole of foot; 2, deepest muscles and tendons of foot; 3, triceps on back of upper arm; 4, muscles of back of scapula and arm; 5, muscles of the front part of upper arm; 6, superficial muscles of lower arm.

PLATE 25. Fig. 1, deep muscles of front part of lower arm; 4, deepest muscles of hand; 2, superficial muscles of back of lower arm; 3, deeper muscles.

PLATE 27. Fig. 1 represents a portion of the heart dissected, and showing by the network of white lines the lymphatics of that organ. 1, right auricle; 2, vena cava; 3, aorta; 4, pulmonary vein; 6, pulmonary artery; 5, coronary artery and vein. Fig. 2, a diagram of the relations of the blood-vessels near the heart, the windpipe, and the nerves 10 and 13. Fig. 3, internal view of 1, right, and 2, left heart; 3, right auricle; 4, pulmonary vein; 5, vena cava; 6, 7, tricuspid valves; 9, muscles attached to the tricuspid; 10, Mitral valve; 11, pulmonary artery open; 12, one semilunar valve; 13, sinus behind another.

PLATE 29. Fig. 1 is a beautiful representation of the thoracic duct, 1, opening into the veins at 3, and receiving in its course the lymphatics and lacteals. Fig. 2 is a magnified portion of the second stomach, showing the folds of its inner surface and the tubular glands between them; 3 shows the same tubular glands, the papillæ standing up, and the Peyer's glands, 1 and 2. Fig. 4 is an exquisitely truthful, very much magnified view of the tubes and papillæ, in some of which the cells forming their surface are seen; from others they are removed, and the capillaries and their connecting artery and vein, 14, 15, are brought into view; the lacteals, 16, are also seen commencing in the centre of the papillæ, as at 17. Fig. 5 represents a portion of the alimentary canal, and the lacteals extending from it toward the thoracic duct.

PLATE 30. Fig. 1 represents a perpendicular section of the eye magnified, (see description of the eye.) Fig. 2 is a section of the nose, mouth, throat, spinal column, and larynx, admirably showing many things to the glance of the eye. It is a delightful study, worthy of the frequent reference made to it in the body of the work; 10, opening of the Eustachian tube; 23, amygdaloid glands; 30, epiglottis, etc.

*₄* A fuller description of the Plates will be given in succeeding editions, if found desirable. Teachers will please indicate their opinions.

PLATE 1.

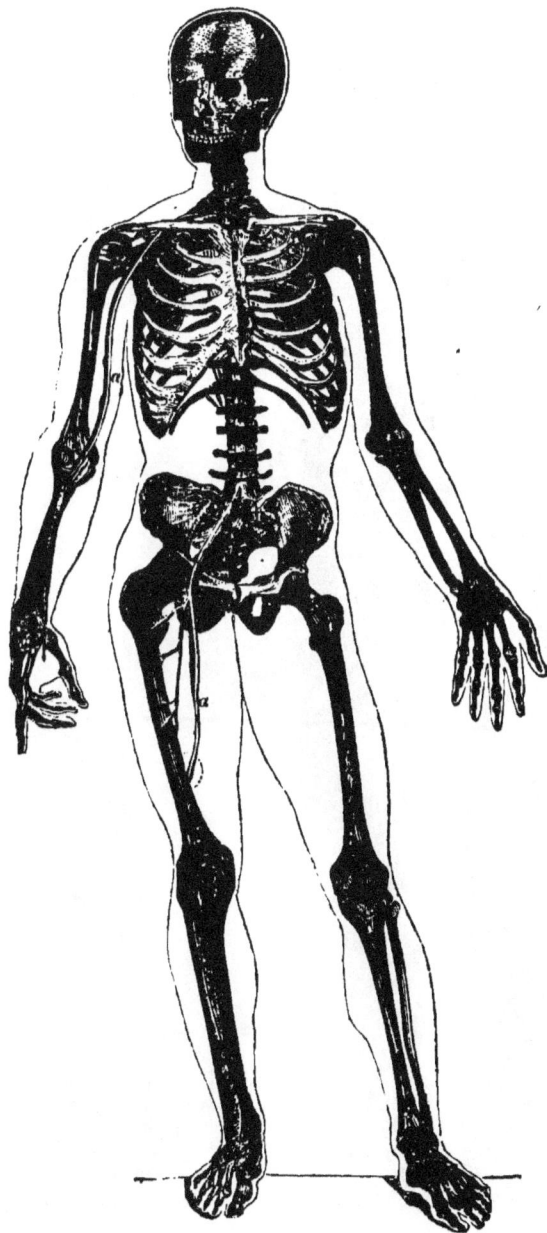

PLATE 2.

# PLATE 3.

Fig. 1.

Fig. 2.

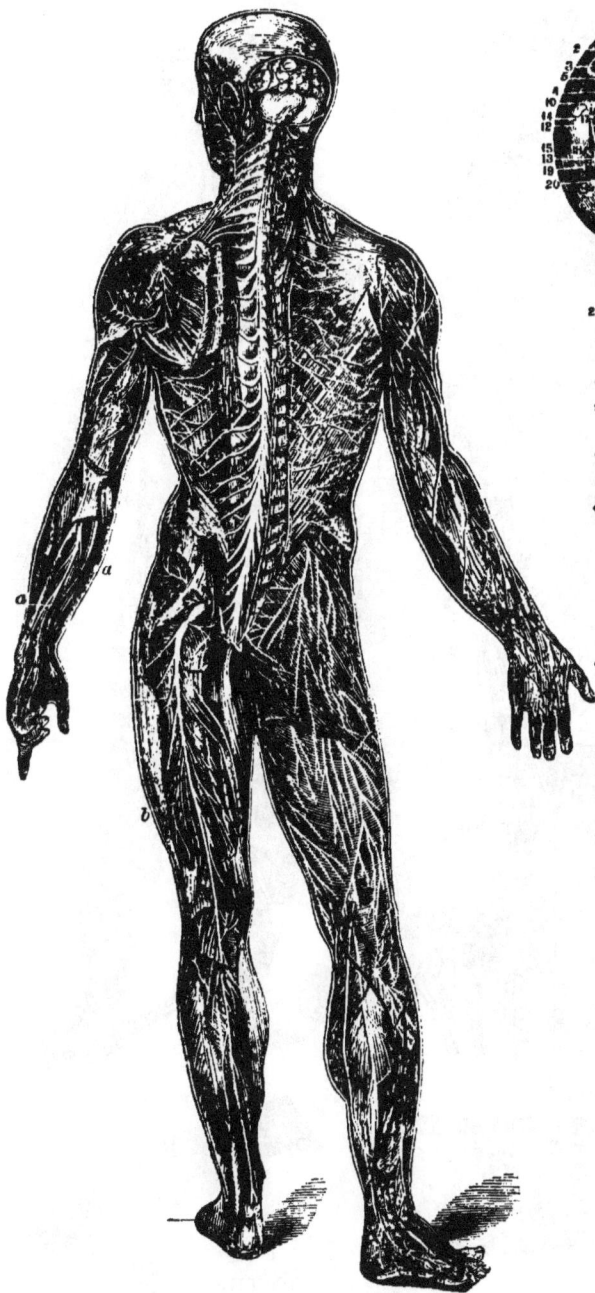

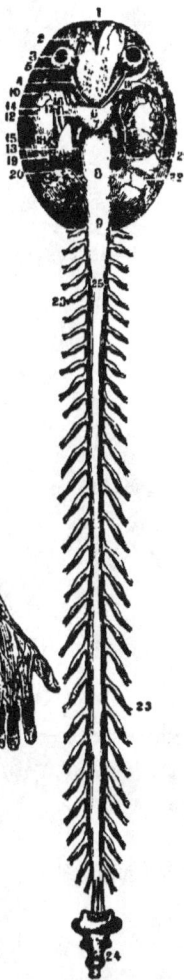

# PLATE 3* or 26.

### Fig. 1.

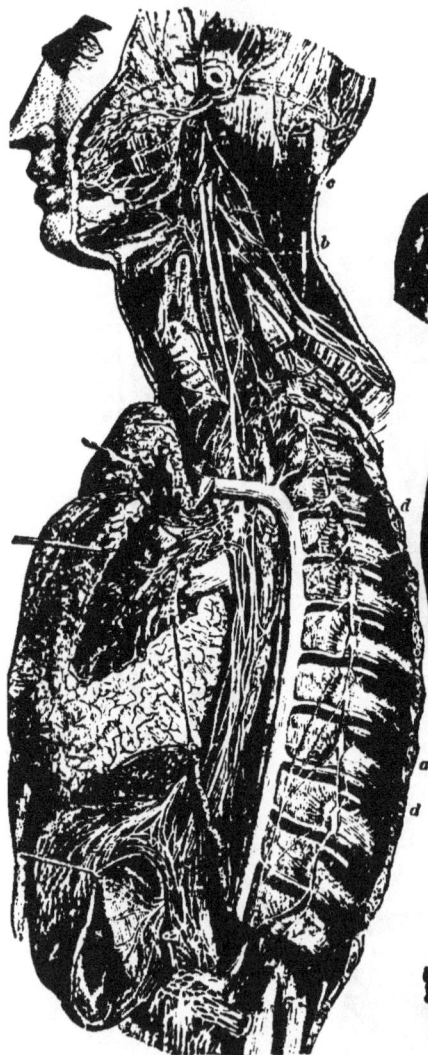

### Fig. 2.

### Fig. 3.

### Fig. 4.

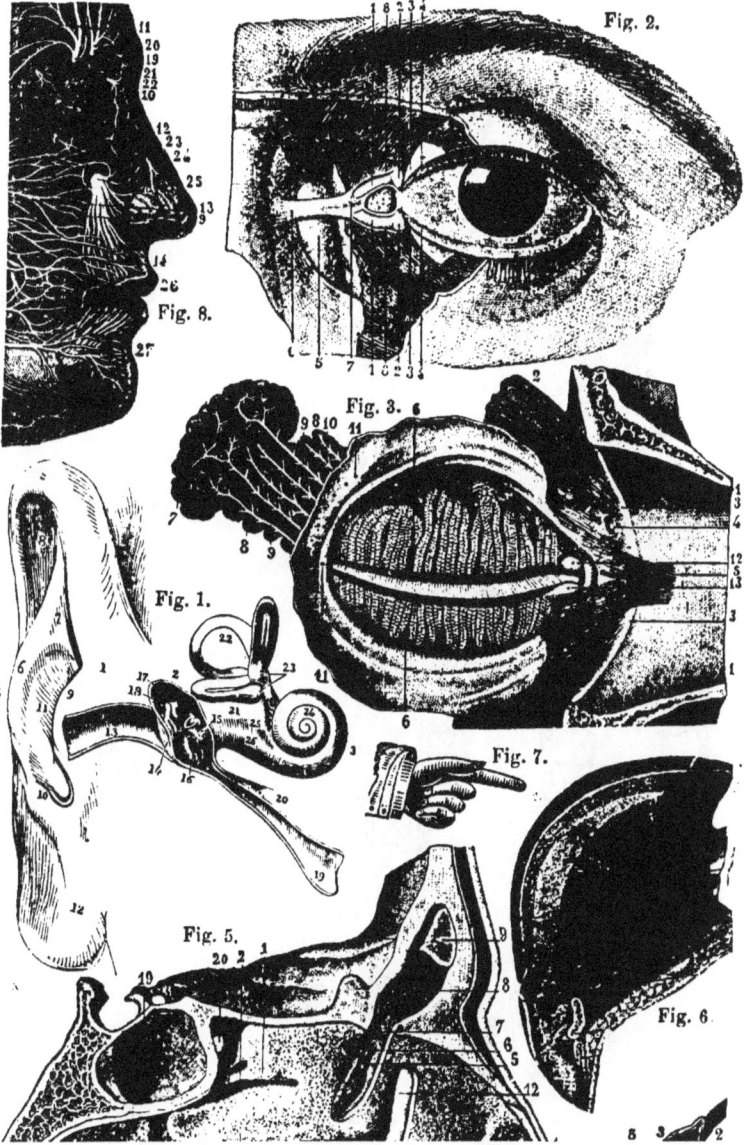

PLATE 4.

Fig. 2.

Fig. 8.

Fig. 3.

Fig. 1.

Fig. 7.

Fig. 5.

Fig. 6.

PLATE 5.

Fig. 1.

Fig. 2.

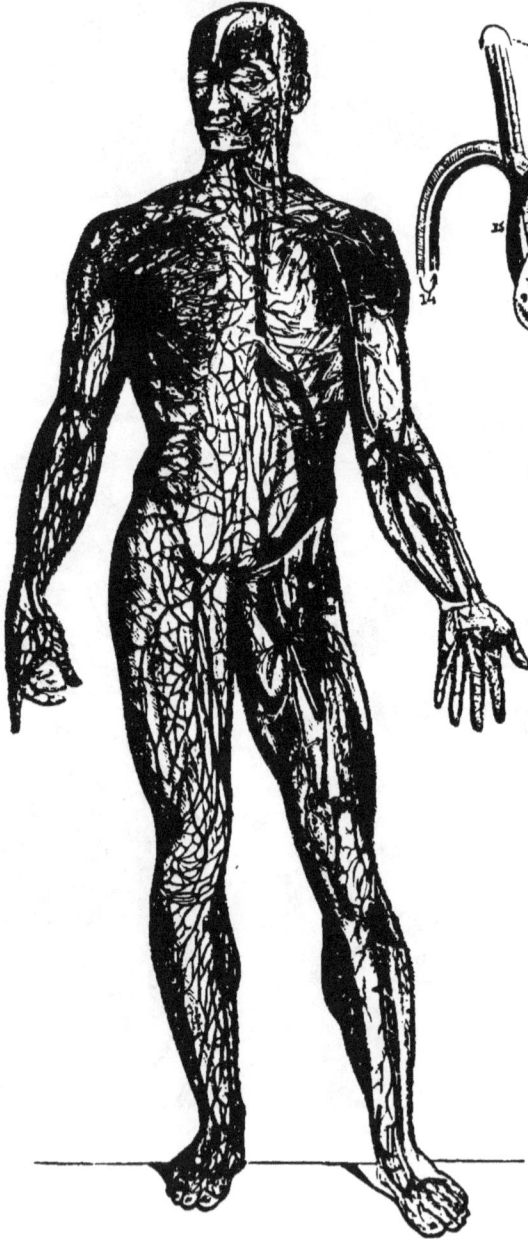

PLATE 5* or 28.

Fig. 1.

Fig. 2.

Fig. 4.

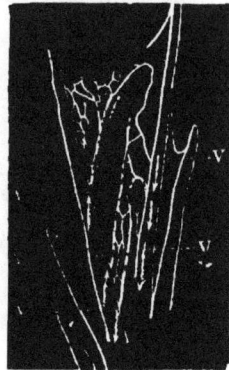

Fig. 3.

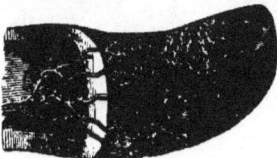

PLATE 6.

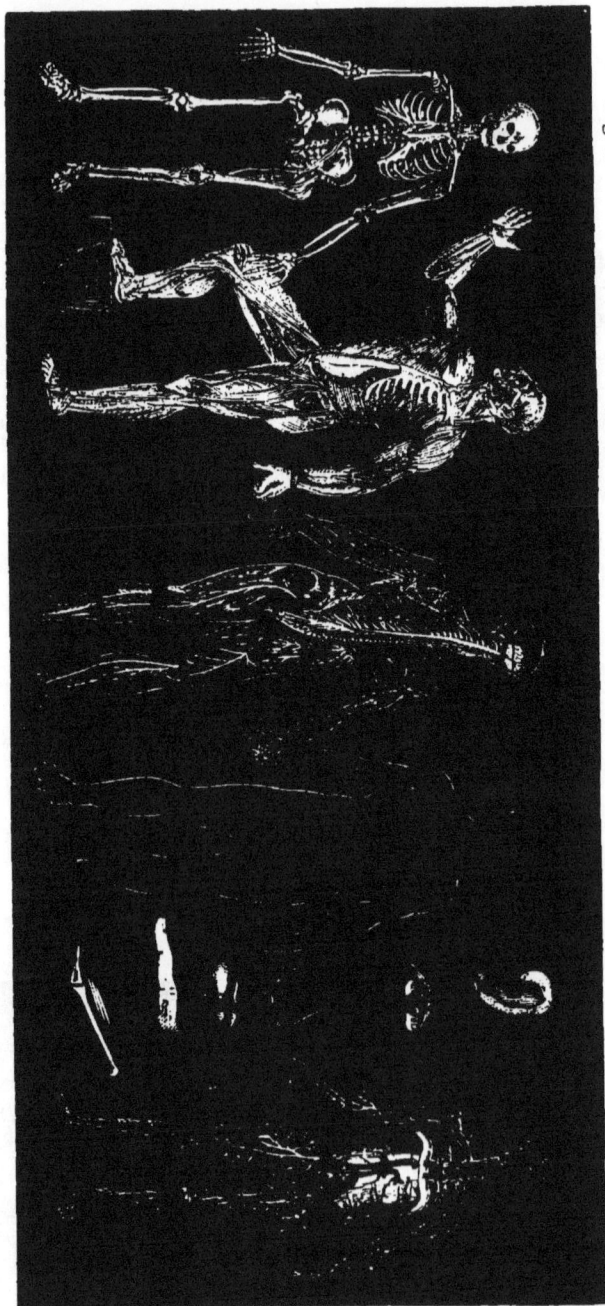

Fig. 1.

Fig. 2.

Fig. 3.

Fig. 4.

Fig. 5.

Fig. 6.

PLATE 7.

Fig. 1.

Fig. 2

Fig. 3.

PLATE 9.

Fig. 1.

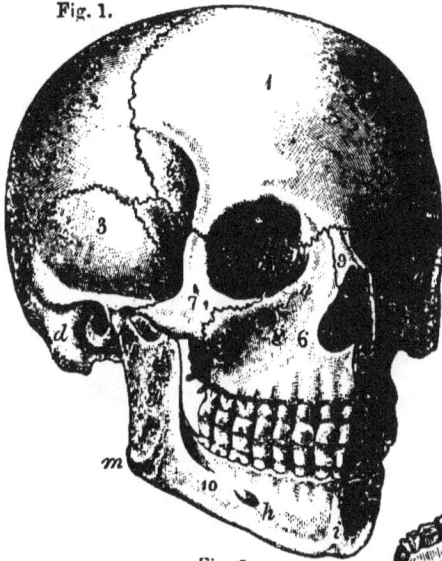

Fig. 2.

Fig. 3.

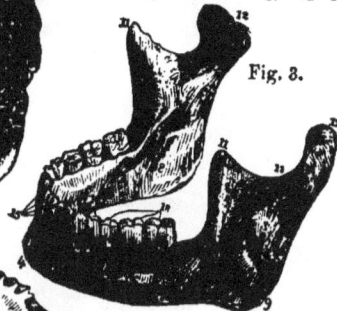

Fig. 5.

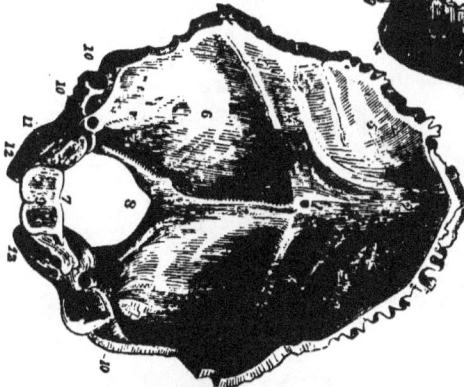

Fig. 4.

Fig. 6.

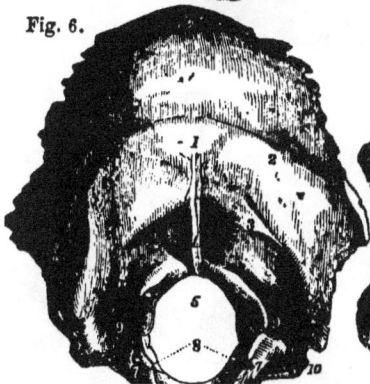

Fig. 7.

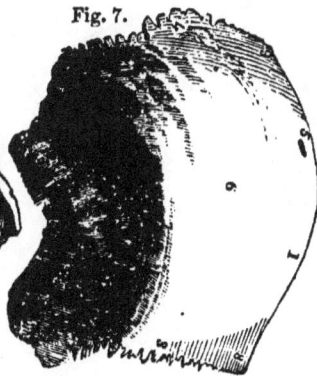

PLATE 10.

Fig. 1.

Fig. 2.

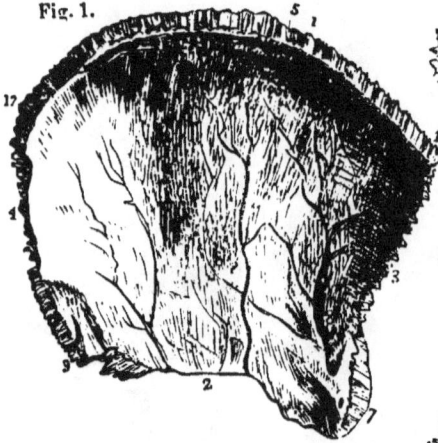

Fig. 4.

Fig. 8.

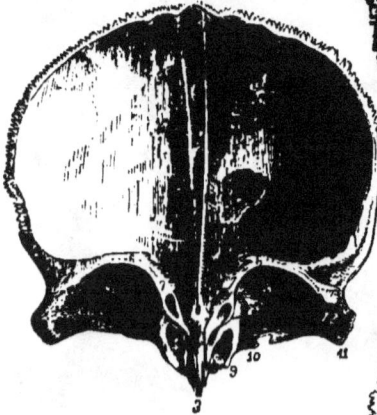

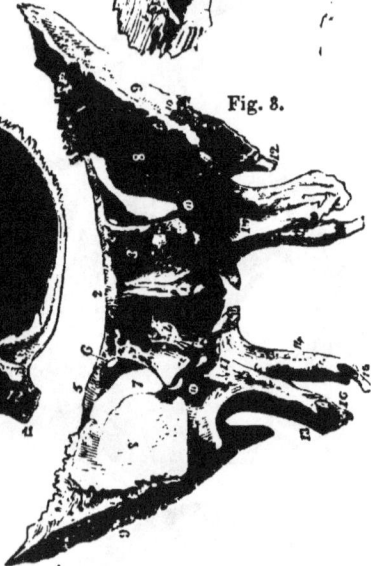

Fig. 5.

Fig. 6.

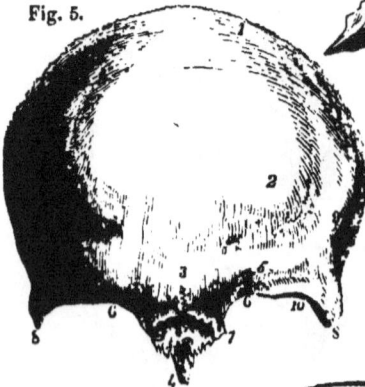

# PLATE 11

Fig. 1.

Fig. 2.

Fig. 3.

Fig. 4.

Fig. 5.

Fig. 6.

PLATE 12.

Fig. 2.

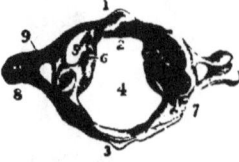

Fig. 3.

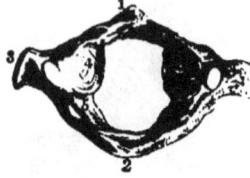

Fig. 1.

Fig. 9.

Fig. 4.

Fig. 8.

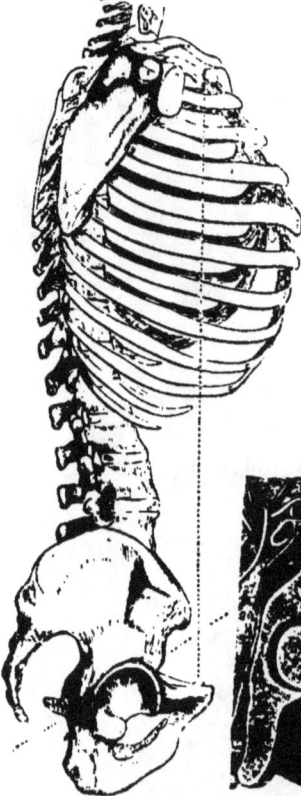

Fig. 5.

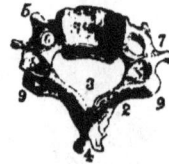

Fig. 6.

Fig. 7.

PLATE 13.

Fig. 1.　Fig. 2.　Fig. 3.

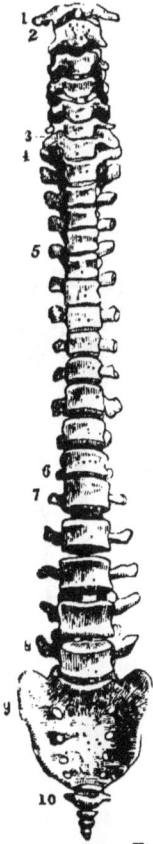

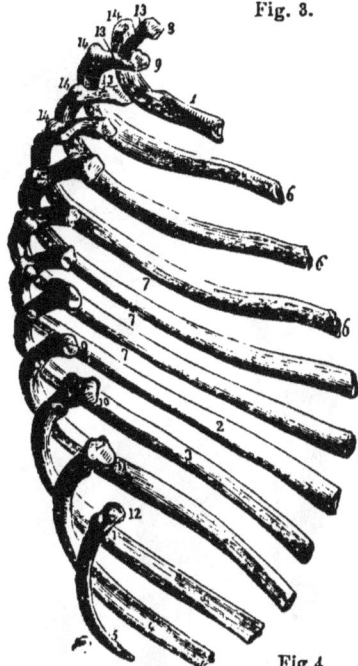

Fig 4.

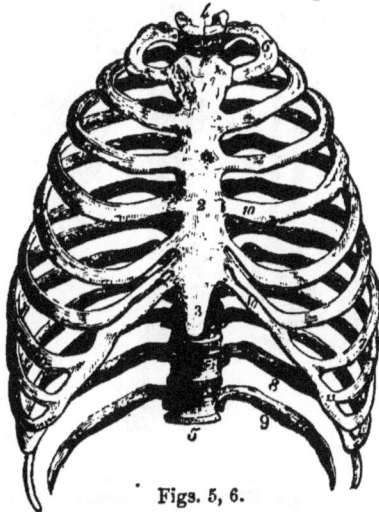

Fig. 7

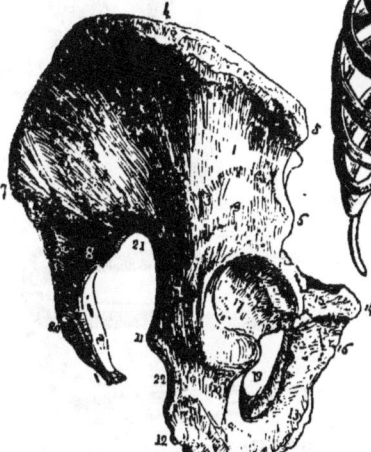

Figs. 5, 6.

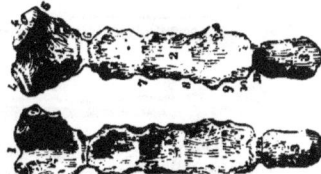

PLATE 14.

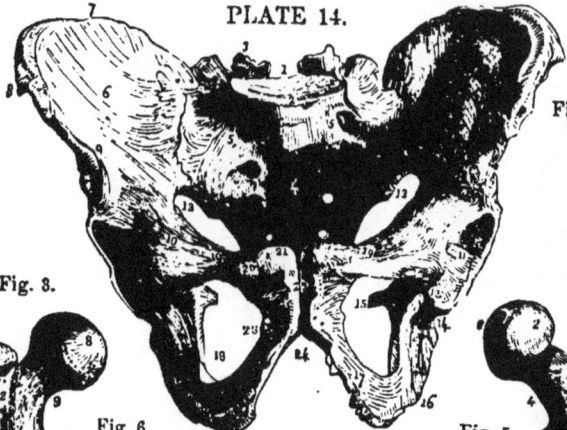

Fig. 1.

Fig. 3.

Fig. 4.

Fig. 6.

Fig. 10.

Fig. 5.

Fig. 2.

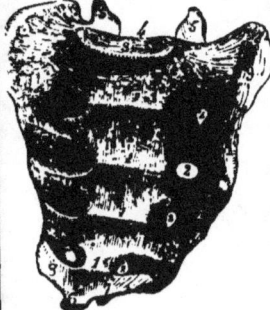

Fig. 7.

Fig. 8.

PLATE 15.

Fig. 1.          Fig. 2.          Fig. 3.

Figs. 4, 5.          Fig. 7.     Fig. 8.     Fig. 9.

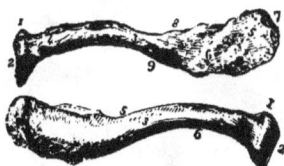

Fig. 6.

PLATE 16.

Fig. 1.

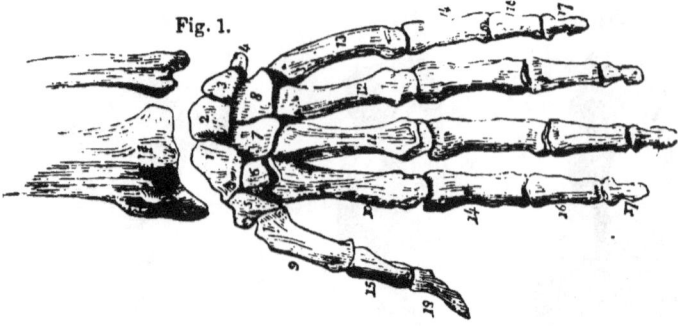

Fig. 2.

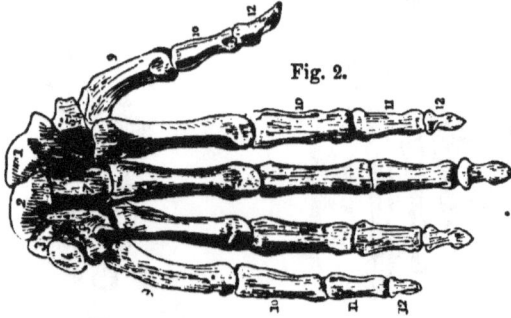

Fig. 3.

Fig. 5.

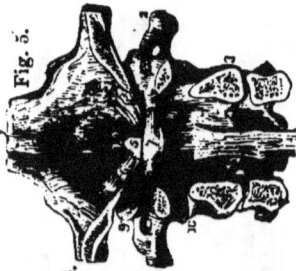

Fig. 4.

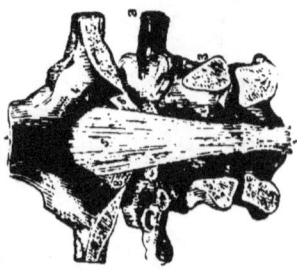

Fig. 6.

Fig. 8.

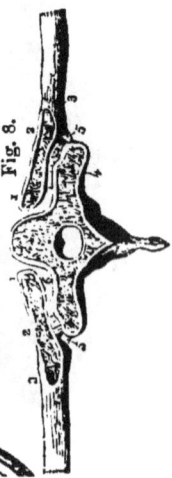

Fig. 7.

# PLATE 17.

Fig. 1.

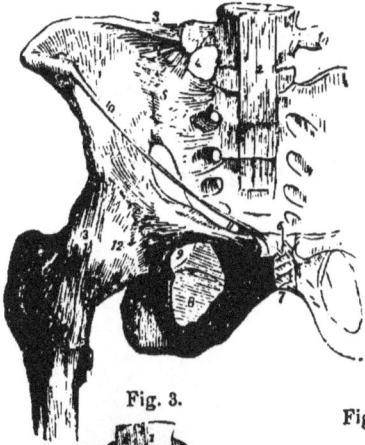

Fig. 2.

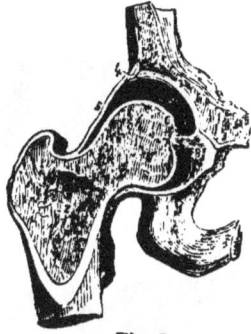

Fig. 3.

Fig. 4.

Fig. 5.

Fig. 6.

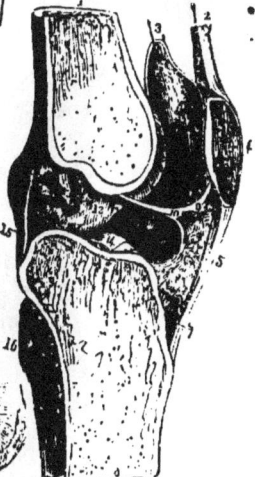

Fig. 8.

Fig. 7.

## PLATE 18.

### Fig. 1.

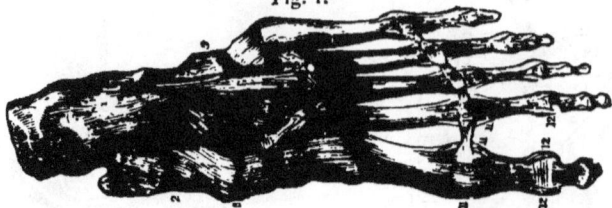

### Fig. 2.

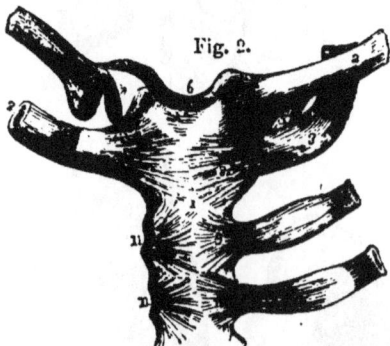

### Fig. 6.

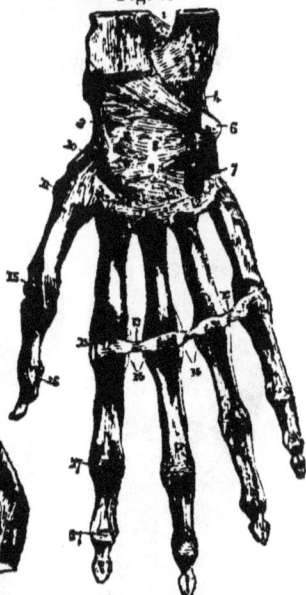

### Fig. 3.

### Fig. 5.

### Fig. 4.

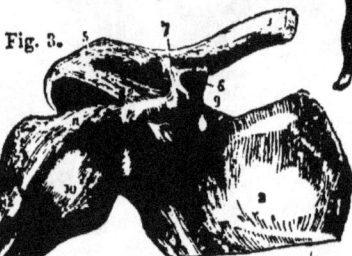

### Fig. 7.

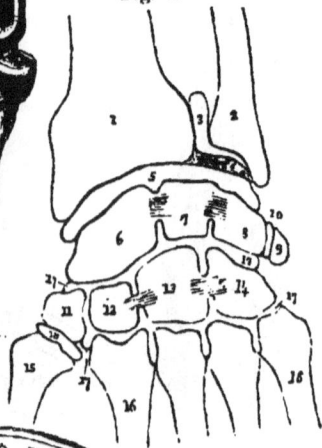

PLATE 19.

Fig. 1.

Fig. 2.

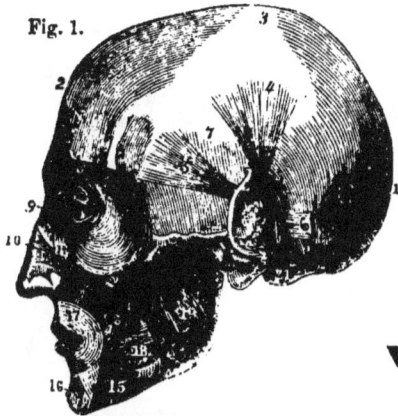

Fig. 3.

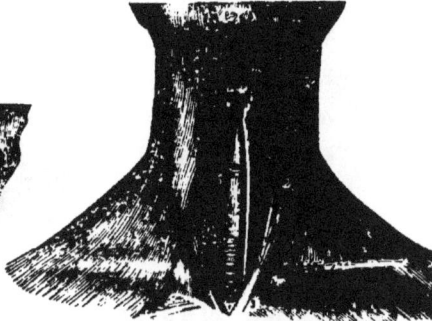

Fig. 4.

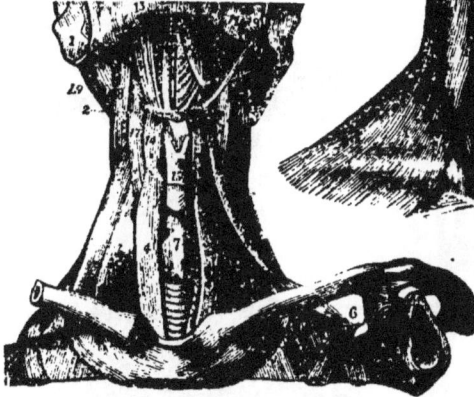

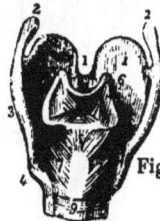

Fig. 8.

Fig. 5.

Fig. 6.

Fig. 7.

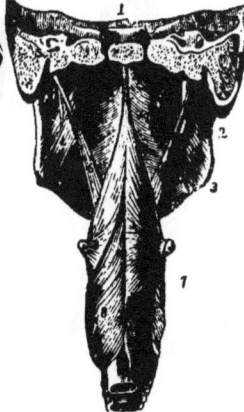

PLATE 20.

Fig. 1.

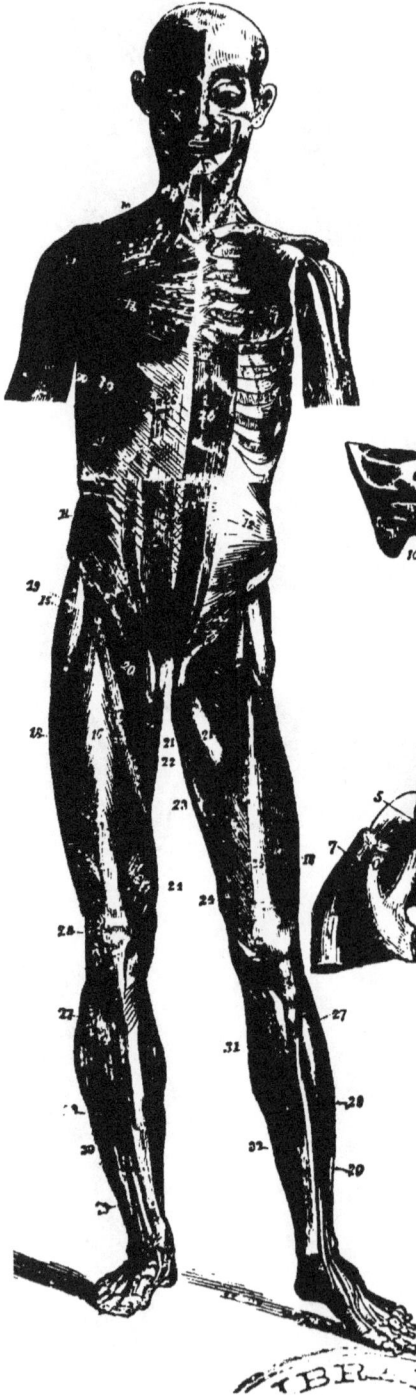

Fig. 2.

Fig. 3.

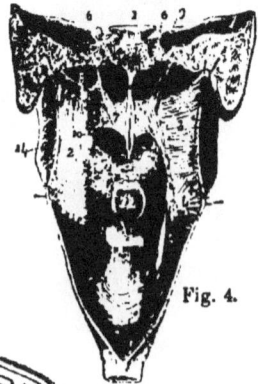

Fig. 4.

PLATE 21.

Fig. 1.

Fig. 3.

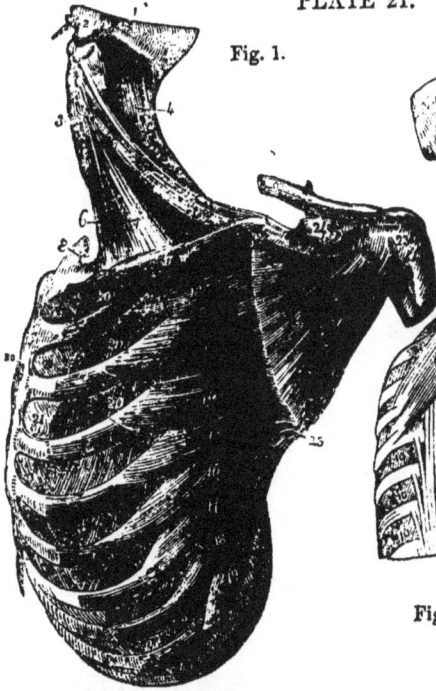

Fig. 4.

Fig. 2.

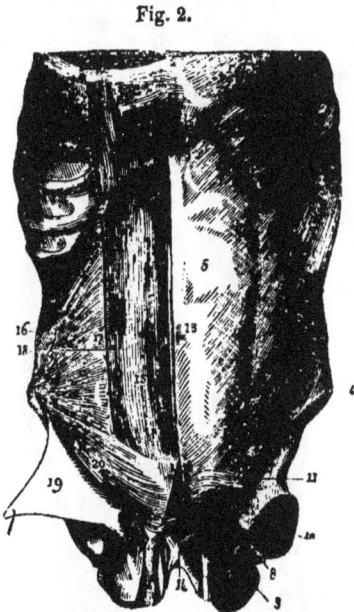

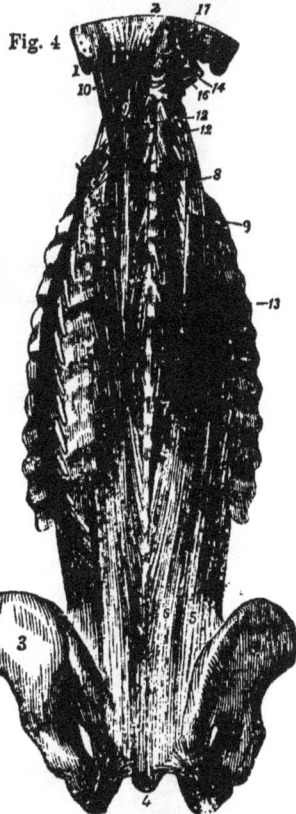

PLATE 22.

Fig. 1.

Fig.3.

Fig. 2.

Fig. 4.

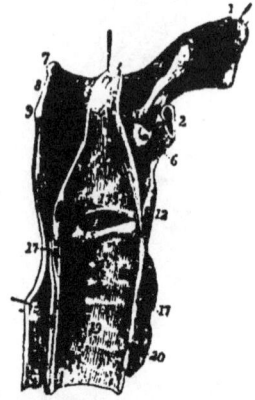

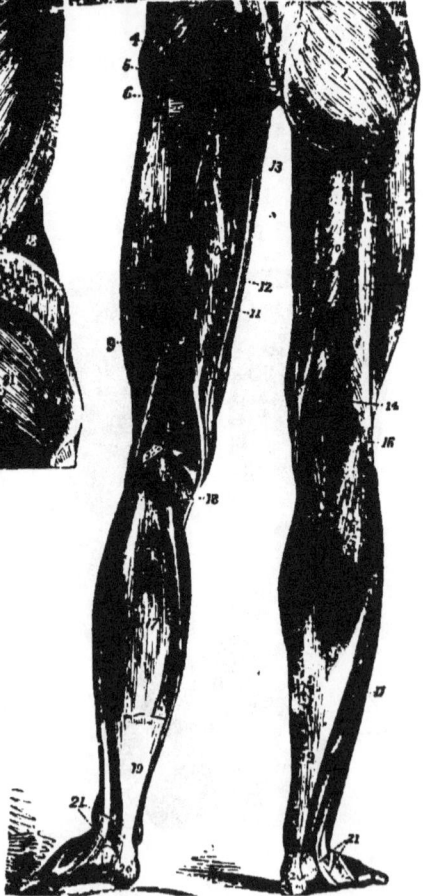

PLATE 23.

Fig. 1.

Fig. 2.

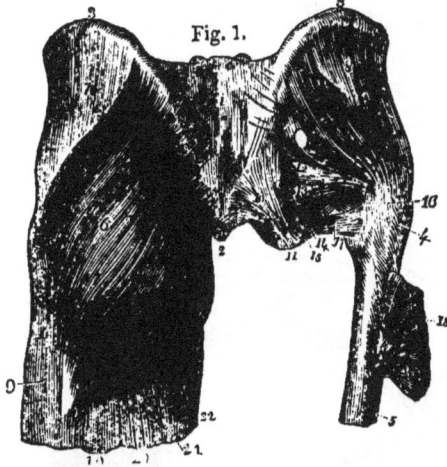

Fig. 3.

Fig. 4.

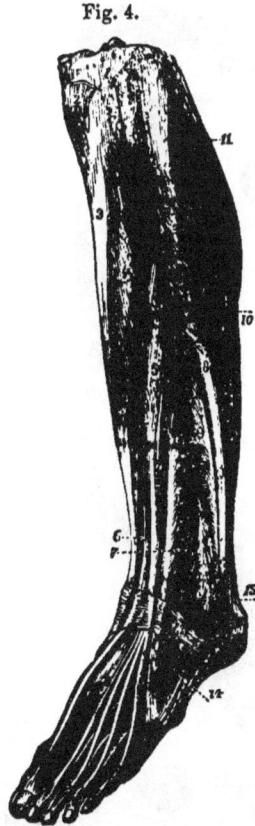

Fig. 5.

PLATE 24.

Fig. 1.

Fig. 2.

Fig. 3.

Fig. 6.

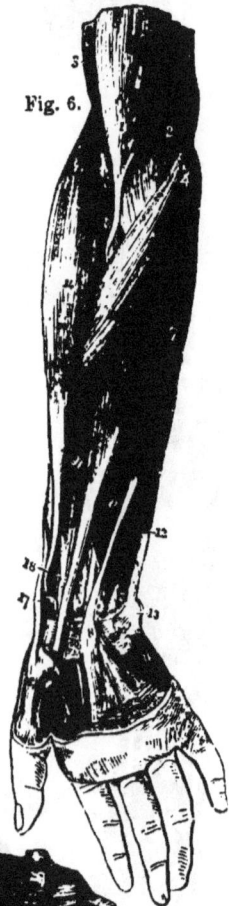

Fig. 4.

Fig. 5.

PLATE 25.

Fig. 1.

Fig. 2.

Fig. 3.

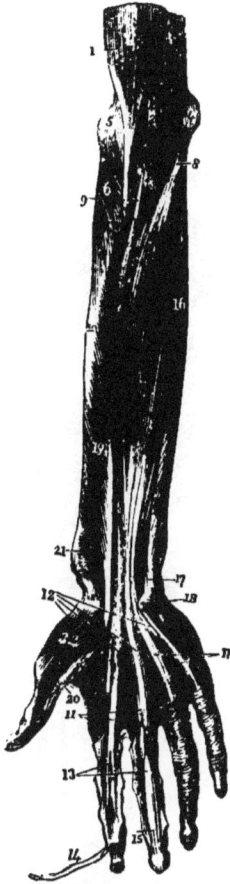

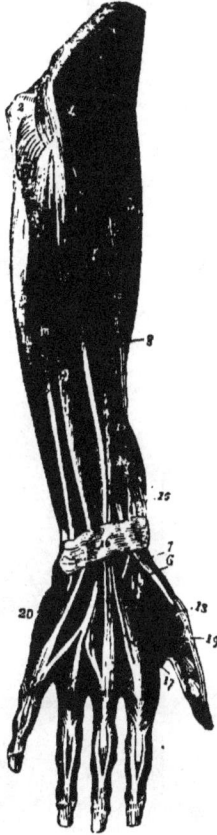

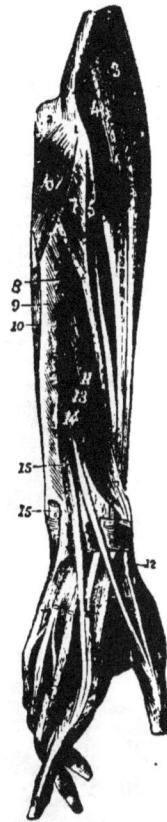

Fig. 4.

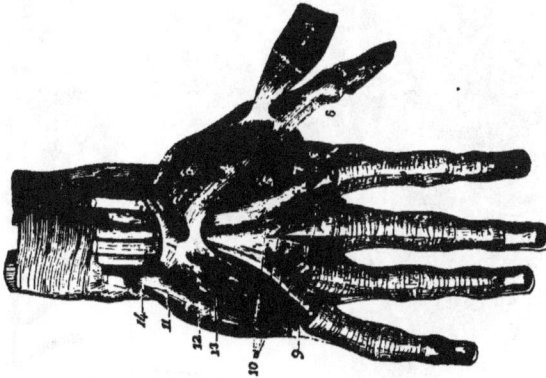

PLATE 27.

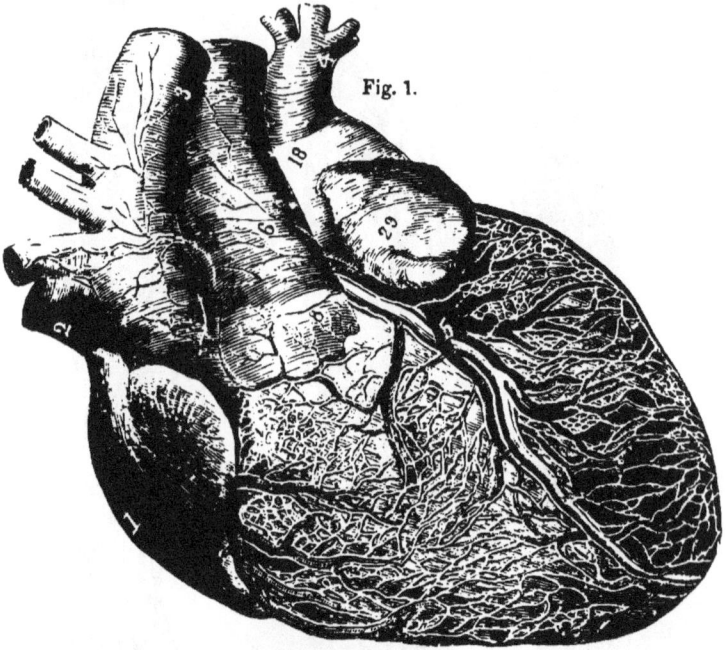

Fig. 1.

Fig. 3.

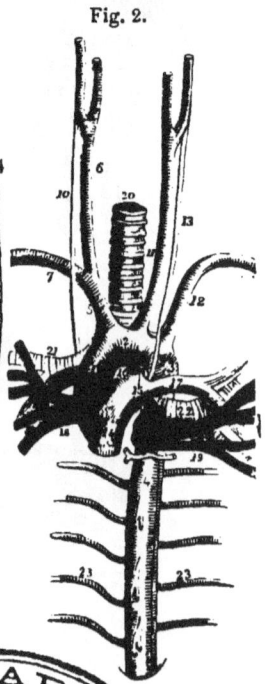

Fig. 2.

PLATE 29.

Fig. 1.

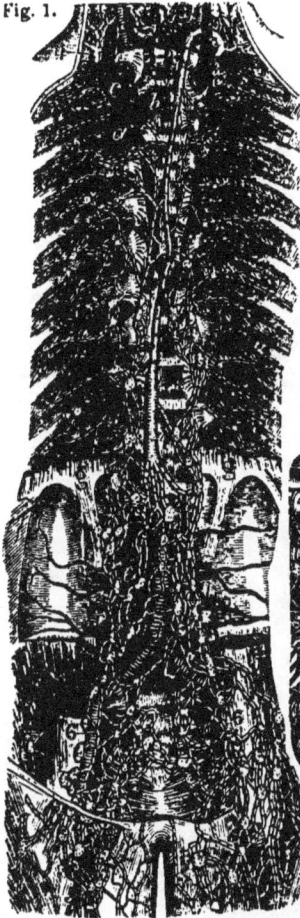

Fig. 2.

Fig. 3.

Fig. 5.

Fig. 4.

PLATE 30.

Fig. 1.

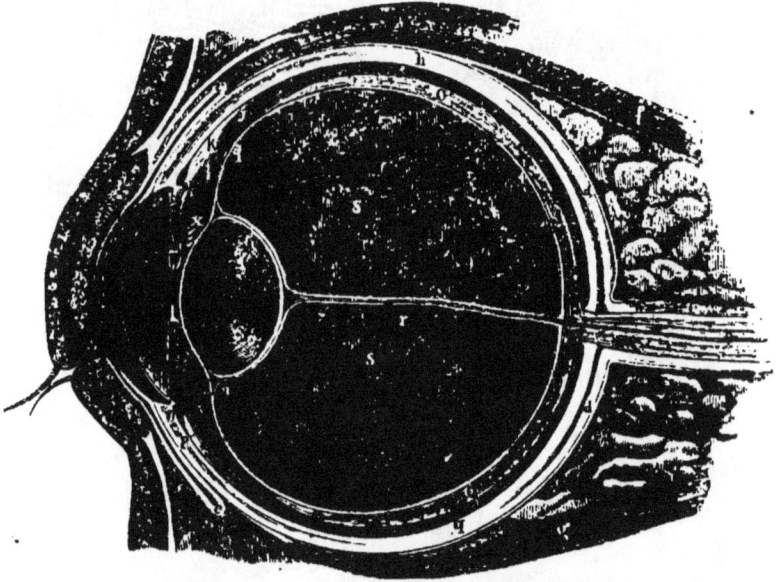

Fig. 2.

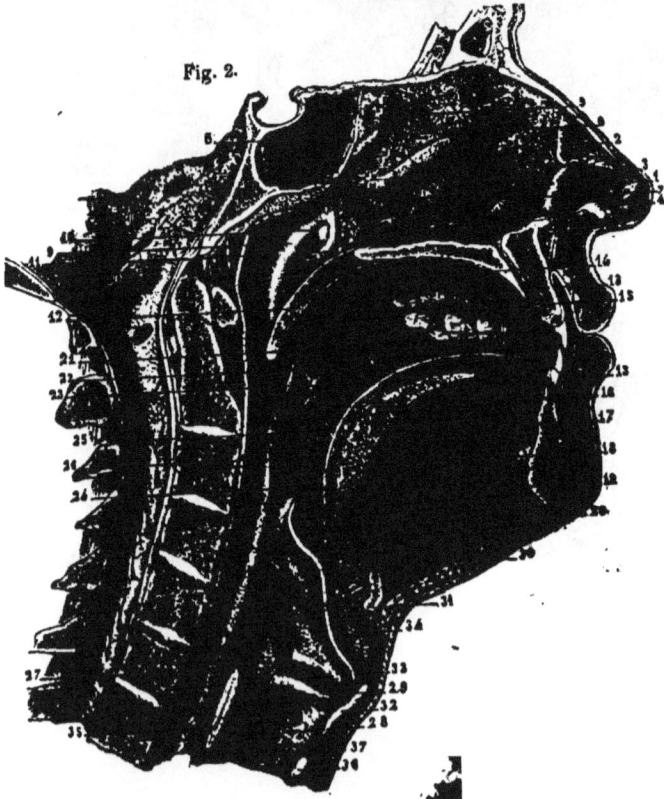

www.ingramcontent.com/pod-product-compliance
Lightning Source LLC
Chambersburg PA
CBHW021342210326
41599CB00011B/723